INTRODUCTORY REAL ANALYSIS

A. N. KOLMOGOROV
S. V. FOMIN

Revised English Edition
Translated and Edited by

Richard A. Silverman

DOVER PUBLICATIONS, INC.
NEW YORK

Published in Canada by General Publishing Company, Ltd., 30 Lesmill Road, Don Mills, Toronto,
Ontario.
Published in the United Kingdom by Constable
and Company, Ltd.

This Dover edition, first published in 1975, is
an unabridged, slightly corrected republication of
the work originally published by Prentice-Hall,
Inc., Englewood Cliffs, N.J., in 1970.

International Standard Book Number: 0-486-61226-0
Library of Congress Catalog Card Number: 74-18669

Manufactured in the United States of America

Dover Publications, Inc.
31 East 2nd Street
Mineola, N.Y. 11501

EDITOR'S PREFACE

The present course is a freely revised and restyled version of pp. 1–380 of the second edition of the Russian original (Moscow, 1968). The result is a comprehensive, but manageably proportioned and entirely elementary introduction to real and functional analysis, from a consistently modern point of view.

As in the other volumes of this series, I have not hesitated to make a number of pedagogical and mathematical improvements that occurred to me in the course of the translation. In particular, more theorems, examples and remarks have been explicitly identified as such. Each of the 37 sections has been equipped with a problem set, leading to a total of some 350 problems. Most of the problem material can be found in one form or another in the Russian original (either explicitly, or in the guise of side comments, or even as text deemed to be of an ancillary character), but quite a few extra problems have been introduced. In the same spirit of free adaptation, the bibliography has been modified to meet the needs of those who cannot easily read languages other than English. These changes have not been made capriciously, but with a view to enhancing the readability and "learnability" of the course.

A theorem or problem cited without giving a page number always refers to the corresponding theorem or problem in the section where the reference is made. The same applies to examples cited in a given subsection.

R. A. S.

CONTENTS

1

SET THEORY

I. Sets and Functions

1.1. Basic definitions. Mathematics habitually deals with "sets" made up of "elements" of various kinds, e.g., the set of faces of a polyhedron, the set of points on a line, the set of all positive integers, and so on. Because of their generality, it is hard to define these concepts in a way that does more than merely replace the word "set" by some equivalent term like "class," "family," "collection," etc. and the word "element" by some equivalent term like "member." We will adopt a "naive" point of view and regard the notions of a set and the elements of a set as primitive and well-understood.

The set concept plays a key role in modern mathematics. This is partly due to the fact that set theory, originally developed towards the end of the nineteenth century, has by now become an extensive subject in its own right. More important, however, is the great influence which set theory has exerted and continues to exert on mathematical thought as a whole. In this chapter, we introduce the basic set-theoretic notions and notation to be used in the rest of the book.

Sets will be denoted by capital letters like $A, B, \ldots$, and elements of sets by small letters like $a, b, \ldots$. The set with elements $a, b, c, \ldots$ is often denoted by $\{a, b, c, \ldots\}$, i.e., by writing the elements of the set between curly brackets. For example, $\{1\}$ is the set whose only member is 1, while $\{1, 2, \ldots, n, \ldots\}$ is the set of all positive integers. The statement "the element a belongs to the set A" is written symbolically as $a \in A$, while $a \notin A$ means that "the element a does not belong to the set A." If every element of a set A also belongs to a set B, we say that A is a *subset* of the set B and write $A \subset B$ or $B \supset A$ (also read as "A is contained in B" or

"*B* contains *A*"). For example, the set of all even numbers is a subset of the set of all real numbers. We say that two sets *A* and *B* are *equal* and write $A = B$ if *A* and *B* consist of precisely the same elements. Note that $A = B$ if and only if $A \subset B$ and $B \subset A$, i.e., if and only if every element of *A* is an element of *B* and every element of *B* is an element of *A*. If $A \subset B$ but $A \neq B$, we call *A* a *proper subset* of *B*.

Sometimes it is not known in advance whether or not a certain set (for example, the set of roots of a given equation) contains any elements at all. Thus it is convenient to introduce the concept of the *empty set*, i.e., the set containing no elements at all. This set will be denoted by the symbol $\varnothing$. The set $\varnothing$ is clearly a subset of every set (why?).

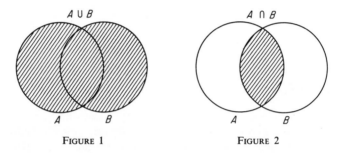

FIGURE 1   FIGURE 2

1.2. Operations on sets. Let *A* and *B* be any two sets. Then by the *sum* or *union* of *A* and *B*, denoted by $A \cup B$, is meant the set consisting of all elements which belong to at least one of the sets *A* and *B* (see Figure 1). More generally, by the sum or union of an *arbitrary* number (finite or infinite) of sets A_α (indexed by some parameter α), we mean the set, denoted by

$$\bigcup_\alpha A_\alpha,$$

of all elements belonging to at least one of the sets A_α.

By the *intersection* $A \cap B$ of two given sets *A* and *B*, we mean the set consisting of all elements which belong to both *A* and *B* (see Figure 2). For example, the intersection of the set of all even numbers and the set of all integers divisible by 3 is the set of all integers divisible by 6. By the intersection of an *arbitrary* number (finite or infinite) of sets A_α, we mean the set, denoted by

$$\bigcap_\alpha A_\alpha,$$

of all elements belonging to every one of the sets A_α. Two sets *A* and *B* are said to be *disjoint* if $A \cap B = \varnothing$, i.e., if they have no elements in common. More generally, let $\mathscr{F}$ be a family of sets such that $A \cap B = \varnothing$ for every pair of sets *A*, *B* in $\mathscr{F}$. Then the sets in $\mathscr{F}$ are said to be *pairwise disjoint*.

It is an immediate consequence of the above definitions that the operations $\cup$ and $\cap$ are commutative and associative, i.e., that

$$A \cup B = B \cup A, \quad (A \cup B) \cup C = A \cup (B \cup C),$$

$$A \cap B = B \cap A, \quad (A \cap B) \cap C = A \cap (B \cap C).$$

Moreover, the operations $\cup$ and $\cap$ obey the following distributive laws:

$$(A \cup B) \cap C = (A \cap C) \cup (B \cap C), \tag{1}$$

$$(A \cap B) \cup C = (A \cup C) \cap (B \cup C). \tag{2}$$

For example, suppose $x \in (A \cup B) \cap C$, so that x belongs to the left-hand

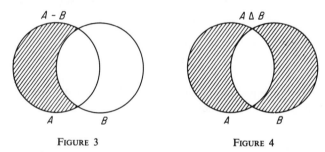

FIGURE 3   FIGURE 4

side of (1). Then x belongs to both C and $A \cup B$, i.e., x belongs to both C and at least one of the sets A and B. But then x belongs to at least one of the sets $A \cap C$ and $B \cap C$, i.e., $x \in (A \cap C) \cup (B \cap C)$, so that x belongs to the right-hand side of (1). Conversely, suppose $x \in (A \cap C) \cup (B \cap C)$. Then x belongs to at least one of the two sets $A \cap C$ and $B \cap C$. It follows that x belongs to both C and at least one of the two sets A and B, i.e., $x \in C$ and $x \in A \cup B$ or equivalently $x \in (A \cup B) \cap C$. This proves (1), and (2) is proved similarly.

By the *difference* $A - B$ between two sets A and B (in that order), we mean the set of all elements of A which do not belong to B (see Figure 3). Note that it is not assumed that $A \supset B$. It is sometimes convenient (e.g., in measure theory) to consider the *symmetric difference* of two sets A and B, denoted by $A \triangle B$ and defined as the union of the two differences $A - B$ and $B - A$ (see Figure 4):

$$A \triangle B = (A - B) \cup (B - A).$$

We will often be concerned later with various sets which are all subsets of some underlying basic set R, for example, various sets of points on the real line. In this case, given a set A, the difference $R - A$ is called the *complement* of A, denoted by CA.

An important role is played in set theory and its applications by the following "duality principle":

$$R - \bigcup_{\alpha} A_{\alpha} = \bigcap_{\alpha} (R - A_{\alpha}), \tag{3}$$

$$R - \bigcap_{\alpha} A_{\alpha} = \bigcup_{\alpha} (R - A_{\alpha}). \tag{4}$$

In words, the complement of a union equals the intersection of the complements, and the complement of an intersection equals the union of the complements. According to the duality principle, any theorem involving a family of subsets of a fixed set R can be converted *automatically* into another, "dual" theorem by replacing all subsets by their complements, all unions by intersections and all intersections by unions. To prove (3), suppose

$$x \in R - \bigcup_{\alpha} A_{\alpha}. \tag{5}$$

Then x does not belong to the union

$$\bigcup_{\alpha} A_{\alpha}, \tag{6}$$

i.e., x does not belong to any of the sets A_{α}. It follows that x belongs to each of the complements $R - A_{\alpha}$, and hence

$$x \in \bigcap_{\alpha} (R - A_{\alpha}). \tag{7}$$

Conversely, suppose (7) holds, so that x belongs to every set $R - A_{\alpha}$. Then x does not belong to any of the sets A_{α}, i.e., x does not belong to the union (6), or equivalently (5) holds. This proves (3), and (4) is proved similarly (give the details).

Remark. The designation "symmetric difference" for the set $A \triangle B$ is not too apt, since $A \triangle B$ has much in common with the sum $A \cup B$. In fact, in $A \cup B$ the two statements "x belongs to A" and "x belongs to B" are joined by the conjunction "or" used in the "either . . . or . . . or both . . ." sense, while in $A \triangle B$ the same two statements are joined by "or" used in the ordinary "either . . . or . . ." sense (as in "to be or not to be"). In other words, x belongs to $A \cup B$ if and only if x belongs to either A or B or both, while x belongs to $A \triangle B$ if and only if x belongs to either A or B but not both. The set $A \triangle B$ can be regarded as a kind of "modulo-two sum" of the sets A and B, i.e., a sum of the sets A and B in which elements are dropped if they are counted twice (once in A and once in B).

1.3. Functions and mappings. Images and preimages. A rule associating a unique real number $y = f(x)$ with each element of a set of real numbers X is said to define a *(real) function* f on X. The set X is called the *domain (of definition)* of f, and the set Y of all numbers $f(x)$ such that $x \in X$ is called the *range* of f.

More generally, let M and N be two arbitrary sets. Then a rule associating a unique element $b = f(a) \in N$ with each element $a \in M$ is again said to define a *function* f on M (or a function f with *domain* M). In this more general context, f is usually called a *mapping of* M *into* N. By the same token, f is said to *map* M *into* N (and a into b).

If a is an element of M, the corresponding element $b = f(a)$ is called the *image* of a (*under* the mapping f). Every element of M with a given element $b \in N$ as its image is called a *preimage* of b. Note that in general b may have several preimages. Moreover, N may contain elements with no preimages at all. If b has a unique preimage, we denote this preimage by $f^{-1}(b)$.

If A is a subset of M, the set of all elements $f(a) \in N$ such that $a \in A$ is called the *image* of A, denoted by $f(A)$. The set of all elements of M whose images belong to a given set $B \subset N$ is called the *preimage* of B, denoted by $f^{-1}(B)$. If no element of B has a preimage, then $f^{-1}(B) = \varnothing$. A function f is said to map M *into* N if $f(M) \subset N$, as is always the case, and *onto* N if $f(M) = N$.[1] Thus every "onto mapping" is an "into mapping," but not conversely.

Suppose f maps M *onto* N. Then f is said to be *one-to-one* if each element $b \in N$ has a unique preimage $f^{-1}(b)$. In this case, f is said to establish a *one-to-one correspondence* between M and N, and the mapping f^{-1} associating $f^{-1}(b)$ with each $b \in N$ is called the *inverse* of f.

THEOREM 1. *The preimage of the union of two sets is the union of the preimages of the sets*:

$$f^{-1}(A \cup B) = f^{-1}(A) \cup f^{-1}(B).$$

Proof. If $x \in f^{-1}(A \cup B)$, then $f(x) \in A \cup B$, so that $f(x)$ belongs to at least one of the sets A and B. But then x belongs to at least one of the sets $f^{-1}(A)$ and $f^{-1}(B)$, i.e., $x \in f^{-1}(A) \cup f^{-1}(B)$.

Conversely, if $x \in f^{-1}(A) \cup f^{-1}(B)$, then x belongs to at least one of the sets $f^{-1}(A)$ and $f^{-1}(B)$. Therefore $f(x)$ belongs to at least one of the sets A and B, i.e., $f(x) \in A \cup B$. But then $x \in f^{-1}(A \cup B)$. ∎[2]

THEOREM 2. *The preimage of the intersection of two sets is the intersection of the preimages of the sets*:

$$f^{-1}(A \cap B) = f^{-1}(A) \cap f^{-1}(B).$$

Proof. If $x \in f^{-1}(A \cap B)$, then $f(x) \in A \cap B$, so that $f(x) \in A$ and $f(x) \in B$. But then $x \in f^{-1}(A)$ and $x \in f^{-1}(B)$, i.e., $x \in f^{-1}(A) \cap f^{-1}(B)$.

Conversely, if $x \in f^{-1}(A) \cap f^{-1}(B)$, then $x \in f^{-1}(A)$ and $x \in f^{-1}(B)$. Therefore $f(x) \in A$ and $f(x) \in B$, i.e., $f(x) \in A \cap B$. But then $x \in f^{-1}(A \cap B)$. ∎

---

[1] As in the case of real functions, the set $f(M)$ is called the *range* of f.

[2] The symbol ∎ stands for Q.E.D. and indicates the end of a proof.

THEOREM 3. *The image of the union of two sets equals the union of the images of the sets*:
$$f(A \cup B) = f(A) \cup f(B).$$

Proof. If $y \in f(A \cup B)$, then $y = f(x)$ where x belongs to at least one of the sets A and B. Therefore $y = f(x)$ belongs to at least one of the sets $f(A)$ and $f(B)$, i.e., $y \in f(A) \cup f(B)$.

Conversely, if $y \in f(A) \cup f(B)$, then $y = f(x)$ where x belongs to at least one of the sets A and B, i.e., $x \in A \cup B$ and hence $y = f(x) \in f(A \cup B)$. ∎

Remark 1. Surprisingly enough, the image of the intersection of two sets does not necessarily equal the intersection of the images of the sets. For example, suppose the mapping f projects the xy-plane onto the x-axis, carrying the point (x, y) into the $(x, 0)$. Then the segments $0 \leqslant x \leqslant 1$, $y = 0$ and $0 \leqslant x \leqslant 1, y = 1$ do not intersect, although their images coincide.

Remark 2. Theorems 1–3 continue to hold for unions and intersections of an *arbitrary* number (finite or infinite) of sets A_α:

$$f^{-1}\left(\bigcup_\alpha A_\alpha\right) = \bigcup_\alpha f^{-1}(A_\alpha),$$

$$f^{-1}\left(\bigcap_\alpha A_\alpha\right) = \bigcap_\alpha f^{-1}(A_\alpha),$$

$$f\left(\bigcup_\alpha A_\alpha\right) = \bigcup_\alpha f(A_\alpha).$$

1.4. Decomposition of a set into classes. Equivalence relations. Decom⁻ positions of a given set into pairwise disjoint subsets play an important role in a great variety of problems. For example, the plane (regarded as a point set) can be decomposed into lines parallel to the x-axis, three-dimensional space can be decomposed into concentric spheres, the inhabitants of a given city can be decomposed into different age groups, and so on. Any such representation of a given set M as the union of a family of pairwise disjoint subsets of M is called a *decomposition* or *partition* of M into *classes*.

A decomposition is usually made on the basis of some criterion, allowing us to assign the elements of M to one class or another. For example, the set of all triangles in the plane can be decomposed into classes of congruent triangles or into classes of triangles of equal area, the set of all functions of x can be decomposed into classes of functions all taking the same value at a given point x, and so on. Despite the great variety of such criteria, they are not completely arbitrary. For example, it is obviously impossible to partition all real numbers into classes by assigning the number b to the same class as the number a if and only if $b > a$. In fact, if $b > a$, b must be assigned to the same class as a, but then a cannot be assigned to the same class as b, since $a < b$. Moreover, since a is not greater than itself, a cannot even be assigned to the class containing itself! As another example, it is impossible to partition the points of the plane into classes by assigning two points to the same class if and only if the distance between them is less than 1. In fact, if the distance between a and b is less than 1 and if the distance between b and c is less than 1, it does not follow that the distance between a and c is less than 1. Thus, by assigning a to the same class as b and b to the same class as c, we may well find that two points fall in the same class even though the distance between them is greater than 1!

These examples suggest conditions which must be satisfied by any criterion if it is to be used as the basis for partitioning a given set into classes. Let M be a set, and let certain ordered pairs (a, b) of elements of M be called "labelled." If (a, b) is a labelled pair, we say that a is *related* to b by the *(binary) relation R* and write aRb.[3] For example, if a and b are real numbers, aRb might mean $a < b$, while if a and b are triangles, aRb might mean that a and b have the same area. A relation between elements of M is called a relation *on M* if there is at least one labelled pair (a, b) for every $a \in M$. A relation R on M is called an *equivalence relation* (on M) if it satisfies the following three conditions:

1) *Reflexivity*: aRa for every $a \in M$;
2) *Symmetry*: If aRb, then bRa;
3) *Transitivity*: If aRb and bRc, then aRc.

THEOREM 4. *A set M can be partitioned into classes by a relation R (acting as a criterion for assigning two elements to the same class) if and only if R is an equivalence relation on M.*

Proof. Every partition of M determines a binary relation on M, where aRb means that "a belongs to the same class as b." It is then obvious that R must be reflexive, symmetric and transitive, i.e., that R is an equivalence relation on M.

Conversely, let R be an equivalence relation on M, and let K_a be the set of all elements $x \in M$ such that xRa (clearly $a \in K_a$, since R is reflexive). Then two classes K_a and K_b are either identical or disjoint. In fact, suppose an element c belongs to both K_a and K_b, so that cRa and cRb. Then aRc by the symmetry, and hence

$$aRb \qquad (8)$$

---

[3] Put somewhat differently, let M^2 be the set of all ordered pairs (a, b) with $a, b \in M$, and let $\mathscr{R}$ be the subset of M^2 consisting of all labelled pairs. Then aRb if and only if $(a, b) \in \mathscr{R}$, i.e., a binary relation is essentially just a subset of M^2. As an exercise, state the three conditions for R to be an equivalence relation in terms of ordered pairs and the set $\mathscr{R}$.

by the transitivity. If now $x \in K_a$, then xRa and hence xRb by (8) and the transitivity, i.e., $x \in K_b$. Virtually the same argument shows that $x \in K_b$ implies $x \in K_a$. Therefore $K_a = K_b$ if K_a and K_b have an element in common. Therefore the distinct sets K_a form a partition of M into classes. ∎

Remark. Because of Theorem 4, one often talks about the decomposition of M into *equivalence* classes.

There is an intimate connection between mappings and partitions into classes, as shown by the following examples:

Example 1. Let f be a mapping of a set A into a set B and partition A into sets, each consisting of all elements with the same image $b = f(a) \in B$. This gives a partition of A into classes. For example, suppose f projects the xy-plane onto the x-axis, by mapping the point (x, y) into the point $(x, 0)$. Then the preimages of the points of the x-axis are vertical lines, and the representation of the plane as the union of these lines is the decomposition into classes corresponding to f.

Example 2. Given any partition of a set A into classes, let B be the set of these classes and associate each element $a \in A$ with the class (i.e., element of B) to which it belongs. This gives a mapping of A into B. For example, suppose we partition three-dimensional space into classes by assigning to the same class all points which are equidistant from the origin of coordinates. Then every class is a sphere of a certain radius. The set of all these classes can be identified with the set of points on the half-line $[0, \infty)$, each point corresponding to a possible value of the radius. In this sense, the decomposition of space into concentric spheres corresponds to the mapping of space into the half-line $[0, \infty)$.

Example 3. Suppose we assign all real numbers with the same fractional part[4] to the same class. Then the mapping corresponding to this partition has the effect of "winding" the real line onto a circle of unit circumference.

Problem 1. Prove that if $A \cup B = A$ and $A \cap B = A$, then $A = B$.

Problem 2. Show that in general $(A - B) \cup B \neq A$.

Problem 3. Let $A = \{2, 4, \ldots, 2n, \ldots\}$ and $B = \{3, 6, \ldots, 3n, \ldots\}$. Find $A \cap B$ and $A - B$.

---

[4] The largest integer $\leqslant x$ is called the *integral part* of x, denoted by $[x]$, and the quantity $x - [x]$ is called the *fractional part* of x.

Problem 4. Prove that

a) $(A - B) \cap C = (A \cap C) - (B \cap C)$;

b) $A \bigtriangleup B = (A \cup B) - (A \cap B)$.

Problem 5. Prove that

$$\bigcup_{\alpha} A_{\alpha} - \bigcup_{\alpha} B_{\alpha} \subset \bigcup_{\alpha} (A_{\alpha} - B_{\alpha}).$$

Problem 6. Let A_n be the set of all positive integers divisible by n. Find the sets

a) $\displaystyle\bigcup_{n=2}^{\infty} A_n$;   b) $\displaystyle\bigcap_{n=2}^{\infty} A_n$.

Problem 7. Find

a) $\displaystyle\bigcup_{n=1}^{\infty} \left[a + \frac{1}{n}, b - \frac{1}{n} \right]$;   b) $\displaystyle\bigcap_{n=1}^{\infty} \left(a - \frac{1}{n}, b + \frac{1}{n} \right)$.

Problem 8. Let A_{α} be the set of points lying on the curve

$$y = \frac{1}{x^{\alpha}} \qquad (0 < x < \infty).$$

What is

$$\bigcap_{\alpha \geqslant 1} A_{\alpha}?$$

Problem 9. Let $y = f(x) = \langle x \rangle$ for all real x, where $\langle x \rangle$ is the fractional part of x. Prove that every closed interval of length 1 has the same image under f. What is this image? Is f one-to-one? What is the preimage of the interval $\frac{1}{4} < y < \frac{3}{4}$? Partition the real line into classes of points with the same image.

Problem 10. Given a set M, let $\mathcal{R}$ be the set of all ordered pairs on the form (a, a) with $a \in M$, and let aRb if and only if $(a, b) \in \mathcal{R}$. Interpret the relation R.

Problem 11. Give an example of a binary relation which is

a) Reflexive and symmetric, but not transitive;
b) Reflexive, but neither symmetric nor transitive;
c) Symmetric, but neither reflexive nor transitive;
d) Transitive, but neither reflexive nor symmetric.

2. Equivalence of Sets. The Power of a Set

2.1. Finite and infinite sets. The set of all vertices of a given polyhedron, the set of all prime numbers less than a given number, and the set of all residents of New York City (at a given time) have a certain property in common, namely, each set has a definite number of elements which can be found in principle, if not in practice. Accordingly, these sets are all said to be *finite*. Clearly, we can be sure that a set is finite without knowing the number of elements in it. On the other hand, the set of all positive integers, the set of all points on the line, the set of all circles in the plane, and the set of all polynomials with rational coefficients have a different property in common, namely, if we remove one element from each set, then remove two elements, three elements, and so on, there will still be elements left in the set at each stage. Accordingly, sets of this kind are said to be *infinite*.

Given two finite sets, we can always decide whether or not they have the same number of elements, and if not, we can always determine which set has more elements than the other. It is natural to ask whether the same is true of infinite sets. In other words, does it make sense to ask, for example, whether there are more circles in the plane than rational points on the line, or more functions defined in the interval [0, 1] than lines in space? As will soon be apparent, questions of this kind can indeed be answered.

To compare two finite sets A and B, we can count the number of elements in each set and then compare the two numbers, but alternatively, we can try to establish a *one-to-one correspondence* between (the elements of) A and B, i.e., a correspondence such that each element in A corresponds to one and only one element in B and vice verse. It is clear that a one-to-one correspondence between two finite sets can be set up if and only if the two sets have the same number of elements. For example, to ascertain whether or not the number of students in an assembly is the same as the number of seats in the auditorium, there is no need to count the number of students and the number of seats. We need merely observe whether or not there are empty seats or students with no place to sit down. If the students can all be seated with no empty seats left, i.e., if there is a one-to-one correspondence between the set of students and the set of seats, then these two sets obviously have the same number of elements. The important point here is that the first method (counting elements) works only for finite sets, while the second method (setting up a one-to-one correspondence) works for infinite sets as well as for finite sets.

2.2. Countable sets. The simplest infinite set is the set Z_+ of all positive integers. An infinite set is called *countable* if its elements can be put in one-to-one correspondence with those of Z_+. In other words, a countable set is a set whose elements can be numbered $a_1, a_2, \ldots, a_n, \ldots$. By an *uncountable* set we mean, of course, an infinite set which is not countable.

We now give some examples of countable sets:

Example 1. The set Z of all integers, positive, negative or zero, is countable. In fact, we can set up the following one-to-one correspondence between Z and the set Z_+ of all positive integers:

$$0, \quad -1, \quad 1, \quad -2, \quad 2, \ldots$$
$$1, \quad 2, \quad 3, \quad 4, \quad 5, \ldots$$

More explicitly, we associate the nonnegative integer $n \geqslant 0$ with the odd number $2n + 1$, and the negative integer $n < 0$ with the even number $2|n|$, i.e.,

$$n \leftrightarrow 2n + 1 \quad \text{if} \quad n \geqslant 0,$$
$$n \leftrightarrow 2|n| \quad \quad \text{if} \quad n < 0$$

(the symbol $\leftrightarrow$ denotes a one-to-one correspondence).

Example 2. The set of all positive even numbers is countable, as shown by the obvious correspondence $n \leftrightarrow 2n$.

Example 3. The set $2, 4, 8, \ldots, 2^n, \ldots$ of powers of 2 is countable, as shown by the obvious correspondence $n \leftrightarrow 2^n$.

Example 4. The set Q of all rational numbers is countable. To see this, we first note that every rational number α can be written as a fraction p/q, $q > 0$ in lowest terms with a positive denominator. Call the sum $|p| + q$ the "height" of the rational number α. For example,

$$\frac{0}{1} = 0$$

is the only rational number of height 0,

$$\frac{-1}{1}, \quad \frac{1}{1}$$

are the only rational numbers of height 2,

$$\frac{-2}{1}, \quad \frac{-1}{2}, \quad \frac{1}{2}, \quad \frac{2}{1}$$

are the only rational numbers of height 3, and so on. We can now arrange all rational numbers in order of increasing height (with the numerators increasing in each set of rational numbers of the same height). In other words, we first count the rational numbers of height 1, then those of height 2 (suitably arranged), those of height 3, and so on. In this way, we assign every rational number a unique positive integer, i.e., we set up a one-to-one correspondence between the set Q of all rational numbers and the set Z_+ of all positive integers.

Next we prove some elementary theorems involving countable sets:

THEOREM 1. *Every subset of a countable set is countable.*

Proof. Let A be countable, with elements $a_1, a_2, \ldots$, and let B be a subset of A. Among the elements $a_1, a_2, \ldots$, let $a_{n_1}, a_{n_2}, \ldots$ be those in the set B. If the set of numbers $n_1, n_2, \ldots$ has a largest number, then B is finite. Otherwise B is countable (consider the correspondence $i \leftrightarrow a_{n_i}$). ∎

THEOREM 2. *The union of a finite or countable number of countable sets $A_1, A_2, \ldots$ is itself countable.*

Proof. We can assume that no two of the sets $A_1, A_2, \ldots$ have elements in common, since otherwise we could consider the sets

$$A_1, \quad A_2 - A_1, \quad A_3 - (A_1 \cup A_2), \ldots$$

instead, which are countable by Theorem 1 and have the same union as the original sets. Suppose we write the elements of $A_1, A_2, \ldots$ in the form of an infinite table

$$
\begin{array}{llll}
a_{11} & a_{12} & a_{13} & a_{14} \cdots \\
a_{21} & a_{22} & a_{23} & a_{24} \cdots \\
a_{31} & a_{32} & a_{33} & a_{34} \cdots \\
a_{41} & a_{42} & a_{43} & a_{44} \cdots \\
\multicolumn{4}{c}{\cdots \cdots \cdots \cdots}
\end{array}
\tag{1}
$$

where the elements of the set A_1 appear in the first row, the elements of the set A_2 appear in the second row, and so on. We now count all the elements in (1) "diagonally," i.e., first we choose a_{11}, then a_{12}, then a_{21}, and so on, moving in the way shown in the following table:[5]

$$
\begin{array}{llll}
a_{11} \rightarrow a_{12} & a_{13} \rightarrow a_{14} \cdots \\
\swarrow \quad \nearrow \quad \swarrow \\
a_{21} \quad a_{22} \quad a_{23} \quad a_{24} \cdots \\
\downarrow \nearrow \quad \swarrow \\
a_{31} \quad a_{32} \quad a_{33} \quad a_{34} \cdots \\
\swarrow \\
a_{41} \quad a_{42} \quad a_{43} \quad a_{44} \cdots
\end{array}
\tag{2}
$$

It is clear that this procedure associates a unique number to each element in each of the sets $A_1, A_2, \ldots$, thereby establishing a one-to-one correspondence between the union of the sets $A_1, A_2, \ldots$ and the set Z_+ of all positive integers. ∎

THEOREM 3. *Every infinite set has a countable subset.*

---

[5] Discuss the obvious modifications of (1) and (2) in the case of only a finite number of sets $A_1, A_2, \ldots$.

Proof. Let M be an infinite set and a_1 any element of M. Being infinite, M contains an element a_2 distinct from a_1, an element a_3 distinct from both a_1 and a_2, and so on. Continuing this process (which can never terminate due to a "shortage" of elements, since M is infinite), we get a countable subset

$$A = \{a_1, a_2, \ldots, a_n, \ldots\}$$

of the set M. ∎

Remark. Theorem 3 shows that countable sets are the "smallest" infinite sets. The question of whether there exist uncountable (infinite) sets will be considered below.

2.3. Equivalence of sets. We arrived at the notion of a countable set M by considering one-to-one correspondences between M and the set Z_+ of all positive integers. More generally, we can consider one-to-one correspondences between any two sets M and N:

DEFINITION. *Two sets M and N are said to be **equivalent** (written $M \sim N$) if there is a one-to-one correspondence between the elements of M and the elements of N.*

The concept of equivalence[6] is applicable to both finite and infinite sets. Two finite sets are equivalent if and only if they have the same number of elements. We can now define a countable set as a set equivalent to the set Z_+ of all positive integers. It is clear that two sets which are equivalent to a third set are equivalent to each other, and in particular that any two countable sets are equivalent.

Example 1. The sets of points in any two closed intervals $[a, b]$ and $[c, d]$ are equivalent, and Figure 5 shows how to set up a one-to-one correspondence between them. Here two points p and q correspond to each other if and only if they lie on the same ray emanating from the point O in which the extensions of the line segments ac and bd intersect.

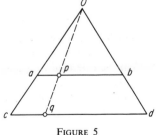

FIGURE 5

Example 2. The set of all points z in the complex plane is equivalent to the set of all

---

[6] Not to be confused with our previous use of the word in the phrase "equivalence relation." However, note that set equivalence is an equivalence relation in the sense of Sec. 1.4, being obviously reflexive, symmetric and transitive. Hence any family of sets can be partitioned into classes of equivalent sets.

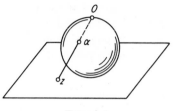

FIGURE 6

points α on a sphere. In fact, a one-to-one correspondence $z \leftrightarrow \alpha$ between the points of the two sets can be established by using stereographic projection, as shown in Figure 6 (O is the north pole of the sphere).

Example 3. The set of all points x in the open unit interval $(0, 1)$ is equivalent to the set of all points y on the whole real line. For example, the formula

$$y = \frac{1}{\pi} \arctan x + \frac{1}{2}$$

establishes a one-to-one correspondence between these two sets.

The last example and the examples in Sec. 2.2 show that an infinite set is sometimes equivalent to one of its proper subsets. For example, there are "as many" positive integers as integers of arbitrary sign, there are "as many" points in the interval $(0, 1)$ as on the whole real line, and so on. This fact is characteristic of all infinite sets (and can be used to define such sets), as shown by

THEOREM 4. *Every infinite set is equivalent to one of its proper subsets.*

Proof. According to Theorem 3, any infinite set M contains a countable subset. Let this subset be

$$A = \{a_1, a_2, \ldots, a_n, \ldots\},$$

and partition A into two countable subsets

$$A_1 = \{a_1, a_3, a_5, \ldots\}, \qquad A_2 = \{a_2, a_4, a_6, \ldots\}.$$

Obviously, we can establish a one-to-one correspondence between the countable sets A and A_1 (merely let $a_n \leftrightarrow a_{2n-1}$). This correspondence can be extended to a one-to-one correspondence between the sets $A \cup (M - A) = M$ and $A_1 \cup (M - A) = M - A_2$ by simply assigning x itself to each element $x \in M - A$. But $M - A_2$ is a proper subset of M. ∎

2.4. Uncountability of the real numbers. Several examples of countable sets were given in Sec. 2.2, and many more examples of such sets could be given. In fact, according to Theorem 2, the union of a finite or countable number of countable sets is itself countable. It is now natural to ask whether there exist infinite sets which are uncountable. The existence of such sets is shown by

THEOREM 5. *The set of real numbers in the closed unit interval* [0, 1] *is uncountable.*

Proof. Suppose we have somehow managed to count some or all of the real numbers in [0, 1], arranging them in a list

$$\begin{aligned}
\alpha_1 &= 0.a_{11}a_{12} \ldots a_{1n} \ldots, \\
\alpha_2 &= 0.a_{21}a_{22} \ldots a_{2n} \ldots, \\
&\cdot\cdot\cdot\cdot\cdot\cdot\cdot\cdot\cdot\cdot\cdot\cdot\cdot\cdot \\
\alpha_n &= 0.a_{n1}a_{n2} \ldots a_{nn} \ldots, \\
&\cdot\cdot\cdot\cdot\cdot\cdot\cdot\cdot\cdot\cdot\cdot\cdot\cdot\cdot
\end{aligned} \tag{3}$$

where a_{ik} is the kth digit in the decimal expansion of the number α_i. Consider the decimal

$$\beta = 0.b_1b_2 \ldots b_n \ldots \tag{4}$$

constructed as follows: For b_1 choose any digit (from 0 to 9) different from a_{11}, for b_2 any digit different from a_{22}, and so on, and in general for b_n any digit different from a_{nn}. Then the decimal (4) cannot coincide with any decimal in the list (3). In fact, β differs from α_1 in at least the first digit, from α_2 in at least the second digit, and so on, since in general $b_n \neq a_{nn}$ for all n. Thus no list of real numbers in the interval [0, 1] can include all the real numbers in [0, 1].

The above argument must be refined slightly since certain numbers, namely those of the form $p/10^q$, can be written as decimals in two ways, either with an infinite run of zeros or an infinite run of nines. For example,

$$\tfrac{1}{2} = \tfrac{5}{10} = 0.5000 \ldots = 0.4999 \ldots,$$

so that the fact that two decimals are distinct does not necessarily mean that they represent distinct real numbers. However, this difficulty disappears if in constructing β, we require that β contain neither zeros nor nines, for example by setting $b_n = 2$ if $a_{nn} = 1$ and $b_n = 1$ if $a_{nn} \neq 1$. ∎

Thus the set [0, 1] is uncountable. Other examples of uncountable sets equivalent to [0, 1] are

1) The set of points in any closed interval $[a, b]$;
2) The set of points on the real line;
3) The set of points in any open interval (a, b);
4) The set of all points in the plane or in space;
5) The set of all points on a sphere or inside a sphere;
6) The set of all lines in the plane;
7) The set of all *continuous* real functions of one or several variables.

The fact that the sets 1) and 2) are equivalent to [0, 1] is proved as in Examples 1 and 3, pp. 13 and 14, while the fact that the sets 3)–7) are equivalent to [0, 1] is best proved *indirectly* (cf. Problems 7 and 9).

2.5. The power of a set. Given any two sets M and N, suppose M and N are equivalent. Then M and N are said to have the same *power*. Roughly speaking, "power" is something shared by equivalent sets. If M and N are finite, then M and N have the same number of elements, and the concept of the power of a set reduces to the usual notion of the number of elements in a set. The power of the set Z_+ of all positive integers, and hence the power of any countable set, is denoted by the symbol $\aleph_0$, read "aleph null." A set equivalent to the set of real numbers in the interval [0, 1], and hence to the set of *all* real numbers, is said to have the power of the *continuum*, denoted by c (or often by $\aleph$).

For the powers of finite sets, i.e., for the positive integers, we have the notions of "greater than" and "less than," as well as the notion of equality. We now show how these concepts are extended to the case of infinite sets.

Let A and B be any two sets, with powers $m(A)$ and $m(B)$, respectively. If A is equivalent to B, then $m(A) = m(B)$ by definition. If A is equivalent to a subset of B and if no subset of A is equivalent to B, then, by analogy with the finite case, it is natural to regard $m(A)$ as less than $m(B)$ or $m(B)$ as greater than $m(A)$. Logically, however, there are two further possibilities:

    a) B has a subset equivalent to A, and A has a subset equivalent to B;

    b) A and B are not equivalent, and neither has a subset equivalent to the other.

In case a), A and B are equivalent and hence have the same power, as shown by the Cantor-Bernstein theorem (Theorem 7 below). Case b) would obviously show the existence of powers that cannot be compared, but it follows from the well-ordering theorem (see Sec. 3.7) that this case is actually impossible. Therefore, taking both of these theorems on faith, we see that any two sets A and B either have the same power or else satisfy one of the relations $m(A) < m(B)$ or $m(A) > m(B)$. For example, it is clear that $\aleph_0 < c$ (why?).

Remark. The very deep problem of the existence of powers between $\aleph_0$ and c is touched upon in Sec. 3.9. As a rule, however, the infinite sets encountered in analysis are either countable or else have the power of the continuum.

We have already noted that countable sets are the "smallest" infinite sets. It has also been shown that there are infinite sets of power greater than that of a countable set, namely sets with the power of the continuum. It is natural to ask whether there are infinite sets of power greater than that of the continuum or, more generally, whether there is a "largest" power. These questions are answered by

THEOREM 6. *Given any set M, let $\mathscr{M}$ be the set whose elements are all possible subsets of M. Then the power of $\mathscr{M}$ is greater than the power of the original set M.*

Proof. Clearly, the power μ of the set $\mathscr{M}$ cannot be less than the power m of the original set M, since the "single-element subsets" (or "singletons") of M form a subset of $\mathscr{M}$ equivalent to M. Thus we need only show that m and μ do not coincide. Suppose a one-to-one correspondence

$$a \leftrightarrow A, \qquad b \leftrightarrow B, \ldots$$

has been established between the elements $a, b, \ldots$ of M and certain elements $A, B, \ldots$ of $\mathscr{M}$ (i.e., certain subsets of M). Then $A, B, \ldots$ do not exhaust all the elements of $\mathscr{M}$, i.e., all the subsets of M. To see this, let X be the set of elements of M which do not belong to their "associated subsets." More exactly, if $a \leftrightarrow A$ we assign a to X if $a \notin A$, but not if $a \in A$. Clearly, X is a subset of M and hence an element of $\mathscr{M}$. Suppose there is an element $x \in M$ such that $x \leftrightarrow X$, and consider whether or not x belongs to X. Suppose $x \notin X$. Then $x \in X$, since, by definition, X contains *every* element not contained in its associated subset. On the other hand, suppose $x \notin X$. Then $x \in X$, since X consists precisely of those elements which do not belong to their associated subsets. In any event, the element x corresponding to the subset X must simultaneously belong to X and not belong to X. But this is impossible! It follows that there is no such element x. Therefore no one-to-one correspondence can be established between the sets M and $\mathscr{M}$, i.e., $m \neq \mu$. ∎

Thus, given any set M, there is a set $\mathscr{M}$ of larger power, a set $\mathscr{M}^*$ of still larger power, and so on indefinitely. In particular, there is no set of "largest" power.

2.6. The Cantor-Bernstein theorem. Next we prove an important theorem already used in the preceding section:

THEOREM 7 (*Cantor-Bernstein*). *Given any two sets A and B, suppose A contains a subset A_1 equivalent to B, while B contains a subset B_1 equivalent to A. Then A and B are equivalent.*

Proof. By hypothesis, there is a one-to-one function f mapping A into B_1 and a one-to-one function g mapping B into A_1:

$$f(A) = B_1 \subset B, \qquad g(B) = A_1 \subset A.$$

Therefore

$$A_2 = gf(A) = g(f(A)) = g(B_1)$$

is a subset of A_1 equivalent to all of A. Similarly,

$$B_2 = fg(B) = f(g(B)) = f(A_1)$$

is a subset of B_1 equivalent to B. Let A_3 be the subset of A into which the mapping gf carries the set A_1, and let A_4 be the subset of A into which gf carries A_2. More generally, let A_{k+2} be the set into which A_k ($k = 1, 2, \ldots$) is carried by gf. Then clearly

$$A \supset A_1 \supset A_2 \supset \cdots \supset A_k \supset A_{k+1} \supset \cdots$$

Setting

$$D = \bigcap_{k=1}^{\infty} A_k,$$

we can represent A as the following union of pairwise disjoint sets:

$$A = (A - A_1) \cup (A_1 - A_2) \cup (A_2 - A_3) \cup \cdots$$
$$\cup (A_k - A_{k+1}) \cup \cdots \cup D. \quad (5)$$

Similarly, we can write A_1 in the form

$$A_1 = (A_1 - A_2) \cup (A_2 - A_3) \cup \cdots \cup (A_k - A_{k+1}) \cup \cdots \cup D. \quad (6)$$

Clearly, (5) and (6) can be rewritten as

$$A = D \cup M \cup N, \quad (5')$$

$$A_1 = D \cup M \cup N_1, \quad (6')$$

where

$$M = (A_1 - A_2) \cup (A_3 - A_4) \cup \cdots,$$
$$N = (A - A_1) \cup (A_2 - A_3) \cup \cdots,$$
$$N_1 = (A_2 - A_3) \cup (A_4 - A_5) \cup \cdots.$$

But $A - A_1$ is equivalent to $A_2 - A_3$ (the former is carried into the latter by the one-to-one function gf), $A_2 - A_3$ is equivalent to $A_4 - A_5$, and so on. Therefore N is equivalent to N_1. It follows from the representations (5') and (6') that a one-to-one correspondence can be set up between the sets A and A_1. But A_1 is equivalent to B, by hypothesis. Therefore A is equivalent to B. ∎

Remark. Here we can even "afford the unnecessary luxury" of explicitly writing down a one-to-one function carrying A into B, i.e.,

$$\varphi(a) = \begin{cases} g^{-1}(a) & \text{if} \quad a \in D \cup M, \\ f(a) & \text{if} \quad a \in D \cup N \end{cases}$$

(see Figure 7).

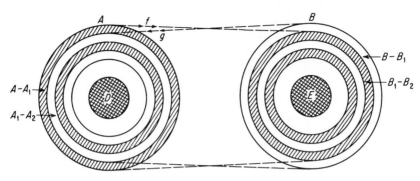

FIGURE 7

Problem 1. Prove that a set with an uncountable subset is itself uncountable.

Problem 2. Let M be any infinite set and A any countable set. Prove that $M \sim M \cup A$.

Problem 3. Prove that each of the following sets is countable:

a) The set of all numbers with two distinct decimal expansions (like $0.5000\ldots$ and $0.4999\ldots$);
b) The set of all rational points in the plane (i.e., points with rational coordinates);
c) The set of all rational intervals (i.e., intervals with rational end points);
d) The set of all polynomials with rational coefficients.

Problem 4. A number α is called *algebraic* if it is a root of a polynomial equation with rational coefficients. Prove that the set of all algebraic numbers is countable.

Problem 5. Prove the existence of uncountably many *transcendental* numbers, i.e., numbers which are not algebraic.

Hint. Use Theorems 2 and 5.

Problem 6. Prove that the set of all real functions (more generally, functions taking values in a set containing at least two elements) defined on a set M is of power greater than the power of M. In particular, prove that the power of the set of *all* real functions (continuous and discontinuous) defined in the interval $[0, 1]$ is greater than c.

Hint. Use the fact that the set of all characteristic functions (i.e., functions taking only the values 0 and 1) on M is equivalent to the set of all subsets of M.

Problem 7. Give an indirect proof of the equivalence of the closed interval $[a, b]$, the open interval (a, b) and the half-open interval $[a, b)$ or $(a, b]$.

Hint. Use Theorem 7.

Problem 8. Prove that the union of a finite or countable number of sets each of power c is itself of power c.

Problem 9. Prove that each of the following sets has the power of the continuum:

   a) The set of all infinite sequences of positive integers;
   b) The set of all ordered n-tuples of real numbers;
   c) The set of all infinite sequences of real numbers.

Problem 10. Develop a contradiction inherent in the notion of the "set of all sets which are not members of themselves."

Hint. Is this set a member of itself?

Comment. Thus we will be careful to avoid sets which are "too big," like the "set of all sets."

3. Ordered Sets and Ordinal Numbers

3.1. Partially ordered sets. A binary relation R on a set M is said to be a *partial ordering* (and the set M itself is said to be *partially ordered*) if

   1) R is reflexive (aRa for every $a \in M$);
   2) R is transitive (aRb and bRc together imply aRc);
   3) R is *antisymmetric* in the sense that aRb and bRa together imply $a = b$.

For example, if M is the set of all real numbers and aRb means $a \leqslant b$, then R is a partial ordering. This suggests writing $a \leqslant b$ (or equivalently $b \geqslant a$) instead of aRb whenever R is a partial ordering, and we will do so from now on. Similarly, we write $a < b$ if $a \leqslant b$, $a \neq b$ and $b > a$ if $b \geqslant a$, $b \neq a$.

The following examples give some idea of the generality of the concept of a partial ordering:

Example 1. Any set M can be partially ordered in a trivial way by setting $a \leqslant b$ if and only if $a = b$.

Example 2. Let M be the set of all continuous functions $f, g, \ldots$ defined in a closed interval $[\alpha, \beta]$. Then we get a partial ordering by setting $f \leqslant g$ if and only if $f(t) \leqslant g(t)$ for every $t \in [\alpha, \beta]$.

Example 3. The set of all subsets $M_1, M_2, \ldots$ is partially ordered if $M_1 \leqslant M_2$ means that $M_1 \subset M_2$.

Example 4. The set of all integers greater than 1 is partially ordered if $a \leqslant b$ means that "b is divisible by a."

An element a of a partially ordered set is said to be *maximal* if $a \leqslant b$ implies $b = a$ and *minimal* if $b \leqslant a$ implies $b = a$. Thus in Example 4 every prime number (greater than 1) is a minimal element.

3.2. Order-preserving mappings. Isomorphisms. Let M and M' be any two partially ordered sets, and let f be a one-to-one mapping of M onto M'. Then f is said to be *order-preserving* if $a \leqslant b$ (where $a, b \in M$) implies $f(a) \leqslant f(b)$ (in M'). An order-preserving mapping f such that $f(a) \leqslant f(b)$ implies $a \leqslant b$ is called an *isomorphism*. In other words, an isomorphism between two partially ordered sets M and M' is a one-to-one mapping of M onto M' such that $f(a) \leqslant f(b)$ if and only if $a \leqslant b$. Two partially ordered sets M and M' are said to be *isomorphic* (to each other) if there exists an isomorphism between them.

Example. Let M be the set of positive integers greater than 1 partially ordered as in Example 4, Sec. 3.1, and let M' be the same set partially ordered in the natural way, i.e., in such a way that $a \leqslant b$ if and only if $b - a$ is nonnegative. Then the mapping of M onto M' carrying every integer n into itself is order-preserving, but not an isomorphism.

Isomorphism between partially ordered sets is an equivalence relation as defined in Sec. 1.4, being obviously reflexive, symmetric and transitive. Hence any given family of partially ordered sets can be partitioned into disjoint classes of isomorphic sets.[7] Clearly, two isomorphic partially ordered sets can be regarded as identical in cases where it is the structure of the partial ordering rather than the specific nature of the elements of the sets that is of interest.

3.3. Ordered sets. Order types. Given two elements a and b of a partially ordered set M, it may turn out that neither of the relations $a \leqslant b$ or $b \leqslant a$ holds. In this case, a and b are said to be *noncomparable*. Thus, in general, the relation $\leqslant$ is defined only for certain pairs of elements, which is why M is said to be *partially* ordered. However, suppose M has no noncomparable elements. Then M is said to be *ordered* (synonymously, *simply* or *linearly ordered*). In other words, a set M is ordered if it is partially ordered and if, given any two distinct elements $a, b \in M$, either $a < b$ or $b < a$. Obviously, any subset of an ordered set is itself ordered.

Each of the sets figuring in Examples 1-4, Sec. 3.1 is partially ordered, but not ordered. Simple examples of ordered sets are the set of all positive integers, the set of all rational numbers, the set of all real numbers in the

---

[7] Note that we avoid talking about the "family of *all* partially ordered sets" (recall Problem 10, p. 20).

interval $[0, 1]$, and so on (with the usual relations of "greater than" and "less than").

Since an ordered set is a special kind of partially ordered set, the concepts of order-preserving mapping and isomorphism apply equally well to ordered sets. Two isomorphic ordered sets are said to have the same (*order*) *type*. Thus "type" is something shared by all isomorphic ordered sets, just as "power" is something shared by all equivalent sets (considered as "plain" sets, without regard for possible orderings).

The simplest example of an ordered set is the set of all positive integers $1, 2, 3, \ldots$ arranged in increasing order, with the usual meaning of the symbol $<$. The order type of this set is denoted by the symbol ω. Two isomorphic ordered sets obviously have the same power (an isomorphism is a *one-to-one* correspondence). Thus it makes sense to talk about the power corresponding to a given order type. For example, the power $\aleph_0$ corresponds to the order type ω. The converse is not true, since a set of a given power can in general be ordered in many different ways. It is only in the finite case that the number of elements in a set uniquely determines its type, designated by the same symbol n as the number of elements in the set. For example, besides the "natural" order type ω of the set of positive integers, there is another order type corresponding to the sequence

$$1, 3, 5, \ldots, 2, 4, 6, \ldots,$$

where odd and even numbers are separately arranged in increasing order, but any odd number precedes any even number. It can be shown that the number of distinct order types of a set of power $\aleph_0$ is infinite and in fact uncountable.

3.4. Ordered sums and products of ordered sets. Let M_1 and M_2 be two ordered sets of types θ_1 and θ_2, respectively. Then we can introduce an ordering in the union $M_1 \cup M_2$ of the two sets by assuming that

1) a and b have the same ordering as in M_1 if $a, b \in M_1$;
2) a and b have the same ordering as in M_2 if $a, b \in M_2$;
3) $a < b$ if $a \in M_1, b \in M_2$

(verify that this is actually an ordering of $M_1 \cup M_2$). The set $M_1 \cup M_2$ ordered in this way is called the *ordered sum* of M_1 and M_2, denoted by $M_1 + M_2$. Note that the order of terms matters here, i.e., in general $M_2 + M_1$ is not isomorphic to $M_1 + M_2$. More generally, we can define the ordered sum of any finite number of ordered sets by writing (cf. Problem 6)

$$M_1 + M_2 + M_3 = (M_1 + M_2) + M_3,$$
$$M_1 + M_2 + M_3 + M_4 = (M_1 + M_2 + M_3) + M_4,$$

and so on. By the *ordered sum* of the types θ_1 and θ_2, denoted by $\theta_1 + \theta_2$, we mean the order type of the set $M_1 + M_2$.

Example. Consider the order types ω and n. It is easy to see that $n + \omega = \omega$. In fact, if finitely many terms are written to the left of the sequence $1, 2, \ldots, k, \ldots$, we again get a set of the same type (why?). On the other hand, the order type $\omega + n$, i.e., the order type of the set[8]

$$\{1, 2, \ldots, k, \ldots, a_1, a_2, \ldots, a_n\},$$

is obviously not equal to ω.

Again let M_1 and M_2 be two ordered sets of types θ_1 and θ_2, respectively. Suppose we replace each element of M_2 by a "replica" of the set M_1. Then the resulting set, denoted by $M_1 \cdot M_2$, is called the *ordered product* of M_1 and M_2. More exactly, $M_1 \cdot M_2$ is the set of all pairs (a, b) where $a \in M_1$, $b \in M_2$, ordered in such a way that

1) $(a_1, b_1) < (a_2, b_2)$ if $b_1 < b_2$ (for arbitrary a_1, a_2);
2) $(a_1, b) < (a_2, b)$ if $a_1 < a_2$.

Note that the order of factors matters here, i.e., in general $M_2 \cdot M_1$ is not isomorphic to $M_1 \cdot M_2$. The ordered product of any finite number of ordered sets can be defined by writing (cf. Problem 6)

$$M_1 \cdot M_2 \cdot M_3 = (M_1 \cdot M_2) \cdot M_3,$$

$$M_1 \cdot M_2 \cdot M_3 \cdot M_4 = (M_1 \cdot M_2 \cdot M_3) \cdot M_4,$$

and so on. By the *ordered product* of the types θ_1 and θ_2, denoted by $\theta_1 \cdot \theta_2$, we mean the order type of the set $M_1 \cdot M_2$.

3.5. Well-ordered sets. Ordinal numbers. A key concept in the theory of ordered sets is given by

DEFINITION 1. *An ordered set M is said to be **well-ordered** if every nonempty subset A of M has a **smallest** (or "**first**") **element**, i.e., an element μ such that $\mu < a$ for every $a \in A$.*

Example 1. Every finite ordered set is obviously well-ordered.

Example 2. Every nonempty subset of a well-ordered set is itself well-ordered.

Example 3. The set M or rational numbers in the interval $[0, 1]$ is ordered but not well-ordered. It is true that M has a smallest element, namely the

---

[8] Here we use the same curly bracket notation as in Sec. 1.1, but the order of terms is now crucial.

number 0, but the subset of M consisting of all *positive* rational numbers has no smallest element.

DEFINITION 2. *The order type of a well-ordered set is called an* **ordinal number** *or simply an* **ordinal.**[9] *If the set is infinite, the ordinal is said to be* **transfinite.**

Example 4. The set of positive integers $1, 2, \ldots, k, \ldots$ arranged in increasing order is well-ordered, and hence its order type ω is a (transfinite) ordinal. The order type $\omega + n$ of the set

$$\{1, 2, \ldots, k, \ldots, a_1, a_2, \ldots, a_n\}$$

is also an ordinal.

Example 5. The set

$$\{\ldots, -k, \ldots, -3, -2, -1\} \tag{1}$$

is ordered but not well-ordered. It is true that any nonempty subset A of (1) has a largest element (i.e., an element ν such that $a < \nu$ for every $a \in A$), but in general A will not have a smallest element. In fact, the set (1) itself has no smallest element. Hence the order type of (1), denoted by ω^*, is not an ordinal number.

THEOREM 1. *The ordered sum of a finite number of well-ordered sets $M_1, M_2, \ldots, M_n$ is itself a well-ordered set.*

Proof. Let M be an arbitrary subset of the ordered sum $M_1 + M_2 + \cdots + M_n$, and let M_k be the first of the sets $M_1, M_2, \ldots, M_n$ (namely the set with smallest index) containing elements of M. Then $M \cap M_k$ is a subset of the well-ordered set M_k, and as such has a smallest element μ. Clearly μ is the smallest element of M itself. ∎

COROLLARY. *The ordered sum of a finite number of ordinal numbers is itself an ordinal number.*

Thus new ordinal numbers can be constructed from any given set of ordinal numbers. For example, starting from the positive integers (i.e., the finite ordinal numbers) and the ordinal number ω, we can construct the new ordinal numbers

$$\omega + n, \quad \omega + \omega, \quad \omega + \omega + n, \quad \omega + \omega + \omega,$$

and so on.

THEOREM 2. *The ordered product of two well-ordered sets M_1 and M_2 is itself a well-ordered set.*

---

[9] This is a good place to point out that the terms "cardinal number" and "power" (of a set) are synonymous.

Proof. Let M be an arbitrary subset of $M_1 \cdot M_2$, so that M is a set of ordered pairs (a, b) with $a \in M_1$, $b \in M_2$. The set of all second elements b of pairs in M is a subset of M_2, and as such has a smallest element since M_2 is well-ordered. Let b_1 denote this smallest element, and consider all pairs of the form (a, b_1) contained in M. The set of all first elements a of these pairs is a subset of M_1, and as such has a smallest element since M_1 is well-ordered. Let a_1 denote this smallest element. Then the pair (a_1, b_1) is clearly the smallest element of M. ∎

COROLLARY 1. *The ordered product of a finite number of well-ordered sets is itself a well-ordered set.*

COROLLARY 2. *The ordered product of a finite number of ordinal numbers is itself an ordinal number.*

Thus it makes sense to talk about the ordinal numbers

$$\omega \cdot n, \ \omega^2, \ \omega^2 \cdot n, \ \omega^3,$$

and so on. It is also possible to define such ordinal numbers as[10]

$$\omega^\omega, \ \omega^{\omega^\omega}, \ \ldots$$

3.6. Comparison of ordinal numbers. If n_1 and n_2 are two finite ordinal numbers, then they either coincide or else one is larger than the other. As we now show, the same is true of transfinite ordinal numbers. We begin by observing that every element a of a well-ordered set M determines an (*initial*) *section P*, the set of all $x \in M$ such that $x < a$, and a *remainder Q*, the set of all $x \in M$ such that $x \geqslant a$. Given any two ordinal numbers α and β, let M and N be well-ordered sets of order type α and β, respectively. Then we say that

1) $\alpha = \beta$ if M and N are isomorphic;
2) $\alpha < \beta$ if M is isomorphic to some section of N;
3) $\alpha > \beta$ if N is isomorphic to some section of M

(note that this definition makes sense for finite α and β).

LEMMA. *Let f be an isomorphism of a well-ordered set A onto some subset $B \subset A$. Then $f(a) \geqslant a$ for all $a \in A$.*

Proof. If there are elements $a \in A$ such that $f(a) < a$, then there is a least such element since A is well-ordered. Let a_0 be this element, and let $b_0 = f(a_0)$. Then $b_0 < a_0$, and hence $f(b_0) < f(a_0) = b_0$ since f is an isomorphism. But then a_0 is not the smallest element such that $f(a) < a$. Contradiction! ∎

---

[10] See e.g., A. A. Fraenkel, *Abstract Set Theory*, third edition, North-Holland Publishing Co., Amsterdam (1966), pp. 202–208.

It follows from the lemma that a well-ordered set A cannot be isomorphic to any of its sections, since if A were isomorphic to the section determined by a, then clearly $f(a) < a$. In other words, the two relations

$$\alpha = \beta, \qquad \alpha < \beta$$

are incompatible, and so are

$$\alpha = \beta, \qquad \alpha > \beta.$$

Moreover, the two relations

$$\alpha < \beta, \qquad \alpha > \beta$$

are incompatible, since otherwise we could use the transitivity to deduce $\alpha < \alpha$, which is impossible by the lemma. Therefore, if one of the three relations

$$\alpha < \beta, \qquad \alpha = \beta, \qquad \alpha > \beta \tag{2}$$

holds, the other two are automatically excluded. We must still show that one of the relations (2) always holds, thereby proving that any two ordinal numbers are comparable.

THEOREM 3. *Two given ordinal numbers α and β satisfy one and only one of the relations*

$$\alpha < \beta, \qquad \alpha = \beta, \qquad \alpha > \beta.$$

Proof. Let $W(\alpha)$ be the set of all ordinals $<\alpha$. Any two numbers γ and γ' in $W(\alpha)$ are comparable[11] and the corresponding ordering of $W(\alpha)$ makes it a well-ordered set of type α. In fact, if a set

$$A = \{\ldots, a, \ldots, b, \ldots\}$$

is of type α, then by definition, the ordinals less than α are the types of well-ordered sets isomorphic to sections of A. Hence the ordinals themselves are in one-to-one correspondence with the elements of A. In other words, the elements of a set of type α can be numbered by using the ordinals less than α:

$$A = \{a_1, a_2, \ldots, a_n, \ldots\}.$$

Now let α and β be any two ordinals. Then $W(\alpha)$ and $W(\beta)$ are well-ordered sets of types α and β, respectively. Moreover, let $C = A \cap B$ be the intersection of the sets A and B, i.e., the set of all ordinals less than both α and β. Then C is well-ordered, of type γ, say. We now show that $\gamma \leqslant \alpha$. If $C = A$, then obviously $\gamma = \alpha$. On the other hand, if $C \neq A$, then C is a section of A and hence $\gamma < \alpha$. In fact, let $\xi \in C$, $\eta \in A - C$. Then ξ and η are comparable, i.e., either $\xi < \eta$ or $\xi > \eta$. But $\eta < \xi < \alpha$

---

[11] Recall the meaning of $\gamma < \alpha$, $\gamma' < \alpha$, and use the fact that a section of a section of a well-ordered set is itself a section of a well-ordered set.

is impossible, since then $\eta \in C$. Therefore $\xi < \eta$ and hence C is a section of A, which implies $\gamma < \alpha$. Moreover, γ is the first element of the set $A - C$. Thus $\gamma \leqslant \alpha$, as asserted, and similarly $\gamma \leqslant \beta$. The case $\gamma < \alpha$, $\gamma < \beta$ is impossible, since then $\gamma \in A - C$, $\gamma \in B - C$. But then $\gamma \notin C$ on the one hand and $\gamma \in A \cap B = C$ on the other hand. It follows that there are only three possibilities

$$\begin{aligned}
\gamma &= \alpha, & \gamma &= \beta, & \alpha &= \beta, \\
\gamma &= \alpha, & \gamma &< \beta, & \alpha &< \beta, \\
\gamma &< \alpha, & \gamma &= \beta, & \alpha &> \beta,
\end{aligned}$$

i.e., α and β are comparable. ∎

THEOREM 4. *Let A and B be well-ordered sets. Then either A is equivalent to B or one of the sets is of greater power than the other, i.e., the powers of A and B are comparable.*

Proof. There is a definite power corresponding to each ordinal. But we have just seen that ordinals are comparable, and so are the corresponding powers (recall the definition of inequality of powers given in Sec. 2.5). ∎

3.7. The well-ordering theorem, the axiom of choice and equivalent assertions. Theorem 4 shows that the powers of two well-ordered sets are always comparable. In 1904, Zermelo succeeded in proving the

WELL-ORDERING THEOREM. *Every set can be well-ordered.*

It follows from the well-ordering theorem and Theorem 5 that the powers of two *arbitrary* sets are always comparable, a fact already used in Sec. 2.5. Zermelo's proof, which will not be given here,[12] rests on the following basic

AXIOM OF CHOICE. *Given any set M, there is a "choice function" f such that $f(A)$ is an element of A for every nonempty subset $A \subset M$.*

We will assume the validity of the axiom of choice without further ado. In fact, without the axiom of choice we would be severely hampered in making set-theoretic constructions. However, it should be noted that from the standpoint of the foundations of set theory, there are still deep and controversial problems associated with the use of the axiom of choice.

There are a number of assertions equivalent to the axiom of choice, i.e., assertions each of which both implies and is implied by the axiom of choice. One of these is the well-ordering theorem, which obviously implies the axiom of choice. In fact, if an arbitrary set M can be well-ordered, then, by merely choosing the "first" element in each subset $A \subset M$, we get the function $f(A)$

---

[12] A. A. Fraenkel, *op. cit.*, pp. 222–227.

figuring in the statement of the axiom of choice. On the other hand, the axiom of choice implies the well-ordering theorem, as already noted without proof.

To state further assertions equivalent to the axiom of choice, we need some more terminology:

DEFINITION 3. *Let M be a partially ordered set, and let A be any subset of M such that a and b are comparable for every a, b ∈ A. Then A is called a* **chain** *(in M). A chain C is said to be* **maximal** *if there is no other chain C' in M containing C as a proper subset.*

DEFINITION 4. *An element a of a partially ordered set M is called an* **upper bound** *of a subset M' ⊂ M if a' ≤ a for every a' ∈ M'.*

We now have the vocabulary needed to state two other assertions equivalent to the axiom of choice:

HAUSDORFF'S MAXIMAL PRINCIPLE. *Every chain in a partially ordered set M is contained in a maximal chain in M.*

ZORN'S LEMMA. *If every chain in a partially ordered set M has an upper bound, then M contains a maximal element.*

For the proof of the equivalence of the axiom of choice, the well-ordering theorem, Hausdorff's maximal principle and Zorn's lemma, we refer the reader elsewhere.[13] Of these various equivalent assertions, Zorn's lemma is perhaps the most useful.

3.8. Transfinite induction. Mathematical propositions are very often proved by using the following familiar

THEOREM 4 (*Mathematical induction*). *Given a proposition $P(n)$ formulated for every positive integer n, suppose that*

1) $P(1)$ *is true;*
2) *The validity of $P(k)$ for all $k \leq n$ implies the validity of $P(n + 1)$.*

Then $P(n)$ is true for all $n = 1, 2, \ldots$

Proof. Suppose $P(n)$ fails to be true for all $n = 1, 2, \ldots$, and let n_1 be the smallest integer for which $P(n)$ is false (the existence of n_1 follows from the well-ordering of the positive integers). Clearly $n_1 > 1$, so that $n_1 - 1$ is a positive integer. Therefore $P(n)$ is valid for all $k \leq n_1 - 1$ but not for n_1. Contradiction! ∎

Replacing the set of all positive integers by an arbitrary well-ordered set,

---

[13] See e.g., G. Birkhoff, *Lattice Theory*, third edition, American Mathematical Society, Providence, R.I. (1967), pp. 205–206.

we get

THEOREM 4′. (*Transfinite induction*). *Given a well ordered set* A,[14] *let* $P(a)$ *be a proposition formulated for every element* $a \in A$. *Suppose that*

1) $P(a)$ *is true for the smallest element of* A;
2) *The validity of* $P(a)$ *for all* $a < a^*$ *implies the validity of* $P(a^*)$.

Then $P(a)$ *is true for all* $a \in A$.

Proof. Suppose $P(a)$ fails to be true tor all $a \in A$. Then $P(a)$ is false for all a in some nonempty subset $A^* \subset A$. By the well-ordering, A^* has a smallest element a^*. Therefore $P(a)$ is valid for all $a < a^*$ but not for a^*. Contradiction! ∎

Remark. Since any set can be well-ordered, by the well-ordering theorem, transfinite induction can in principle be applied to any set M whatsoever. In practice, however, Zorn's lemma is a more useful tool, requiring only that M be partially ordered.

3.9. Historical remarks. Set theory as a branch of mathematics in its own right stems from the pioneer work of Georg Cantor (1845–1918). Originally met with disbelief, Cantor's ideas subsequently became widespread. By now, the set-theoretic point of view has become standard in the most diverse fields of mathematics. Basic concepts, like groups, rings, fields, linear spaces, etc. are habitually defined as sets of elements of an arbitrary kind obeying appropriate axioms.

Further development of set theory led to a number of logical difficulties, which naturally gave rise to attempts to replace "naive" set theory by a more rigorous, *axiomatic* set theory. It turns out that certain set-theoretic questions, which would at first seem to have "yes" or "no" answers, are in fact of a different kind. Thus it was shown by Gödel in 1940 that a negative answer to the question "Is there an uncountable set of power less than that of the continuum" is consistent with set theory (axiomatized in a way we will not discuss here), but it was recently shown by Cohen that an affirmative answer to the question is also consistent in the same sense!

Problem 1. Exhibit both a partial ordering and a simple ordering of the set of all complex numbers.

Problem 2. What is the minimal element of the set of all subsets of a given set X, partially ordered by set inclusion. What is the maximal element?

Problem 3. A partially ordered set M is said to be a *directed set* if, given any two elements $a, b \in M$, there is an element $c \in M$ such that $a \leqslant c, b \leqslant c$. Are the partially ordered sets in Examples 1-4, Sec. 3.1 all directed sets?

---

[14] For example, the set of all transfinite ordinals less than a given ordinal.

Problem 4. By the *greatest lower bound* of two elements a and b of a partially ordered set M, we mean an element $c \in M$ such that $c \leqslant a$, $c \leqslant b$ and there is no element $d \in M$ such that $c < d \leqslant a$, $d \leqslant b$. Similarly, by the *least upper bound* of a and b, we mean an element $c \in M$ such that $a \leqslant c$, $b \leqslant c$ and there is no element $d \in M$ such that $a \leqslant d < c$, $b \leqslant d$. By a *lattice* is meant a partially ordered set any two element of which have both a greatest lower bound and a least upper bound. Prove that the set of all subsets of a given set X, partially ordered by set inclusion, is a lattice. What is the set-theoretic meaning of the greatest lower bound and least upper bound of two elements of this set?

Problem 5. Prove that an order-preserving mapping of one ordered set onto another is automatically an isomorphism.

Problem 6. Prove that ordered sums and products of ordered sets are associative, i.e., prove that if M_1, M_2 and M_3 are ordered sets, then

$$(M_1 + M_2) + M_3 = M_1 + (M_2 + M_3), \quad (M_1 \cdot M_2) \cdot M_3 = M_1 \cdot (M_2 \cdot M_3),$$

where the operations $+$ and $\cdot$ are the same as in Sec. 3.4.

Comment. This allows us to drop parentheses in writing ordered sums and products.

Problem 7. Construct well-ordered sets with ordinals

$$\omega + n, \quad \omega + \omega, \quad \omega + \omega + n, \quad \omega + \omega + \omega, \ldots$$

Show that the sets are all countable.

Problem 8. Construct well-ordered sets with ordinals

$$\omega \cdot n, \quad \omega^2, \quad \omega^2 \cdot n, \quad \omega^3, \ldots$$

Show that the sets are all countable.

Problem 9. Show that

$$\omega + \omega = \omega \cdot 2, \quad \omega + \omega + \omega = \omega \cdot 3, \ldots$$

Problem 10. Prove that the set $W(\alpha)$ of all ordinals less than a given ordinal α is well-ordered.

Problem 11. Prove that any nonempty set of ordinals is well-ordered.

Problem 12. Prove that the set M of all ordinals corresponding to a countable set is itself uncountable.

Problem 13. Let $\aleph_1$ be the power of the set M in the preceding problem. Prove that there is no power m such that $\aleph_0 < m < \aleph_1$.

4. Systems of Sets[15]

4.1. Rings of sets. By a *system of sets* we mean any set whose elements are themselves sets. Unless the contrary is explicitly stated, the elements of a given system of sets will be assumed to be certain subsets of some fixed set X. Systems of sets will usually be denoted by capital script letters like $\mathscr{R}$, $\mathscr{S}$, etc. Our chief interest will be systems of sets which have certain closure properties under the operations introduced in Sec. 1.1.

DEFINITION 1. *A nonempty system of sets $\mathscr{R}$ is called a* **ring** *(of sets) if $A \bigtriangleup B \in \mathscr{R}$ and $A \cap B \in \mathscr{R}$ whenever $A \in \mathscr{R}$, $B \in \mathscr{R}$.*

Since

$$A \cup B = (A \bigtriangleup B) \bigtriangleup (A \cap B),$$

$$A - B = A \bigtriangleup (A \cap B),$$

we also have $A \cup B \in \mathscr{R}$ and $A - B \in \mathscr{R}$ whenever $A \in \mathscr{R}$, $B \in \mathscr{R}$. Thus a ring of sets is a system of sets closed under the operations of taking unions, intersections, differences, and symmetric differences. Clearly, a ring of sets is also closed under the operations of taking finite unions and intersections:

$$\bigcup_{k=1}^{n} A_k, \qquad \bigcap_{k=1}^{n} A_k.$$

A ring of sets must contain the empty set $\varnothing$, since $A - A = \varnothing$.

A set E is called the *unit* of a system of sets $\mathscr{S}$ if $E \in \mathscr{S}$ and

$$A \cap E = A$$

for every $A \in \mathscr{S}$. Clearly E is unique (why?). Thus the unit of $\mathscr{S}$ is just the maximal set of $\mathscr{S}$, i.e., the set containing all other sets of $\mathscr{S}$. A ring of sets with a unit is called an *algebra (of sets)*.

Example 1. Given a set A, the system $\mathscr{M}(A)$ of all subsets of A is an algebra of sets, with unit $E = A$.

Example 2. The system $\{\varnothing, A\}$ consisting of the empty set $\varnothing$ and any nonempty set A is an algebra of sets, with $E = A$.

Example 3. The system of all *finite* subsets of a given set A is a ring of sets. This ring is an algebra if and only if A itself is finite.

Example 4. The system of all bounded subsets of the real line is a ring of sets, which does not contain a unit.

---

[15] The material in this section need not be read now, since it will not be needed until Chapter 7.

THEOREM 1. *The intersection*

$$\mathscr{R} = \bigcap_{\alpha} \mathscr{R}_{\alpha}$$

of any set of rings is itself a ring.

Proof. An immediate consequence of Definition 1.   ∎

THEOREM 2. *Given any nonempty system of sets $\mathscr{S}$, there is a unique ring $\mathscr{P}$ containing $\mathscr{S}$ and contained in every ring containing $\mathscr{S}$.*

Proof. If $\mathscr{P}$ exists, then clearly $\mathscr{P}$ is unique (why?). To prove the existence of $\mathscr{P}$, consider the union

$$X = \bigcup_{A \in \mathscr{S}} A$$

of all sets A belonging to $\mathscr{S}$ and the ring $\mathscr{M}(X)$ of all subsets of X. Let Σ be the set of all rings of sets contained in $\mathscr{M}(X)$ and containing $\mathscr{S}$. Then the intersection

$$\mathscr{P} = \bigcap_{\mathscr{R} \in \Sigma} \mathscr{R}$$

of all these rings clearly has the desired properties. In fact, $\mathscr{P}$ obviously contains $\mathscr{S}$. Moreover, if $\mathscr{R}^*$ is any ring containing $\mathscr{S}$, then the intersection $\mathscr{R} = \mathscr{R}^* \cap \mathscr{M}(X)$ is a ring in Σ and hence $\mathscr{P} \subset \mathscr{R} \subset \mathscr{R}^*$, as required. The ring $\mathscr{P}$ is called the *minimal ring* generated by the system $\mathscr{S}$, and will henceforth be denoted by $\mathscr{R}(\mathscr{S})$.   ∎

Remark. The set $\mathscr{M}(X)$ containing $\mathscr{R}(\mathscr{S})$ has been introduced to avoid talking about the "set of all rings containing $\mathscr{S}$." Such concepts as "the set of all sets," "the set of all rings," etc. are inherently contradictory and should be avoided (recall Problem 10, p. 20).

4.2. Semirings of sets. The following notion is more general than that of a ring of sets and plays an important role in a number of problems (particularly in measure theory):

DEFINITION 2. *A system of sets $\mathscr{S}$ is called a **semiring** (**of sets**) if*

1) $\mathscr{S}$ *contains the empty set $\varnothing$;*
2) $A \cap B \in \mathscr{S}$ *whenever $A \in \mathscr{S}$, $B \in \mathscr{S}$;*
3) *If $\mathscr{S}$ contains the sets A and $A_1 \subset A$, then A can be represented as a finite union*

$$A = \bigcup_{k=1}^{n} A_k \tag{1}$$

of pairwise disjoint sets of $\mathscr{S}$, with the given set A_1 as its first term.

Remark. The representation (1) is called a *finite expansion* of A, with respect to the sets $A_1, A_2, \ldots, A_n$.

Example 1. Every ring of sets $\mathscr{R}$ is a semiring, since if $\mathscr{R}$ contains A and $A_1 \subset A$, then $A = A_1 \cup A_2$ where $A_2 = A - A_1 \in \mathscr{R}$.

Example 2. The set $\mathscr{S}$ of all open intervals (a, b), closed intervals $[a, b]$ and half-open intervals $[a, b)$, $(a, b]$, including the "empty interval" $(a, a) = \varnothing$ and the single-element sets $[a, a] = \{a\}$, is a semiring but not a ring.

LEMMA 1. *Suppose the sets A, $A_1, \ldots, A_n$, where $A_1, \ldots, A_n$ are pairwise disjoint subsets of A, all belong to a semiring $\mathscr{S}$. Then there is a finite expansion*

$$A = \bigcup_{k=1}^{s} A_k \qquad (s \geqslant n)$$

with $A_1, \ldots, A_n$ as its first n terms, where $A_k \in \mathscr{S}$, $A_k \cap A_l = \varnothing$ for all $k, l = 1, \ldots, n$.

Proof. The lemma holds for $n = 1$, by the definition of a semiring. Suppose the lemma holds for $n = m$, and consider $m + 1$ sets $A_1, \ldots, A_m, A_{m+1}$ satisfying the conditions of the lemma. By hypothesis,

$$A = A_1 \cup \cdots \cup A_m \cup B_1 \cup \cdots \cup B_p,$$

where the sets $A_1, \ldots, A_m, B_1, \ldots, B_p$ are pairwise disjoint subsets of A, all belonging to $\mathscr{S}$. Let

$$B_{q1} = A_{m+1} \cap B_q.$$

By the definition of a semiring,

$$B_q = B_{q1} \cup \cdots \cup B_{qr_q},$$

where the sets B_{qj} $(j = 1, \ldots, r_q)$ are pairwise disjoint subsets of B_q, all belonging to $\mathscr{S}$. But then it is easy to see that

$$A = A_1 \cup \cdots \cup A_m \cup A_{m+1} \cup \bigcup_{q=1}^{p} \left(\bigcup_{j=2}^{r_q} B_{qj} \right),$$

i.e., the lemma is true for $n = m + 1$. The proof now follows by mathematical induction. ∎

LEMMA 2. *Given any finite system of sets $A_1, \ldots, A_n$ belonging to a semiring $\mathscr{S}$, there is a finite system of pairwise disjoint sets $B_1, \ldots, B_t$ belonging to $\mathscr{S}$ such that every A_k has a finite expansion*

$$A_k = \bigcup_{s \in M_k} B_s \qquad (k = 1, \ldots, n)$$

with respect to certain of the sets B_s.[16]

---

[16] Here M_k denotes some subset of the set $\{1, 2, \ldots, t\}$, depending on the choice of k.

Proof. The lemma is trivial for $n = 1$, since we need only set $t = 1$, $B_1 = A_1$ Suppose the lemma is true for $n = m$, and consider a system of sets $A_1, \ldots, A_m, A_{m+1}$ in $\mathscr{S}$. Let $B_1, \ldots, B_t$ be sets of $\mathscr{S}$ satisfying the conditions of the lemma with respect to $A_1, \ldots, A_m$, and let

$$B_{s1} = A_{m+1} \cap B_s.$$

Then, by Lemma 1, there is an expansion

$$A_{m+1} = \left(\bigcup_{s=1}^{t} B_{s1} \right) \cup \left(\bigcup_{p=1}^{q} B_p' \right) \qquad (B_p' \in \mathscr{S}),$$

while, by the very definition of a semiring, there is an expansion

$$B_s = B_{s1} \cup B_{s2} \cup \cdots \cup B_{sr_s} \qquad (B_{sj} \in \mathscr{S}).$$

It is easy to see that

$$A_k = \bigcup_{s \in M_k} \left(\bigcup_{j=1}^{r_s} B_{sj} \right) \qquad (k = 1, \ldots, m)$$

for some suitable M_k. Moreover, the sets B_{sj}, B_p' are pairwise disjoint. Hence the sets B_{sj}, B_p' satisfy the conditions of the lemma with respect to $A_1, \ldots, A_m, A_{m+1}$. The proof now follows by mathematical induction. ∎

4.3. The ring generated by a semiring. According to Theorem 1, there is a unique minimal ring $\mathscr{R}(\mathscr{S})$ generated by a given system of sets $\mathscr{S}$. The actual construction of $\mathscr{R}(\mathscr{S})$ is quite complicated for arbitrary $\mathscr{S}$. However, the construction is completely straightforward if $\mathscr{S}$ is a semiring, as shown by

THEOREM 3. *If $\mathscr{S}$ is a semiring, then $\mathscr{R}(\mathscr{S})$ coincides with the system $\mathscr{Z}$ of all sets A which have finite expansions*

$$A = \bigcup_{k=1}^{n} A_k$$

with respect to the sets $A_k \in \mathscr{S}$.

Proof. First we prove that $\mathscr{Z}$ is a ring. Let A and B be any two sets in $\mathscr{Z}$. Then there are expansions

$$A = \bigcup_{i=1}^{m} A_i \qquad (A_i \in \mathscr{S}),$$

$$B = \bigcup_{j=1}^{n} B_j \qquad (B_j \in \mathscr{S}).$$

Since $\mathscr{S}$ is a semiring, the sets

$$C_{ij} = A_i \cap B_j$$

also belong to $\mathscr{S}$. By Lemma 1, there are expansions

$$A_i = \left(\bigcup_{j=1}^{n} C_{ij}\right) \cup \left(\bigcup_{k=1}^{r_i} D_{ik}\right) \quad (D_{ik} \in \mathscr{S}),$$

$$B_j = \left(\bigcup_{i=1}^{m} C_{ij}\right) \cup \left(\bigcup_{l=1}^{s_j} E_{jl}\right) \quad (E_{jl} \in \mathscr{S}). \tag{2}$$

It follows from (2) that $A \cap B$ and $A \triangle B$ have the expansions

$$A \cap B = \bigcup_{i,j} C_{ij},$$

$$A \triangle B = \left(\bigcup_{i,k} D_{ik}\right) \cup \left(\bigcup_{j,l} E_{jl}\right),$$

and hence belong to $\mathscr{Z}$. Therefore $\mathscr{Z}$ is a ring. The fact that $\mathscr{Z}$ is the *minimal* ring generated by $\mathscr{S}$ is obvious. ∎

4.4. Borel algebras. There are many problems (particularly in measure theory) involving unions and intersections not only of a finite number of sets, but also of a *countable* number of sets. This motivates the following concepts:

DEFINITION 3. *A ring of sets is called a σ-ring if it contains the union*

$$S = \bigcup_{n=1}^{\infty} A_n$$

whenever it contains the sets $A_1, A_2, \ldots, A_n, \ldots$. A σ-ring with a unit E is called a σ-algebra.

DEFINITION 4. *A ring of sets is called a δ-ring if it contains the intersection*

$$D = \bigcap_{n=1}^{\infty} A_n$$

whenever it contains the sets $A_1, A_2, \ldots, A_n, \ldots$. A δ-ring with a unit E is called a δ-algebra.

THEOREM 4. *Every σ-algebra is a δ-algebra and conversely.*

Proof. An immediate consequence of the "dual" formulas

$$\bigcup_{n} A_n = E - \bigcap_{n} (E - A_n),$$

$$\bigcap_{n} A_n = E - \bigcup_{n} (E - A_n). \quad ∎$$

The term *Borel algebra* (or briefly, *B-algebra*) is often used to denote a σ-algebra (equivalently, a δ-algebra). The simplest example of a B-algebra is the set of all subsets of a given set A.

Given any system of sets $\mathscr{S}$, there always exists at least one B-algebra containing $\mathscr{S}$. In fact, let

$$X = \bigcup_{A \in \mathscr{S}} A.$$

Then the system $\mathscr{B}$ of all subsets of X is clearly a B-algebra containing $\mathscr{S}$.

If $\mathscr{B}$ is any B-algebra containing $\mathscr{S}$ and if E is its unit, then every $A \in \mathscr{S}$ is contained in E and hence

$$X = \bigcup_{A \in \mathscr{S}} A \subset E.$$

A B-algebra $\mathscr{B}$ is called *irreducible* (with respect to the system $\mathscr{S}$) if $X = E$, i.e., an irreducible B-algebra is a B-algebra containing no points that do not belong to one of the sets $A \in \mathscr{S}$. In every case, it will be enough to consider only irreducible B-algebras.

Theorem 2 has the following analogue for irreducible B-algebras:

THEOREM 5. *Given any nonempty system of sets $\mathscr{S}$, there is a unique irreducible[17] B-algebra $\mathscr{B}(\mathscr{S})$ containing $\mathscr{S}$ and contained in every B-algebra containing $\mathscr{S}$.*

Proof. The proof is virtually identical with that of Theorem 2. The B-algebra $\mathscr{B}(\mathscr{S})$ is called the *minimal B-algebra* generated by the system $\mathscr{S}$ or the *Borel closure* of $\mathscr{S}$. ∎

Remark. An important role is played in analysis by *Borel sets* or *B-sets*. These are the subsets of the real line belonging to the minimal B-algebra generated by the set of all closed intervals $[a, b]$.

Problem 1. Let X be an uncountable set, and let $\mathscr{R}$ be the ring consisting of all finite subsets of X and their complements. Is $\mathscr{R}$ a σ-ring?

Problem 2. Are open intervals Borel sets?

Problem 3. Let $y = f(x)$ be a function defined on a set M and taking values in a set N. Let $\mathscr{M}$ be a system of subsets of M, and let $f(\mathscr{M})$ denote the system of all images $f(A)$ of sets $A \in \mathscr{M}$. Moreover, let $\mathscr{N}$ be a system of subsets of N, and let $f^{-1}(\mathscr{N})$ denote the system of all preimages $f^{-1}(B)$ of sets $B \in \mathscr{N}$. Prove that

a) If $\mathscr{N}$ is a ring, so is $f^{-1}(\mathscr{N})$;
b) If $\mathscr{N}$ is an algebra, so is $f^{-1}(\mathscr{N})$;
c) If $\mathscr{N}$ is a B-algebra, so is $f^{-1}(\mathscr{N})$;
d) $\mathscr{R}(f^{-1}(\mathscr{N})) = f^{-1}(\mathscr{R}(\mathscr{N}))$;
e) $\mathscr{B}(f^{-1}(\mathscr{N})) = f^{-1}(\mathscr{B}(\mathscr{N}))$.

Which of these assertions remain true if $\mathscr{N}$ is replaced by $\mathscr{M}$ and f^{-1} by f?

[17] More exactly, irreducible with respect to $\mathscr{S}$.

2

METRIC SPACES

---

5. Basic Concepts

5.1. Definitions and examples. One of the most important operations in mathematical analysis is the taking of limits. Here what matters is not so much the algebraic nature of the real numbers,[1] but rather the fact that distance from one point to another on the real line (or in two or three-dimensional space) is well-defined and has certain properties. Roughly speaking, a *metric space* is a set equipped with a distance (or "metric") which has these same properties. More exactly, we have

DEFINITION 1. *By a **metric space** is meant a pair (X, ρ) consisting of a set X and a **distance** ρ, i.e., a single-valued, nonnegative, real function $\rho(x, y)$ defined for all $x, y \in X$ which has the following three properties*:

1) $\rho(x, y) = 0$ *if and only if* $x = y$;
2) *Symmetry*: $\rho(x, y) = \rho(y, x)$;
3) *Triangle inequality*: $\rho(x, z) \leqslant \rho(x, y) + \rho(y, z)$.

We will often refer to the set X as a "space" and its elements $x, y, \ldots$ as "points." Metric spaces are usually denoted by a single letter, like

$$R = (X, \rho),$$

or even by the same letter X as used for the underlying space, in cases where there is no possibility of confusion.

---

[1] I.e., the fact that the real numbers form a *field*.

37

Example 1. Setting

$$\rho(x, y) = \begin{cases} 0 & \text{if } x = y, \\ 1 & \text{if } x \neq y, \end{cases}$$

where x and y are elements of an arbitrary set X, we obviously get a metric space, which might be called a "discrete space" or a "space of isolated points."

Example 2. The set of all real numbers with distance

$$\rho(x, y) = |x - y|$$

is a metric space, which we denote by R^1.

Example 3. The set of all ordered n-tuples

$$x = (x_1, x_2, \ldots, x_n)$$

of real numbers $x_1, x_2, \ldots, x_n$, with distance

$$\rho(x, y) = \sqrt{\sum_{k=1}^{n} (x_k - y_k)^2}, \tag{1}$$

is a metric space denoted by R^n and called *n-dimensional Euclidean space* (or simply *Euclidean n-space*). The distance (1) obviously has properties 1) and 2) in Definition 1. Moreover, it is easy to see that (1) satisfies the triangle inequality. In fact, let

$$x = (x_1, x_2, \ldots, x_n), \quad y = (y_1, y_2, \ldots, y_n), \quad z = (z_1, z_2, \ldots, z_n)$$

be three points in R^n, and let

$$a_k = x_k - y_k, \quad b_k = y_k - z_k \qquad (k = 1, \ldots, n).$$

Then the triangle inequality takes the form

$$\sqrt{\sum_{k=1}^{n} (x_k - z_k)^2} \leqslant \sqrt{\sum_{k=1}^{n} (x_k - y_k)^2} + \sqrt{\sum_{k=1}^{n} (y_k - z_k)^2}, \tag{2}$$

or equivalently

$$\sqrt{\sum_{k=1}^{n} (a_k + b_k)^2} \leqslant \sqrt{\sum_{k=1}^{n} a_k^2} + \sqrt{\sum_{k=1}^{n} b_k^2}. \tag{2'}$$

It follows from the *Cauchy-Schwarz inequality*

$$\left(\sum_{k=1}^{n} a_k b_k \right)^2 \leqslant \sum_{k=1}^{n} a_k^2 \sum_{k=1}^{n} b_k^2 \tag{3}$$

(see Problem 2) that

$$\sum_{k=1}^{n} (a_k + b_k)^2 = \sum_{k=1}^{n} a_k^2 + 2 \sum_{k=1}^{n} a_k b_k + \sum_{k=1}^{n} b_k^2$$

$$\leqslant \sum_{k=1}^{n} a_k^2 + 2 \sqrt{\sum_{k=1}^{n} a_k^2 \sum_{k=1}^{n} b_k^2} + \sum_{k=1}^{n} b_k^2 = \left(\sqrt{\sum_{k=1}^{n} a_k^2} + \sqrt{\sum_{k=1}^{n} b_k^2} \right)^2$$

Taking square roots, we get (2') and hence (2).

Example 4. Take the same set of ordered n-tuples $x = (x_1, \ldots, x_n)$ as in the preceding example, but this time define the distance by the function

$$\rho_1(x, y) = \sum_{k=1}^{n} |x_k - y_k|. \tag{4}$$

It is clear that (4) has all three properties of a distance figuring in Definition 1. The corresponding metric space will be denoted by R_1^n.

Example 5. Take the same set as in Examples 3 and 4, but this time define distance between two points $x = (x_1, \ldots, x_n)$ and $y = (y_1, \ldots, y_n)$ by the formula

$$\rho_0(x, y) = \max_{1 \leqslant k \leqslant n} |x_k - y_k|. \tag{5}$$

Then we again get a metric space (verify all three properties of the distance). This space, denoted by R_0^n, is often as useful as the Euclidean space R^n.

Remark. The last three examples show that it is sometimes important to use a different notation for a metric space than for the underlying set of points in the space, since the latter can be "metrized" in a variety of different ways.

Example 6. The set $C_{[a,b]}$ of all continuous functions defined on the closed interval $[a, b]$, with distance

$$\rho(f, g) = \max_{a \leqslant t \leqslant b} |f(t) - g(t)| \tag{6}$$

is a metric space of great importance in analysis (again verify the three properties of distance). This metric space and the underlying set of "points" will both be denoted by the symbol $C_{[a,b]}$. Instead of $C_{[0,1]}$, we will often write just C. A space like $C_{[a,b]}$ is often called a "function space," to emphasize that its elements are functions.

Example 7. Let l_2 be the set of all infinite sequences[2]

$$x = (x_1, x_2, \ldots, x_k, \ldots)$$

of real numbers $x_1, x_2, \ldots, x_k, \ldots$ satisfying the convergence condition

$$\sum_{k=1}^{\infty} x_k^2 < \infty,$$

---

[2] The infinite sequence with general term x_k can be written as $\{x_k\}$ or simply as $x_1, x_2, \ldots, x_k, \ldots$ (this notation is familiar from calculus). It can also be written in "point notation" as $x = (x_1, x_2, \ldots, x_k, \ldots)$, i.e., as an "ordered ∞-tuple" generalizing the notion of an ordered n-tuple. (In writing $\{x_k\}$ we have another use of curly brackets, but the context will always prevent any confusion between the *sequence* $\{x_k\}$ and the set whose only element is x_k.)

where distance between points is defined by

$$\rho(x, y) = \sqrt{\sum_{k=1}^{\infty} (x_k - y_k)^2}. \tag{7}$$

Clearly (7) makes sense for all $x, y \in l_2$, since it follows from the elementary inequality

$$(x_k \pm y_k)^2 \leqslant 2(x_k^2 + y_k^2)$$

that convergence of the two series

$$\sum_{k=1}^{\infty} x_k^2, \quad \sum_{k=1}^{\infty} y_k^2$$

implies that of the series

$$\sum_{k=1}^{\infty} (x_k - y_k)^2.$$

At the same time, we find that if the points $(x_1, x_2, \ldots, x_k, \ldots)$ and $(y_1, y_2, \ldots, y_k, \ldots)$ both belong to l_2, then so does the point

$$(x_1 + y_1, x_2 + y_2, \ldots, x_k + y_k, \ldots).$$

The function (7) obviously has the first two defining properties of a distance. To verify the triangle inequality, which takes the form

$$\sqrt{\sum_{k=1}^{\infty} (x_k - z_k)^2} \leqslant \sqrt{\sum_{k=1}^{\infty} (x_k - y_k)^2} + \sqrt{\sum_{k=1}^{\infty} (y_k - z_k)^2} \tag{8}$$

for the metric (7), we first note that all three series converge, for the reason just given. Moreover, the inequality

$$\sqrt{\sum_{k=1}^{n} (x_k - z_k)^2} \leqslant \sqrt{\sum_{k=1}^{n} (x_k - y_k)^2} + \sqrt{\sum_{k=1}^{n} (y_k - z_k)^2} \tag{9}$$

holds for all n, as shown in Example 3. Taking the limit as $n \to \infty$ in (9), we get (8), thereby verifying the triangle inequality in l_2. Therefore l_2 is a metric space.

Example 8. As in Example 6, consider the set of all functions continuous on the interval $[a, b]$, but this time define distance by the formula

$$\rho(x, y) = \left(\int_a^b [x(t) - y(t)]^2 \, dt \right)^{1/2}, \tag{10}$$

instead of (6). The resulting metric space will be denoted by $C_{[a,b]}^2$. The first two properties of the metric are obvious, and the fact that (10) satisfies the triangle inequality is an immediate consequence of *Schwarz's inequality*

$$\left(\int_a^b x(t) y(t) \, dt \right)^2 \leqslant \int_a^b x^2(t) \, dt \int_a^b y^2(t) \, dt \tag{11}$$

(see Problem 3), by the continuous analogue of the argument given in Example 3.

Example 9. Next consider the set of all *bounded* infinite sequences of real numbers $x = (x_1, x_2, \ldots, x_k, \ldots)$, and let[3]

$$\rho(x, y) = \sup_{k} |x_k - y_k|. \tag{12}$$

This gives a metric space which we denote by m. The fact that (12) has the three properties of a metric is almost obvious.

Example 10. As in Example 3, consider the set of all ordered n-tuples $x = (x_1, \ldots, x_n)$ of real numbers, but this time define the distance by the more general formula

$$\rho_p(x, y) = \left(\sum_{k=1}^{n} |x_k - y_k|^p \right)^{1/p}, \tag{13}$$

where p is a fixed number $\geqslant 1$ (Examples 3 and 4 correspond to the cases $p = 2$ and $p = 1$, respectively). This gives a metric space, which we denote by R_p^n. It is obvious that $\rho_p(x, y) = 0$ if and only if $x = y$ and that $\rho_p(x, y) = \rho_p(y, x)$, but verification of the triangle inequality for the metric (13) requires a little work. Let

$$x = (x_1, \ldots, x_n), \quad y = (y_1, \ldots, y_n), \quad z = (z_1, \ldots, z_n)$$

be three points in R_p^n, and let

$$a_k = x_k - y_k, \quad b_k = y_k - z_k \qquad (k = 1, \ldots, n),$$

just as in Example 3. Then the triangle inequality

$$\rho_p(x, z) \leqslant \rho_p(x, z) + \rho_p(y, z)$$

takes the form of *Minkowski's inequality*

$$\left(\sum_{k=1}^{n} |a_k + b_k|^p \right)^{1/p} \leqslant \left(\sum_{k=1}^{n} |a_k|^p \right)^{1/p} + \left(\sum_{k=1}^{n} |b_k|^p \right)^{1/p}. \tag{14}$$

The inequality is obvious for $p = 1$, and hence we can confine ourselves to the case $p > 1$.

The proof of (14) for $p > 1$ is in turn based on *Hölder's inequality*

$$\sum_{k=1}^{n} |a_k b_k| \leqslant \left(\sum_{k=1}^{n} |a_k|^p \right)^{1/p} \left(\sum_{k=1}^{n} |b_k|^q \right)^{1/q}, \tag{15}$$

where the numbers $p > 1$ and $q > 1$ satisfy the condition

$$\frac{1}{p} + \frac{1}{q} = 1. \tag{16}$$

---

[3] The least upper bound or supremum of a sequence of real numbers $a_1, a_2, \ldots, a_k, \ldots$ is denoted by $\sup_{k} a_k$.

We begin by observing that the inequality (15) is *homogeneous*, i.e., if it holds for two points $(a_1, \ldots, a_n)$ and $(b_1, \ldots, b_n)$, then it holds for any two points $(\lambda a_1, \ldots, \lambda a_n)$ and $(\mu b_1, \ldots, \mu b_n)$ where λ and μ are arbitrary real numbers. Therefore we need only prove (15) for the case

$$\sum_{k=1}^{n} |a_k|^p = \sum_{k=1}^{n} |b_k|^p = 1. \tag{17}$$

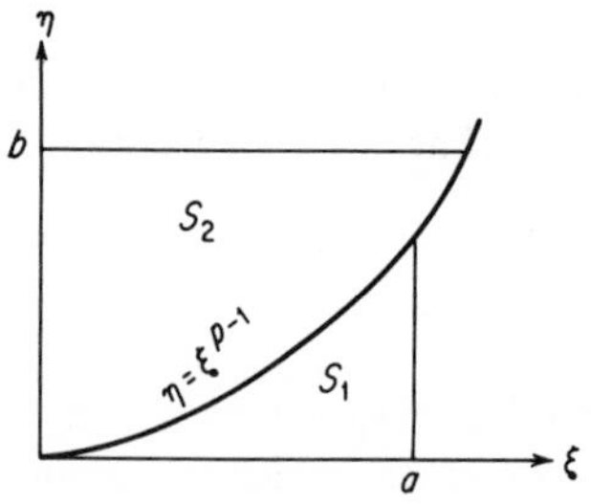

Thus, assuming that (17) holds, we now prove that

$$\sum_{k=1}^{n} |a_k b_k| \leqslant 1. \tag{18}$$

Consider the two areas S_1 and S_2 shown in Figure 8, associated with the curve in the $\xi\eta$-plane defined by the equation

$$\eta = \xi^{p-1},$$

FIGURE 8

or equivalently by the equation

$$\xi = \eta^{q-1}.$$

Then clearly

$$S_1 = \int_0^a \xi^{p-1} \, d\xi = \frac{a^p}{p}, \qquad S_2 = \int_0^b \eta^{q-1} \, d\eta = \frac{b^q}{q}.$$

Moreover, it is apparent from the figure that

$$S_1 + S_2 \geqslant ab$$

for arbitrary positive a and b. It follows that

$$ab \leqslant \frac{a^p}{p} + \frac{b^q}{q}. \tag{19}$$

Setting $a = |a_k|$, $b = |b_k|$, summing over k from 1 to n, and taking account of (16) and (17), we get the desired inequality (18). This proves Hölder's inequality (15). Note that (15) reduces to Schwarz's inequality if $p = 2$.

It is now an easy matter to prove Minkowski's inequality (14), starting from the identity

$$(|a| + |b|)^p = (|a| + |b|)^{p-1} |a| + (|a| + |b|)^{p-1} |b|.$$

In fact, setting $a = a_k$, $b = b_k$ and summing over k from 1 to n, we obtain

$$\sum_{k=1}^{n} (|a_k| + |b_k|)^p = \sum_{k=1}^{n} (|a_k| + |b_k|)^{p-1} |a_k| + \sum_{k=1}^{n} (|a_k| + |b_k|)^{p-1} |b_k|.$$

Next we apply Hölder's inequality (15) to both sums on the right, bearing in mind that $(p - 1)q = p$:

$$\sum_{k=1}^{n}(|a_k| + |b_k|)^p \leqslant \left(\sum_{k=1}^{n}(|a_k| + |b_k|)^p\right)^{1/q}\left(\left[\sum_{k=1}^{n}|a_k|^p\right]^{1/p} + \left[\sum_{k=1}^{n}|b_k|^p\right]^{1/p}\right)$$

Dividing both sides of this inequality by

$$\left(\sum_{k=1}^{n}(|a_k| + |b_k|)^p\right)^{1/q},$$

we get

$$\left(\sum_{k=1}^{n}(|a_k| + |b_k|)^p\right)^{1/p} \leqslant \left(\sum_{k=1}^{n}|a_k|^p\right)^{1/p} + \left(\sum_{k=1}^{n}|b_k|^p\right)^{1/p},$$

which immediately implies (14), thereby proving the triangle inequality in R_p^n.

Example 11. Finally let l_p be the set of all infinite sequences

$$x = (x_1, x_2, \ldots, x_k, \ldots)$$

of real numbers satisfying the convergence condition

$$\sum_{k=1}^{\infty}x_k^p < \infty$$

for some fixed number $p \geqslant 1$, where distance between points is defined by

$$\rho(x, y) = \left(\sum_{k=1}^{\infty}|x_k - y_k|^p\right)^{1/p} \tag{20}$$

(the case $p = 2$ has already been considered in Example 7). It follows from Minkowski's inequality (14) that

$$\left(\sum_{k=1}^{n}|x_k - y_k|^p\right)^{1/p} \leqslant \left(\sum_{k=1}^{n}|x_k|^p\right)^{1/p} + \left(\sum_{k=1}^{n}|y_k|^p\right)^{1/p} \tag{21}$$

for any n. Since the series

$$\sum_{k=1}^{\infty}|x_k|^p, \qquad \sum_{k=1}^{\infty}|y_k|^p$$

converge, by hypothesis, we can take the limit as $n \to \infty$ in (21), obtaining

$$\left(\sum_{k=1}^{\infty}|x_k - y_k|^p\right)^{1/p} \leqslant \left(\sum_{k=1}^{\infty}|x_k|^p\right)^{1/p} + \left(\sum_{k=1}^{\infty}|y_k|^p\right)^{1/p} < \infty.$$

This shows that (20) actually makes sense for arbitrary $x, y \in l_p$. At the same time, we have verified that the triangle inequality holds in l_p (the other two properties of a metric are obviously satisfied). Therefore l_p is a metric space.

Remark. If $R = (X, \rho)$ is a metric space and M is any subset of X, then obviously $R^* = (M, \rho)$ is again a metric space, called a *subspace* of the original metric space R. This device gives us infinitely more examples of metric spaces.

5.2. Continuous mappings and homeomorphisms. Isometric spaces. Let f be a mapping of one metric space X into another metric space Y, so that f associates an element $y = f(x) \in Y$ with each element $x \in X$. Then f is said to be *continuous at the point* $x_0 \in X$ if, given any $\varepsilon > 0$, there exists a $\delta > 0$ such that

$$\rho'(f(x), f(x_0)) < \varepsilon$$

whenever

$$\rho(x, x_0) < \delta$$

(here ρ is the metric in X and ρ' the metric in Y). The mapping f is said to be *continuous on* X if it is continuous at every point $x \in X$.

Remark. This definition reduces to the usual definition of continuity familiar from calculus if X and Y are both numerical sets, i.e., if f is a real function defined on some subset of the real line.

Given two metric spaces X and Y, let f be one-to-one mapping of X onto Y, and suppose f and f^{-1} are both continuous. Then f is called a *homeomorphic mapping*, or simply a *homeomorphism* (between X and Y). Two spaces X and Y are said to be *homeomorphic* if there exists a homeomorphism between them.

Example. The function

$$y = f(x) = \frac{2}{\pi} \arctan x$$

establishes a homeomorphism between the whole real line $(-\infty, \infty)$ and the open interval $(-1, 1)$.

DEFINITION 2. *A one-to-one mapping f of one metric space $R = (X, \rho)$ onto another metric space $R' = (Y, \rho')$ is said to be an **isometric mapping** (or simply an **isometry**) if*

$$\rho(x_1, x_2) = \rho'(f(x_1), f(x_2))$$

for all $x_1, x_2 \in R$. Correspondingly, the spaces R and R' are said to be **isometric** *(to each other).*

Thus if R and R' are isometric, the "metric relations" between the elements of R are the same as those between the elements of R', i.e., R and R' differ only in the explicit nature of their elements (this distinction is unimportant from the standpoint of metric space theory). From now on, we will not distinguish between isometric spaces, regarding them simply as identical.

Remark. We will discuss continuity and homeomorphisms from a more general point of view in Sec. 9.6.

Problem 1. Given a metric space (X, ρ), prove that

a) $|\rho(x, z) - \rho(y, u)| \leqslant \rho(x, y) + \rho(z, u)$ $(x, y, z, u \in X)$;

b) $|\rho(x, z) - \rho(y, z)| \leqslant \rho(x, y)$ $(x, y, z \in X)$.

Problem 2. Verify that

$$\left(\sum_{k=1}^{n} a_k b_k\right)^2 = \sum_{k=1}^{n} a_k^2 \sum_{k=1}^{n} b_k^2 - \frac{1}{2} \sum_{i=1}^{n} \sum_{j=1}^{n} (a_i b_j - b_i a_j)^2.$$

Deduce the Cauchy-Schwarz inequality (3) from this identity.

Problem 3. Verify that

$$\left(\int_a^b x(t) y(t)\, dt\right)^2 = \int_a^b x^2(t)\, dt \int_a^b y^2(t)\, dt - \frac{1}{2} \int_a^b \int_a^b [x(s) y(t) - y(s) x(t)]^2\, ds\, dt.$$

Deduce Schwarz's inequality (11) from this identity.

Problem 4. What goes wrong in Example 10, p. 41 if $p < 1$?

Hint. Show that Minkowski's inequality fails for $p < 1$.

Problem 5. Prove that the metric (5) is the limiting case of the metric (13) in the sense that

$$\rho_0(x, y) = \max_{1 \leqslant k \leqslant n} |x_k - y_k| = \lim_{p \to \infty} \left(\sum_{k=1}^{n} |x_k - y_k|^p\right)^{1/p}.$$

Problem 6. Starting from the inequality (19), deduce *Hölder's integral inequality*

$$\int_a^b x(t) y(t)\, dt \leqslant \left(\int_a^b |x(t)|^p\, dt\right)^{1/p} \left(\int_a^b |y(t)|^q\right)^{1/q} \quad \left(\frac{1}{p} + \frac{1}{q} = 1\right),$$

valid for any functions $x(t)$ and $y(t)$ such that the integrals on the right exist.

Problem 7. Use Hölder's integral inequality to prove *Minkowski's integral inequality*

$$\left(\int_a^b |x(t) + y(t)|^p\, dt\right)^{1/p} \leqslant \left(\int_a^b |x(t)|^p\right)^{1/p} + \left(\int_a^b |y(t)|^p\, dt\right)^{1/p} \quad (p \geqslant 1).$$

Problem 8. Exhibit an isometry between the spaces $C_{[0,1]}$ and $C_{[1,2]}$.

6. Convergence. Open and Closed Sets

6.1. Closure of a set. Limit points. By the *open sphere* (or *open ball*) $S(x_0, r)$ in a metric space R we mean the set of points $x \in R$ satisfying the inequality

$$\rho(x_0, x) < r$$

(ρ is the metric of R).[4] The fixed point x_0 is called the *center* of the sphere, and the number r is called its *radius*. By the *closed sphere* (or *closed ball*) $S[x_0, r]$ with center x_0 and radius r we mean the set of points $x \in R$ satisfying the inequality

$$\rho(x_0, x) \leqslant r.$$

An open sphere of radius ε with center x_0 will also be called an ε-*neighborhood* of x_0, denoted by $O_\varepsilon(x_0)$.

A point $x \in R$ is called a *contact point* of a set $M \subset R$ if every neighborhood of x contains at least one point of M. The set of all contact points of a set M is denoted by $[M]$ and is called the *closure* of M. Obviously $M \subset [M]$, since every point of M is a contact point of M. By the *closure operator* in a metric space R, we mean the mapping of R into R carrying each set $M \subset R$ into its closure $[M]$.

THEOREM 1. *The closure operator has the following properties*:

1) *If $M \subset N$, then $[M] \subset [N]$*;
2) $[[M]] = [M]$;
3) $[M \cup N] = [M] \cup [N]$;
4) $[\varnothing] = \varnothing$.

Proof. Property 1) is obvious. To prove property 2), let $x \in [[M]]$. Then any given neighborhood $O_\varepsilon(x)$ contains a point $x_1 \in [M]$. Consider the sphere $O_{\varepsilon_1}(x_1)$ of radius

$$\varepsilon_1 = \varepsilon - \rho(x, x_1).$$

Clearly $O_{\varepsilon_1}(x_1)$ is contained in $O_\varepsilon(x)$. In fact, if $z \in O_{\varepsilon_1}(x_1)$, then $\rho(z, x_1) < \varepsilon_1$ and hence, since $\rho(x, x_1) = \varepsilon - \varepsilon_1$, it follows from the triangle inequality that

$$\rho(z, x) < \varepsilon_1 + (\varepsilon - \varepsilon_1) = \varepsilon,$$

i.e., $z \in O_\varepsilon(x)$. Since $x_1 \in [M]$, there is a point $x_2 \in M$ in $O_{\varepsilon_1}(x)$. But then $x_2 \in O_\varepsilon(x)$ and hence $x \in [M]$, since $O_\varepsilon(x)$ is an arbitrary neighborhood of x. Therefore $[[M]] \subset [M]$. But obviously $[M] \subset [[M]]$ and hence $[[M]] = [M]$, as required.

To prove property 3), let $x \in [M \cup N]$ and suppose $x \notin [M] \cup [N]$. Then $x \notin [M]$ and $x \notin [N]$. But then there exist neighborhoods $O_{\varepsilon_1}(x)$ and $O_{\varepsilon_2}(x)$ such that $O_{\varepsilon_1}(x)$ contains no points of M while $O_{\varepsilon_2}(x)$ contains

---

[4] Any confusion between "sphere" meant in the sense of spherical surface and "sphere" meant in the sense of a solid sphere (or ball) will always be avoided by judicious use of the adjectives "open" or "closed."

no points of N. It follows that the neighborhood $O_\varepsilon(x)$, where $\varepsilon = \min\{\varepsilon_1, \varepsilon_2\}$, contains no points of either M or N, and hence no points of $M \cup N$, contrary to the assumption that $x \in [M \cup N]$. Therefore $x \in [M] \cup [N]$, and hence

$$[M \cup N] \subset [M] \cup [N], \tag{1}$$

since x is an arbitrary point of $[M \cup N]$. On the other hand, since $M \subset M \cup N$ and $N \subset M \cup N$, it follows from property 1) that $[M] \subset [M \cup N]$ and $[N] \subset [M \cup N]$. But then

$$[M] \cup [N] \subset [M \cup N],$$

which together with (1) implies $[M \cup N] = [M] \cup [N]$.

Finally, to prove property 4), we observe that given any $M \subset R$,

$$[M] = [M \cup \varnothing] = [M] \cup [\varnothing],$$

by property 3). It follows that $[\varnothing] \subset [M]$. But this is possible for arbitrary M only if $[\varnothing] = \varnothing$. (Alternatively, the set with no elements can have no contact points!) $\blacksquare$

A point $x \in R$ is called a *limit point* of a set $M \subset R$ if every neighborhood of x contains infinitely many points of M. The limit point may or may not belong to M. For example, if M is the set of rational numbers in the interval $[0, 1]$, then every point of $[0, 1]$, rational or not, is a limit point of M.

A point x *belonging* to a set M is called an *isolated point* of M if there is a ("sufficiently small") neighborhood of x containing no points of M other than x itself.

6.2. Convergence and limits. A sequence of points $\{x_n\} = x_1, x_2, \ldots,$ $x_n, \ldots$ in a metric space R is said to *converge* to a point $x \in R$ if every neighborhood $O_\varepsilon(x)$ of x contains all points x_n starting from a certain index (more exactly, if, given any $\varepsilon > 0$, there is an integer N_ε such that $O_\varepsilon(x)$ contains all points x_n with $n > N_\varepsilon$). The point x is called the *limit* of the sequence $\{x_n\}$, and we write $x_n \to x$ (as $n \to \infty$). Clearly, $\{x_n\}$ converges to x if and only if

$$\lim_{n \to \infty} \rho(x, x_n) = 0.$$

It is an immediate consequence of the definition of a limit that

1) No sequence can have two distinct limits;
2) If a sequence $\{x_n\}$ converges to a point x, then so does every subsequence of $\{x_n\}$

(give the details).

THEOREM 2. *A necessary and sufficient condition for a point x to be a contact point of a set M is that there exist a sequence $\{x_n\}$ of points of M converging to x.*

Proof. The condition is necessary, since if x is a contact point of M, then every neighborhood $O_{1/n}(x)$ contains at least one point $x_n \in M$, and these points form a sequence $\{x_n\}$ converging to M. The sufficiency is obvious. ∎

THEOREM 2′. *A necessary and sufficient condition for a point x to be a limit point of a set M is that there exist a sequence $\{x_n\}$ of distinct points of M converging to x.*

Proof. Clearly, if x is a limit point of M, then the points $x_n \in O_{1/n}(x) \cap M$ figuring in the proof of Theorem 2 can be chosen to be distinct. This proves the necessity, and the sufficiency is again obvious. ∎

6.3. Dense subsets. Separable spaces. Let A and B be two subsets of a metric space R. Then A is said to be *dense* in B if $[A] \supset B$. In particular, A is said to be *everywhere dense* (in R) if $[A] = R$. A set A is said to be *nowhere dense* if it is dense in no (open) sphere at all.

Example 1. The set of all rational points is dense in the real line R^1.

Example 2. The set of all points $x = (x_1, x_2, \ldots, x_n)$ with rational coordinates is dense in each of the spaces R^n, R_0^n and R_1^n introduced in Examples 3–5, pp. 38–39.

Example 3. The set of all points $x = (x_1, x_2, \ldots, x_k, \ldots)$ with only finitely many nonzero coordinates, each a rational number, is dense in the space l_2 introduced in Example 7, p. 39.

Example 4. The set of all polynomials with rational coefficients is dense in both spaces $C_{[a,b]}$ and $C_{[a,b]}^2$ introduced in Examples 6 and 8, pp. 39 and 40.

DEFINITION. *A metric space is said to be **separable** if it has a countable everywhere dense subset.*

Example 5. The spaces R^1, R^n, R_0^n, R_1^n, l_2, $C_{[a,b]}$, and $C_{[a,b]}^2$ are all separable, since the sets in Examples 1–4 above are all countable.

Example 6. The "discrete space" M described in Example 1, p. 38 contains a countable everywhere dense subset and hence is separable if and only if it is itself a countable set, since clearly $[M] = M$ in this case.

Example 7. There is no countable everywhere dense set in the space m of all bounded sequences, introduced in Example 9, p. 41. In fact, consider the set E of all sequences consisting exclusively of zeros and ones. Clearly, E has the power of the continuum (recall Theorem 6, Sec. 2.5), since there is a one-to-one correspondence between E and the set of all subsets of the set $Z_+ = \{1, 2, \ldots, n, \ldots\}$ (describe the correspondence). According to formula (12), p. 41, the distance between any two points of E equals 1. Suppose we surround each point of E by an open sphere of radius $\frac{1}{2}$, thereby obtaining an uncountably infinite family of pairwise disjoint spheres. Then if some set M is everywhere dense in m, there must be at least one point of M in each of the spheres. It follows that M cannot be countable and hence that m cannot be separable.

6.4. Closed sets. We say that a subset M of a metric space R is *closed* if it coincides with its own closure, i.e., if $[M] = M$. In other words, a set is called closed if it contains all its limit points (see Problem 2).

Example 1. The empty set $\varnothing$ and the whole space R are closed sets.

Example 2. Every closed interval $[a, b]$ on the real line is a closed set.

Example 3. Every closed sphere in a metric space is a closed set. In particular, the set of all functions f in the space $C_{[a,b]}$ such that $|f(t)| \leqslant K$ (where K is a constant) is closed.

Example 4. The set of all functions f in $C_{[a,b]}$ such that $|f(t)| < K$ (an open sphere) is not closed. The closure of this set is the closed sphere in the preceding example.

Example 5. Any set consisting of a finite number of points is closed.

THEOREM 3. *The intersection of an arbitrary number of closed sets is closed. The union of a **finite** number of closed sets is closed.*

Proof. Given arbitrary sets F_α indexed by a parameter α, let x be a limit point of the intersection

$$F = \bigcap_\alpha F_\alpha.$$

Then any neighborhood $O_\varepsilon(x)$ contains infinitely many points of F, and hence infinitely many points of each F_α. Therefore x is a limit point of each F_α and hence belongs to each F_α, since the sets F_α are all closed. It follows that $x \in F$, and hence that F itself is closed.

Next let

$$F = \bigcup_{k=1}^{n} F_k$$

be the union of a finite number of closed sets F_k, and suppose x does not belong to F. Then x does not belong to any of the sets F_k, and hence cannot be a limit point of any of them. But then, for every k, there is a neighborhood $O_{\varepsilon_k}(x)$ containing no more than a finite number of points of F_k. Choosing

$$\varepsilon = \min \{\varepsilon_1, \ldots, \varepsilon_n\},$$

we get a neighborhood $O_\varepsilon(x)$ containing no more than a finite number of points of F, so that x cannot be a limit point of F. This proves that a point $x \notin F$ cannot be a limit point of F. Therefore F is closed. ∎

6.5. Open sets. A point x is called an *interior point* of a set M if x has a neighborhood $O_\varepsilon(x) \subset M$, i.e., a neighborhood consisting entirely of points of M. A set is said to be *open* if its points are all interior points.

Example 1. Every open interval (a, b) on the real line is an open set. In fact, if $a < x < b$, choose $\varepsilon = \min \{x - a, b - x\}$. Then clearly $O_\varepsilon(x) \subset (a, b)$.

Example 2. Every open sphere $S(a, r)$ in a metric space is an open set. In fact, $x \in S(a, r)$ implies $\rho(a, x) < r$. Hence, choosing $\varepsilon = r - \rho(a, x)$, we have $O_\varepsilon(x) = S(x, \varepsilon) \subset S(a, r)$.

Example 3. Let M be the set of all functions f in $C_{[a,b]}$ such that $f(t) < g(t)$, where g is a fixed function in $C_{[a,b]}$. Then M is an open subset of $C_{[a,b]}$.

THEOREM 4. *A subset M of a metric space R is open if and only if its complement $R - M$ is closed.*

Proof. If M is open, then every point $x \in M$ has a neighborhood (entirely) contained in M. Therefore no point $x \in M$ can be a contact point of $R - M$. In other words, if x is a contact point of $R - M$, then $x \in R - M$, i.e., $R - M$ is closed.

Conversely, if $R - M$ is closed, then any point $x \in M$ must have a neighborhood contained in M, since otherwise every neighborhood of x would contain points of $R - M$, i.e., x would be a contact point of $R - M$ not in $R - M$. Therefore M is open. ∎

COROLLARY. *The empty set $\varnothing$ and the whole space R are open sets.*

Proof. An immediate consequence of Theorem 4 and Example 1, Sec. 6.4. ∎

THEOREM 5. *The union of an arbitrary number of open sets is open. The intersection of a finite number of open sets is open.*

Proof. This is the "dual" of Theorem 3. The proof is an immediate consequence of Theorem 4 and formulas (3)–(4), p. 4. ∎

6.6. Open and closed sets on the real line. The structure of open and closed sets in a given metric space can be quite complicated. This is true even for open and closed sets in a Euclidean space of two or more dimensions (R^n, $n \geqslant 2$). In the one-dimensional case, however, it is an easy matter to give a complete description of all open sets (and hence of all closed sets):

THEOREM 6. *Every open set G on the real line is the union of a finite or countable system of pairwise disjoint open intervals.*[5]

Proof. Let x be an arbitrary point of G. By the definition of an open set, there is at least one open interval containing x and contained in G. Let I_x be the union of all such open intervals. Then, as we now show, I_x is itself an open interval. In fact, let[6]

$$a = \inf I_x, \qquad b = \sup I_x$$

(where we allow the cases $a = -\infty$ and $b = +\infty$). Then obviously

$$I_x \subset (a, b). \tag{2}$$

Moreover, suppose y is an arbitrary point of (a, b) distinct from x, where, to be explicit, we assume that $a < y < x$. Then there is a point $y' \in I_x$ such that $a < y' < y$ (why?). Hence G contains an open interval containing the points y' and x. But then this interval also contains y, i.e., $y \in I_x$. (The case $y > x$ is treated similarly.) Moreover, the point x belongs to I_x, by hypothesis. It follows that $I_x \supset (a, b)$, and hence by (2) that $I_x = (a, b)$. Thus I_x is itself an open interval, as asserted, in fact the open interval (a, b).

By its very construction, the interval (a, b) is contained in G and is not a subset of a larger interval contained in G. Moreover, it is clear that two intervals I_x and $I_{x'}$ corresponding to distinct points x and x' either coincide or else are disjoint (otherwise I_x and $I_{x'}$ would both be contained in a larger interval $I_x \cup I_{x'} = I \subset G$. There are no more than countably many such pairwise disjoint intervals I_x. In fact, choosing an arbitrary rational point in each I_x, we establish a one-to-one correspondence between the intervals I_x and a subset of the rational numbers. Finally, it is obvious that

$$G = \bigcup_x I_x. \quad \blacksquare$$

COROLLARY. *Every closed set on the real line can be obtained by deleting a finite or countable system of pairwise disjoint intervals from the line.*

---

[5] The infinite intervals $(-\infty, \infty)$, (a, ∞), and $(-\infty, b)$ are regarded as open.

[6] Given a set of real numbers E, inf E denotes the greatest lower bound or infimum of E, while sup E denotes the least upper bound or supremum of E.

Proof. An immediate consequence of Theorems 4 and 6. ∎

Example 1. Every closed interval $[a, b]$ is a closed set (here a and b are necessarily finite).

Example 2. Every single-element set $\{x_0\}$ is closed.

Example 3. The union of a finite number of closed intervals and single-element sets is a closed set.

Example 4 (The Cantor set). A more interesting example of a closed set on the line can be constructed as follows: Delete the open interval $(\frac{1}{3}, \frac{2}{3})$ from the closed interval $F_0 = [0, 1]$, and let F_1 denote the remaining closed set, consisting of two closed intervals. Then delete the open intervals $(\frac{1}{9}, \frac{2}{9})$ and $(\frac{7}{9}, \frac{8}{9})$ from F_1, and let F_2 denote the remaining closed set, consisting of four closed intervals. Then delete the "middle third" from each of these four intervals, getting a new closed set F_3, and so on (see Figure 9). Continuing this process indefinitely, we get a sequence of closed sets F_n such that

$$F_0 \supset F_1 \supset F_2 \supset \cdots \supset F_n \supset \cdots$$

(such a sequence is said to be *decreasing*). The intersection

$$F = \bigcap_{n=0}^{\infty} F_n$$

of all these sets is called the *Cantor set*. Clearly F is closed, by Theorem 3, and is obtained from the unit interval $[0, 1]$ by deleting a countable number of open intervals. In fact, at the nth stage of the construction, we delete 2^{n-1} intervals, each of length $1/3^n$.

To describe the structure of the set F, we first note that F contains the points

$$0, 1, \tfrac{1}{3}, \tfrac{2}{3}, \tfrac{1}{9}, \tfrac{2}{9}, \tfrac{7}{9}, \tfrac{8}{9}, \ldots, \tag{3}$$

i.e., the end points of the deleted intervals (together with the points 0 and 1).

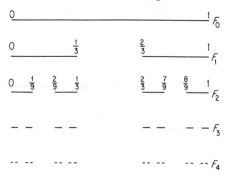

FIGURE 9

However F contains many other points. In fact, given any $x \in [0, 1]$, suppose we write x in ternary notation, representing x as a series

$$x = \frac{a_1}{3} + \frac{a_2}{3^2} + \cdots + \frac{a_n}{3^n} + \cdots, \tag{4}$$

where each of the numbers $a_1, a_2, \ldots, a_n, \ldots$ can only take one of the three values 0, 1, 2. Then it is easy to see that x belongs to F if and only if x has a representation (4) such that none of the numbers $a_1, a_2, \ldots, a_n, \ldots$ equals 1 (think things through).[7]

Remarkably enough, the set F has the power of the continuum, i.e., there are as many points in F as in the whole interval [0, 1], despite the fact that the sum of the lengths of the deleted intervals equals

$$\tfrac{1}{3} + \tfrac{2}{9} + \tfrac{4}{27} + \cdots = 1.$$

To see this, we associate a new point

$$y = \frac{b_1}{2} + \frac{b_2}{2^2} + \cdots + \frac{b_n}{2^n} + \cdots$$

with each point (4), where[8]

$$b_n = \begin{cases} 0 & \text{if} \quad a_n = 0, \\ 1 & \text{if} \quad a_n = 2. \end{cases}$$

In this way, we set up a one-to-one correspondence between F and the whole interval [0, 1]. It follows that F has the power of the continuum, as asserted. Let A_1 be the set of points (3). Then $F = A_1 \cup A_2$, where the set $A_2 = F - A_1$ is uncountable, since A_1 is countable and F itself is not. The points of A_1 are often called "points (of F) of the first kind," while those of A_2 are called "points of the second kind."

Problem 1. Give an example of a metric space R and two open spheres $S(x, r_1)$ and $S(y, r_2)$ in R such that $S(x, r_1) \subset S(y, r_2)$ although $r_1 > r_2$.

Problem 2. Prove that every contact point of a set M is either a limit point of M or an isolated point of M.

---

[7] Just as in the case of ordinary decimals, certain numbers can be written in two distinct ways. For example,

$$\frac{1}{3} = \frac{1}{3} + \frac{0}{3^2} + \frac{0}{3^3} + \cdots + \frac{0}{3^n} + \cdots = \frac{0}{3} + \frac{2}{3^2} + \frac{2}{3^3} + \cdots + \frac{2}{3^n} + \cdots.$$

Since none of the numerators in the second representation equals 1 the point $\tfrac{1}{3}$ belongs to F (this is already obvious from the construction of F).

[8] If x has two representations of the form (4), then one and only one of them has no numerators $a_1, a_2, \ldots, a_n, \ldots$ equal to 1. These are the numbers used to define b_n.

Comment. In particular, $[M]$ can only contain points of the following three types:

a) Limit points of M belonging to M;
b) Limit points of M which do not belong to M;
c) Isolated points of M.

Thus $[M]$ is the union of M and the set of all its limit points.

Problem 3. Prove that if $x_n \to x$, $y_n \to y$ as $n \to \infty$, then $\rho(x_n, y_n) \to \rho(x, y)$.

Hint. Use Problem 1a, p. 45.

Problem 4. Let f be a mapping of one metric space X into another metric space Y. Prove that f is continuous at a point x_0 if and only if the sequence $\{y_n\} = \{f(x_n)\}$ converges to $y = f(x_0)$ whenever the sequence $\{x_n\}$ converges to x_0.

Problem 5. Prove that

a) The closure of any set M is a closed set;
b) $[M]$ is the smallest closed set containing M.

Problem 6. Is the union of infinitely many closed sets necessarily closed? How about the intersection of infinitely many open sets? Give examples.

Problem 7. Prove directly that the point $\frac{1}{4}$ belongs to the Cantor set F, although it is not an end point of any of the open intervals deleted in constructing F.

Hint. The point $\frac{1}{4}$ divides the interval $[0, 1]$ in the ratio $1:3$. It also divides the interval $[0, \frac{1}{3}]$ left after deleting $(\frac{1}{3}, \frac{2}{3})$ in the ratio $3:1$, and so on.

Problem 8. Let F be the Cantor set. Prove that

a) The points of the first kind, i.e., the points (3) form an everywhere dense subset of F;
b) The numbers of the form $t_1 + t_2$, where $t_1, t_2 \in F$, fill the whole interval $[0, 2]$.

Problem 9. Given a metric space R, let A be a subset of R and x a point of R. Then the number

$$\rho(A, x) = \inf_{a \in A} \rho(a, x)$$

is called the *distance between A and x*. Prove that

a) $x \in A$ implies $\rho(A, x) = 0$, but not conversely;
b) $\rho(A, x)$ is a continuous function of x (for fixed A);
c) $\rho(A, x) = 0$ if and only if x is a contact point of A;
d) $[A] = A \cup M$, where M is the set of all points x such that $\rho(A, x) = 0$.

Problem 10. Let A and B be two subsets of a metric space R. Then the number

$$\rho(A, B) = \inf_{\substack{a \in A \\ b \in B}} \rho(a, b)$$

is called the *distance between A and B*. Show that $\rho(A, B) = 0$ if $A \cap B \neq \varnothing$, but not conversely.

Problem 11. Let M_K be the set of all functions f in $C_{[a,b]}$ satisfying a *Lipschitz condition*, i.e., the set of all f such that

$$|f(t_1) - f(t_2)| \leqslant K |t_1 - t_2|$$

for all $t_1, t_2 \in [a, b]$, where K is a fixed positive number. Prove that

a) M_K is closed and in fact is the closure of the set of all differentiable functions on $[a, b]$ such that $|f'(t)| < K$;

b) The set

$$M = \bigcup_K M_K$$

of all functions satisfying a Lipschitz condition for some K is not closed;

c) The closure of M is the whole space $C_{[a,b]}$.

Problem 12. An open set G in n-dimensional Euclidean space R^n is said to be *connected* if any points $x, y \in G$ can be joined by a polygonal line[9] lying entirely in G. For example, the (open) disk $x^2 + y^2 < 1$ is connected, but not the union of the two disks

$$x^2 + y^2 < 1, \qquad (x - 2)^2 + y^2 < 1$$

(even though they share a contact point). An open subset of an open set G is called a *component* of G if it is connected and is not contained in a larger connected subset of G. Use Zorn's lemma to prove that every open set G in R^n is the union of no more than countably many pairwise disjoint components.

Comment. In the case $n = 1$ (i.e., on the real line) every connected open set is an open interval, possibility one of the infinite intervals $(-\infty, \infty)$, (a, ∞), $(-\infty, b)$. Thus Theorem 6 on the structure of open sets on the line is tantamount to two assertions:

1) Every open set on the line is the union of a finite or countable number of components;

2) Every open connected set on the line is an open interval.

---

[9] By a *polygonal line* we mean a curve obtained by joining a finite number of straight line segments end to end.

The first assertion holds for open sets in R^n (and in fact is susceptible to further generalizations), while the second assertion pertains specifically to the real line.

7. Complete Metric Spaces

7.1. Definitions and examples. The reader is presumably already familiar with the notion of the completeness of the real line. The real line is, of course, a particularly simple example of a metric space. We now make the natural generalization of the notion of completeness to the case of an arbitrary metric space.

DEFINITION 1. *A sequence $\{x_n\}$ of points in a metric space R with metric ρ is said to satisfy the **Cauchy criterion** if, given any $\varepsilon > 0$, there is an integer N_ε such that $\rho(x_n, x_{n'}) < \varepsilon$ for all $n, n' > N_\varepsilon$.*

DEFINITION 2. *A subsequence $\{x_n\}$ of points in a metric space R is called a **Cauchy sequence** (or a **fundamental sequence**) if it satisfies the Cauchy criterion.*

THEOREM 1. *Every convergent sequence $\{x_n\}$ is fundamental.*

Proof. If $\{x_n\}$ converges to a limit x, then, given any $\varepsilon > 0$, there is an integer N_ε such that

$$\rho(x_n, x) < \frac{\varepsilon}{2}$$

for all $n > N_\varepsilon$. But then

$$\rho(x_n, x_{n'}) \leqslant \rho(x_n, x) + \rho(x_{n'}, x) < \varepsilon$$

for all $n, n' > N_\varepsilon$. ∎

DEFINITION 3. *A metric space R is said to be **complete** if every Cauchy sequence in R converges to an element of R. Otherwise R is said to be **incomplete**.*

Example 1. Let R be the "space of isolated points" considered in Example 1, p. 38. Then the Cauchy sequences in R are just the "stationary sequences," i.e., the sequences $\{x_n\}$ all of whose terms are the same starting from some index n. Every such sequence is obviously convergent to an element of R. Hence R is complete.

Example 2. The completeness of the real line R^1 is familiar from elementary analysis.

Example 3. The completeness of Euclidean n-space R^n follows from that of R^1. In fact, let

$$x^{(p)} = (x_1^{(p)}, \ldots, x_n^{(p)}) \qquad (p = 1, 2, \ldots)$$

be a fundamental sequence of points of R^n. Then, given any $\varepsilon > 0$, there exists an N_ε such that

$$\sum_{k=1}^{n} (x_k^{(p)} - x_k^{(q)})^2 < \varepsilon^2$$

for all $p, q > N_\varepsilon$. It follows that

$$|x_k^{(p)} - x_k^{(q)}| < \varepsilon \qquad (k = 1, \ldots, n)$$

for all $p, q > N_\varepsilon$, i.e., each $\{x_k^{(p)}\}$ is a fundamental sequence in R^1. Let

$$x = (x_1, \ldots, x_n),$$

where

$$x_k = \lim_{p \to \infty} x_k^{(p)}.$$

Then obviously

$$\lim_{p \to \infty} x^{(p)} = x.$$

This proves the completeness of R^n. The completeness of the spaces R_0^n and R_1^n introduced in Examples 4 and 5, p. 39 is proved in almost the same way (give the details).

Example 4. Let $\{x_n(t)\}$ be a Cauchy sequence in the function space $C_{[a,b]}$ considered in Example 6, p. 39. Then, given any $\varepsilon > 0$, there is an N_ε such that

$$|x_n(t) - x_{n'}(t)| < \varepsilon \qquad (1)$$

for all $n, n' > N_\varepsilon$ and all $t \in [a, b]$. It follows that the sequence $\{x_n(t)\}$ is uniformly convergent. But the limit of a uniformly convergent sequence of continuous functions is itself a continuous function (see Problem 1). Taking the limit as $n' \to \infty$ in (1), we find that

$$|x_n(t) - x(t)| \leqslant \varepsilon$$

for all $n > N_\varepsilon$ and all $t \in [a, b]$, i.e., $\{x_n(t)\}$ converges in the metric of $C_{[a,b]}$ to a function $x(t) \in C_{[a,b]}$. Hence $C_{[a,b]}$ is a complete metric space.

Example 5. Next let $x^{(n)}$ be a sequence in the space l_2 considered in Example 7, p. 39, so that

$$x^{(n)} = (x_1^{(n)}, x_2^{(n)}, \ldots, x_k^{(n)}, \ldots),$$

$$\sum_{k=1}^{\infty} (x_k^{(n)})^2 < \infty \qquad (n = 1, 2, \ldots).$$

Suppose further that $\{x^{(n)}\}$ is a Cauchy sequence. Then, given any $\varepsilon > 0$, there is a N_ε such that

$$\rho^2(x^{(n)}, x^{(n')}) = \sum_{k=1}^{\infty}(x_k^{(n)} - x_k^{(n')})^2 < \varepsilon \qquad (2)$$

if $n, n' > N_\varepsilon$. It follows that

$$(x_k^{(n)} - x_k^{(n')})^2 < \varepsilon \qquad (k = 1, 2, \ldots),$$

i.e., for every k the sequence $\{x_k^{(n)}\}$ is fundamental and hence convergent. Let

$$x_k = \lim_{n \to \infty} x_k^{(n)},$$

$$x = (x_1, x_2, \ldots, x_k, \ldots).$$

Then, as we now show, x is itself a point of l_2 and moreover $\{x^{(n)}\}$ converges to x in the l_2 metric, so that l_2 is a complete metric space.

In fact, (2) implies

$$\sum_{k=1}^{M}(x_k^{(n)} - x_k^{(n')})^2 < \varepsilon \qquad (3)$$

for any fixed M. Holding n fixed in (3) and taking the limit as $n' \to \infty$, we get

$$\sum_{k=1}^{M}(x_k^{(n)} - x_k)^2 \leqslant \varepsilon. \qquad (4)$$

Since (4) holds for arbitrary M, we can in turn take the limit of (4) as $M \to \infty$, obtaining

$$\sum_{k=1}^{\infty}(x_k^{(n)} - x_k)^2 \leqslant \varepsilon. \qquad (5)$$

Just as on p. 40, the convergence of the two series

$$\sum_{k=1}^{\infty}(x_k^{(n)})^2, \qquad \sum_{k=1}^{\infty}(x_k^{(n)} - x_k)^2$$

implies that of the series

$$\sum_{k=1}^{\infty}x_k^2.$$

This proves that $x \in l_2$. Moreover, since ε is arbitrarily small, (5) implies

$$\lim_{n \to \infty}\rho(x^{(n)}, x) = \lim_{n \to \infty}\sqrt{\sum_{k=1}^{\infty}(x_k^{(n)} - x_k)^2} = 0,$$

i.e., $\{x^{(n)}\}$ converges to x in the l_2 metric, as asserted.

Example 6. It is easy to show that the space $C^2_{[a,b]}$ of Example 8, p. 40 is *incomplete.* If

$$\varphi_n(t) = \begin{cases} -1 & \text{if } -1 \leqslant t \leqslant -\dfrac{1}{n}, \\[2mm] nt & \text{if } -\dfrac{1}{n} \leqslant t \leqslant \dfrac{1}{n}, \\[2mm] 1 & \text{if } \dfrac{1}{n} \leqslant t \leqslant 1, \end{cases}$$

then $\{\varphi_n(t)\}$ is a fundamental sequence in $C^2_{[-1,1]}$, since

$$\int_{-1}^{1} [\varphi_n(t) - \varphi_{n'}(t)]^2 \, dt < \frac{2}{\min \{n, n'\}}$$

However, $\{\varphi_n(t)\}$ cannot converge to a function in $C^2_{[-1,1]}$. In fact, consider the discontinuous function

$$\psi(t) = \begin{cases} -1 & \text{if } t < 0, \\ 1 & \text{if } t > 0. \end{cases}$$

Then, given any function $f \in C^2_{[-1,1]}$, it follows from Schwarz's inequality (obviously still valid for piecewise continuous functions) that

$$\left(\int_{-1}^{1} [f(t) - \psi(t)]^2\right)^{1/2} \leqslant \left(\int_{-1}^{1} [f(t) - \varphi_n(t)]^2 \, dt\right)^{1/2} + \left(\int_{-1}^{1} [\varphi_n(t) - \psi(t)]^2 \, dt\right)^{1/2}.$$

But the integral on the left is nonzero, by the continuity of f, and moreover it is clear that

$$\lim_{n \to \infty} \int_{-1}^{1} [\varphi_n(t) - \psi(t)]^2 \, dt = 0.$$

Therefore

$$\int_{-1}^{1} [f(t) - \varphi_n(t)]^2 \, dt$$

cannot converge to zero as $n \to \infty$.

7.2. The nested sphere theorem. A sequence of closed spheres

$$S[x_1, r_1], S[x_2, r_2], \ldots, S[x_n, r_n], \ldots$$

in a metric space R is said to be *nested* (or *decreasing*) if

$$S[x_1, r_1] \supset S[x_2, r_2] \supset \cdots \supset S[x_n, r_n] \supset \cdots .$$

Using this concept, we can prove a simple criterion for the completeness of R:

THEOREM 2 (*Nested sphere theorem*). *A metric space R is complete if and only if every nested sequence* $\{S_n\} = \{S[x_n, r_n]\}$ *of closed spheres in R such that* $r_n \to 0$ *as* $n \to \infty$ *has a nonempty intersection*

$$\bigcap_{n=1}^{\infty} S_n.$$

Proof. If R is complete and if $\{S_n\} = \{S[x_n, r_n]\}$ is any nested sequence of closed spheres in R such that $r_n \to 0$ as $n \to \infty$, then the sequence $\{x_n\}$ of centers of the spheres is fundamental, since $\rho(x_n, x_{n'}) < r_n$ for $n' > n$ and $r_n \to 0$ as $n \to \infty$. Therefore $\{x_n\}$ has a limit. Let

$$x = \lim_{n \to \infty} x_n.$$

Then

$$x \in \bigcap_{n=1}^{\infty} S_n.$$

In fact, S_n contains every point of the sequence $\{x_n\}$ except possibly the points $x_1, x_2, \ldots, x_{n-1}$, and hence x is a limit point of every sphere S_n. But S_n is closed, and hence $x \in S_n$ for all n.

Conversely, suppose every nested sequence of closed spheres in R with radii converging to zero has a nonempty intersection, and let $\{x_n\}$ be any fundamental sequence in R. Then x has a limit in R. To see this, use the fact that $\{x_n\}$ is fundamental to choose a term x_{n_1} of the sequence $\{x_n\}$ such that

$$\rho(x_n, x_{n_1}) < \frac{1}{2}$$

for all $n \geqslant n_1$, and let S_1 be the closed sphere of radius 1 with center x_{n_1}. Then choose a term x_{n_2} of $\{x_n\}$ such that $n_2 > n_1$ and

$$\rho(x_n, x_{n_2}) < \frac{1}{2^2}$$

for all $n \geqslant n_2$, and let S_2 be the closed sphere of radius $\frac{1}{2}$ with center x_{n_2}. Continue this construction indefinitely, i.e., once having chosen terms $x_{n_1}, x_{n_2}, \ldots, x_{n_k}$ $(n_1 < n_2 < \cdots < n_k)$, choose a term $x_{n_{k+1}}$ such that $n_{k+1} > n_k$ and

$$\rho(x_n, x_{n_{k+1}}) < \frac{1}{2^{k+1}}$$

for all $n \geqslant n_{k+1}$, let S_{k+1} be the closed sphere of radius $1/2^k$ with center $x_{n_{k+1}}$, and so on. This gives a nested sequence $\{S_n\}$ of closed spheres with radii converging to zero. By hypothesis, these spheres have a nonempty intersection, i.e., there is a point x in all the spheres. This point is obviously the limit of the sequence $\{x_{n_k}\}$. But if a fundamental sequence contains a subsequence converging to x, then the sequence itself must converge to x (why?), i.e.,

$$\lim_{n \to \infty} x_n = x. \quad \blacksquare$$

7.3. Baire's theorem. It will be recalled from Sec. 6.3 that a subset A of a metric space R is said to be *nowhere dense* in R if it is dense in no (open) sphere at all, or equivalently, if every sphere $S \subset R$ contains another sphere S' such that $S' \cap A = \varnothing$ (check the equivalence). This concept plays an important role in

THEOREM 3 (*Baire*). *A complete metric space R cannot be represented as the union of a countable number of nowhere dense sets.*

Proof. Suppose to the contrary that

$$R = \bigcup_{n=1}^{\infty} A_n, \tag{6}$$

where every set A_n is nowhere dense in R. Let $S_0 \subset R$ be a closed sphere of radius 1. Since A_1 is nowhere dense in S_0, being nowhere dense in R, there is a closed sphere S_1 of radius less than $\frac{1}{2}$ such that $S_1 \subset S_0$ and $S_1 \cap A_1 = \varnothing$. Since A_2 is nowhere dense in S_1, being nowhere dense in S_0, there is a closed sphere S_2 of radius less than $\frac{1}{3}$ such that $S_2 \subset S_1$ and $S_2 \cap A_2 = \varnothing$, and so on. In this way, we get a nested sequence of closed spheres $\{S_n\}$ with radii converging to zero such that

$$S_n \cap A_n = \varnothing \qquad (n = 1, 2, \ldots).$$

By the nested sphere theorem, the intersection

$$\bigcap_{n=1}^{\infty} S_n$$

contains a point x. By construction, x cannot belong to any of the sets A_n, i.e.,

$$x \notin \bigcup_{n=1}^{\infty} A_n.$$

It follows that

$$R \neq \bigcup_{n=1}^{\infty} A_n,$$

contrary to (6). Hence the representation (6) is impossible. $\quad \blacksquare$

COROLLARY. *A complete metric space R without isolated points is uncountable.*

Proof. Every single-element set $\{x\}$ is nowhere dense in R. $\quad \blacksquare$

7.4. Completion of a metric space. As we now show, an incomplete metric space can always be enlarged (in an essentially unique way) to give a complete metric space.

DEFINITION 4. *Given a metric space R with closure [R], a complete metric space R* is called a* **completion** *of R if R ⊂ R* and [R] = R*, i.e., if R is a subset of R* everywhere dense in R*.*

Example 1. Clearly $R^* = R$ if R is already complete (see Problem 7).

Example 2. The space of all real numbers is the completion of the space of all rational numbers.

THEOREM 4. *Every metric space R has a completion. This completion is unique to within an isometric mapping carrying every point x ∈ R into itself.*

Proof. The proof is somewhat lengthy, but completely straightforward. First we prove the *uniqueness*, showing that if R^* and R^{**} are two completions of R, then there is a one-to-one mapping $x^{**} = \varphi(x^*)$ of R^* onto R^{**} such that $\varphi(x) = x$ for all $x \in R$ and

$$\rho_1(x^*, y^*) = \rho_2(x^{**}, y^{**}) \tag{7}$$

$(y^{**} = \varphi(y^*))$, where ρ_1 is the distance in R^* and ρ_2 the distance in R^{**}. The required mapping φ is constructed as follows: Let x^* be an arbitrary point of R^*. Then, by the definition of a completion, there is a sequence $\{x_n\}$ of points of R converging to x^*. The points of the sequence $\{x_n\}$ also belong to R^{**}, where they form a fundamental sequence (why?). Therefore $\{x_n\}$ converges to a point $x^{**} \in R^{**}$, since R^{**} is complete. It is clear that x^{**} is independent of the choice of the sequence $\{x_n\}$ converging to the point x^* (why?). If we set $\varphi(x^*) = x^{**}$, then φ is the required mapping. In fact, $\varphi(x) = x$ for all $x \in R$, since if $x_n \to x \in R$, then obviously $x = x^* \in R^*$, $x^{**} = x$. Moreover, suppose $x_n \to x^*$, $y_n \to y^*$ in R^*, while $x_n \to x^{**}$, $y_n \to y^{**}$ in R^{**}. Then, if ρ is the distance in R,

$$\rho_1(x^*, y^*) = \lim_{n \to \infty} \rho_1(x_n, y_n) = \lim_{n \to \infty} \rho(x_n, y_n) \tag{8}$$

(see Problem 3, p. 54), while at the same time

$$\rho_2(x^{**}, y^{**}) = \lim_{n \to \infty} \rho_2(x_n, y_n) = \lim_{n \to \infty} \rho(x_n, y_n). \tag{8'}$$

But (8) and (8') together imply (7).

We must now prove the *existence* of a completion of R. Given an arbitrary metric space R, we say that two Cauchy sequences $\{x_n\}$ and

$\{\tilde{x}_n\}$ in R are *equivalent* and write $\{x_n\} \sim \{\tilde{x}_n\}$ if

$$\lim_{n \to \infty} \rho(x_n, \tilde{x}_n) = 0.$$

As anticipated by the notation and terminology, $\sim$ is reflexive, symmetric and transitive, i.e., $\sim$ is an equivalence relation in the sense of Sec. 1.4. Therefore the set of all Cauchy sequences of points in the space R can be partitioned into classes of equivalent sequences. Let these *classes* be the points of a new space R^*. Then we define the distance between two arbitrary points $x^*, y^* \in R^*$ by the formula

$$\rho_1(x^*, y^*) = \lim_{n \to \infty} \rho(x_n, y_n), \tag{9}$$

where $\{x_n\}$ is any "representative" of x^* (namely, any Cauchy sequence in the class x^*) and $\{y_n\}$ is any representative of y^*.

The next step is to verify that (9) is in fact a distance, i.e., that (9) exists, is independent of the choice of the sequences $\{x_n\} \in x^*$, $\{y_n\} \in y^*$, and satisfies the three properties of a distance figuring in Definition 1, p. 37. Given any $\varepsilon > 0$, it follows from the triangle inequality in R (recall Problem 1b, p. 45) that

$$|\rho(x_n, y_n) - \rho(x_{n'}, y_{n'})|$$

$$= |\rho(x_n, y_n) - \rho(x_{n'}, y_n) + \rho(x_{n'}, y_n) - \rho(x_{n'}, y_{n'})|$$

$$\leqslant |\rho(x_n, y_n) - \rho(x_{n'}, y_n)| + |\rho(x_{n'}, y_n) - \rho(x_{n'}, y_{n'})|$$

$$\leqslant \rho(x_n, x_{n'}) + \rho(y_n, y_{n'}) < \frac{\varepsilon}{2} + \frac{\varepsilon}{2} = \varepsilon \tag{10}$$

for all sufficiently large n and n'. Therefore the sequence of real numbers $\{s_n\} = \{\rho(x_n, y_n)\}$ is fundamental and hence has a limit. This limit is independent of the choice $\{x_n\} \in x^*$, $\{y_n\} \in y^*$. In fact, suppose

$$\{x_n\}, \{\tilde{x}_n\} \in x^*, \qquad \{y_n\}, \{\tilde{y}_n\} \in y^*.$$

Then

$$|\rho(x_n, y_n) - \rho(\tilde{x}_n, \tilde{y}_n)| \leqslant \rho(x_n, \tilde{x}_n) + \rho(y_n, \tilde{y}_n),$$

by a calculation analogous to (10). But

$$\lim_{n \to \infty} \rho(x_n, \tilde{x}_n) = \lim_{n \to \infty} \rho(y_n, \tilde{y}_n) = 0,$$

since $\{x_n\} \sim \{\tilde{x}_n\}$, $\{y_n\} \sim \{\tilde{y}_n\}$, and hence

$$\lim_{n \to \infty} \rho(x_n, y_n) = \lim_{n \to \infty} \rho(\tilde{x}_n, \tilde{y}_n).$$

As for the three properties of a metric, it is obvious that $\rho_1(x^*, y^*) = \rho_1(y^*, x^*)$, and the fact that $\rho_1(x^*, y^*) = 0$ if and only if $x^* = y^*$ is an immediate consequence of the definition of equivalent Cauchy sequences. To verify the triangle inequality in R^*, we start from the triangle inequality

$$\rho(x_n, z_n) \leqslant \rho(x_n, y_n) + \rho(y_n, z_n)$$

in the original space R and then take the limit as $n \to \infty$, obtaining

$$\lim_{n \to \infty} \rho(x_n, z_n) \leqslant \lim_{n \to \infty} \rho(x_n, y_n) + \lim_{n \to \infty} \rho(y_n, z_n),$$

i.e.,

$$\rho_1(x^*, z^*) \leqslant \rho_1(x^*, y^*) + \rho_1(y^*, z^*).$$

We now come to the crucial step of showing that R^* is a completion of R. Suppose that with every point $x \in R$, we associate the class $x^* \in R^*$ of all Cauchy sequences converging to x. Let

$$x = \lim_{n \to \infty} x_n, \qquad y = \lim_{n \to \infty} y_n.$$

Then clearly

$$\rho(x, y) = \lim_{n \to \infty} \rho(x_n, y_n)$$

(recall Problem 3, p. 54), while on the other hand

$$\rho_1(x^*, y^*) = \lim_{n \to \infty} \rho(x_n, y_n),$$

by definition. Therefore

$$\rho(x, y) = \rho_1(x^*, y^*),$$

and hence the mapping of R into R^* carrying x into x^* is *isometric*. Accordingly, we need no longer distinguish between the original space R and its image in R^*, in particular between the two metrics ρ and ρ_1 (recall the relevant comments on p. 44). In other words, R can be regarded as a subset of R^*. The theorem will be proved once we succeed in showing that

1) R is everywhere dense in R^*, i.e., $[R] = R$;
2) R^* is complete.

To this end, given any point $x^* \in R^*$ and any $\varepsilon > 0$, choose a representative of x^*, namely a Cauchy sequence $\{x_n\}$ in the class x^*. Let N be such that $\rho(x_n, x_{n'}) < \varepsilon$ for all $n, n' > N$. Then

$$\rho(x_n, x^*) = \lim_{n' \to \infty} \rho(x_n, x_{n'}) \leqslant \varepsilon$$

if $n > N$, i.e., every neighborhood of the point x^* contains a point of R. It follows that $[R] = R$.

Finally, to show that R^* is complete, we first note that by the very definition of R^*, any Cauchy sequence $\{x_n\}$ consisting of points in R converges to some point in R^*, namely to the point $x^* \in R^*$ defined by $\{x_n\}$. Moreover, since R is dense in R^*, given any Cauchy sequence

$\{x_n^*\}$ consisting of points in R^*, we can find an equivalent sequence $\{x_n\}$ consisting of points in R. In fact, we need only choose x_n to be any point of R such that $\rho(x_n, x_n^*) < 1/n$. The resulting sequence $\{x_n\}$ is fundamental, and, as just shown, converges to a point $x^* \in R^*$. But then the sequence $\{x_n^*\}$ also converges to x^*. ∎

Example. If R is the space of all rational numbers, then R^* is the space of all real numbers, both equipped with the distance $\rho(x, y) = |x - y|$. In this way, we can "construct the real number system." However, there still remains the problem of suitably defining sums and products of real numbers and verifying that the usual axioms of arithmetic are satisfied.

Problem 1. Prove that the limit $f(t)$ of a uniformly convergent sequence of functions $\{f_n(t)\}$ continuous on $[a, b]$ is itself a function continuous on $[a, b]$.

Hint. Clearly

$$|f(t) - f(t_0)| \leqslant |f(t) - f_n(t)| + |f_n(t) - f_n(t_0)| + |f_n(t_0) - f(t_0)|,$$

where $t, t_0 \in [a, b]$. Use the uniform convergence to make the sum of the first and third terms on the right small for sufficiently large n. Then use the continuity of $f_n(t)$ to make the second term small for t sufficiently close to t_0.

Problem 2. Prove that the space m in Example 9, p. 41 is complete.

Problem 3. Prove that if R is complete, then the intersection $\bigcap_{n=1}^{\infty} S_n$ figuring in Theorem 2 consists of a single point.

Problem 4. By the *diameter* of a subset A of a metric space R is meant the number

$$d(A) = \sup_{x, y \in A} \rho(x, y).$$

Suppose R is complete, and let $\{A_n\}$ be a sequence of closed subsets of R *nested* in the sense that

$$A_1 \supset A_2 \supset \cdots \supset A_n \supset \cdots.$$

Suppose further that

$$\lim_{n \to \infty} d(A_n) = 0.$$

Prove that the intersection $\bigcap_{n=1}^{\infty} A_n$ is nonempty.

Problem 5. A subset A of a metric space R is said to be *bounded* if its diameter $d(A)$ is finite. Prove that the union of a finite number of bounded sets is bounded.

Problem 6. Give an example of a complete metric space R and a nested sequence $\{A_n\}$ of closed subsets of R such that

$$\bigcap_{n=1}^{\infty} A_n = \varnothing.$$

Reconcile this example with Problem 4.

Problem 7. Prove that a subspace of a complete metric space R is complete if and only if it is closed.

Problem 8. Prove that the real line equipped with the distance

$$\rho(x, y) = |\text{arc tan } x - \text{arc tan } y|$$

is an incomplete metric space.

Problem 9. Give an example of a complete metric space homeomorphic to an incomplete metric space.

Hint. Consider the example on p. 44.

Comment. Thus homeomorphic metric spaces can have different "metric properties."

Problem 10. Carry out the program discussed in the last sentence of the example on p. 65.

Hint. If $\{x_n\}$ and $\{y_n\}$ are Cauchy sequences of rational numbers serving as "representatives" of real numbers x^* and y^*, respectively, define $x^* + y^*$ as the real number with representative $\{x_n + y_n\}$.

8. Contraction Mappings

8.1. Definition of a contraction mapping. The fixed point theorem. Let A be a mapping of a metric space R into itself. Then x is called a *fixed point* of A if $Ax = x$, i.e., if A carries x into itself. Suppose there exists a number $\alpha < 1$ such that

$$\rho(Ax, Ay) \leqslant \alpha \rho(x, y) \tag{1}$$

for every pair of points $x, y \in R$. Then A is said to be a *contraction mapping*. Every contraction mapping is automatically continuous, since it follows from the "contraction condition" (1) that $Ax_n \to Ax$ whenever $x_n \to x$.

THEOREM 1 (*Fixed point theorem*[10]). *Every contraction mapping A defined on a complete metric space R has a unique fixed point.*

---

[10] Often called the *method of successive approximations* (see the remark following Theorem 1) or the *principle of contraction mappings*.

Proof. Given an arbitrary point x_0 R, let[11]

$$x_1 = Ax_0, \quad x_2 = Ax_1 = A^2x_0, \ldots, \quad x_n = Ax_{n-1} = A^nx_0, \ldots \quad (2)$$

Then the sequence $\{x_n\}$ is fundamental. In fact, assuming to be explicit that $n \leqslant n'$, we have

$$\rho(x_n, x_{n'}) = \rho(A^nx_0, A^{n'}x_0) \leqslant \alpha^n\rho(x_0, x_{n'-n})$$

$$\leqslant \alpha^n[\rho(x_0, x_1) + \rho(x_1, x_2) + \cdots + \rho(x_{n'-n-1}, x_{n'-n})]$$

$$\leqslant \alpha^n\rho(x_0, x_1)[1 + \alpha + \alpha^2 + \cdots + \alpha^{n'-n-1}] < \alpha^n\rho(x_0, x_1)\frac{1}{1-\alpha}.$$

But the expression on the right can be made arbitrarily small for sufficiently large n, since $\alpha < 1$. Since R is complete, the sequence $\{x_n\}$, being fundamental, has a limit

$$x = \lim_{n \to \infty} x_n.$$

Then, by the continuity of A,

$$Ax = A \lim_{n \to \infty} x_n = \lim_{n \to \infty} Ax_n = \lim_{n \to \infty} x_{n+1} = x.$$

This proves the existence of a fixed point x. To prove the uniqueness of x, we note that if

$$Ax = x, \quad Ay = y,$$

(1) becomes

$$\rho(x, y) \leqslant \alpha\rho(x, y).$$

But then $\rho(x, y) = 0$ since $\alpha < 1$, and hence $x = y$. ∎

Remark. The fixed point theorem can be used to prove existence and uniqueness theorems for solutions of equations of various types. Besides showing that an equation of the form $Ax = x$ has a unique solution, the fixed point theorem also gives a practical method for finding the solution, i.e., calculation of the "successive approximations" (2). In fact, as shown in the proof, the approximations (2) actually converge to the solution of the equation $Ax = x$. For this reason, the fixed point theorem is often called the *method of successive approximations.*

Example 1. Let f be a function defined on the closed interval $[a, b]$ which which maps $[a, b]$ into itself and satisfies a Lipschitz condition

$$|f(x_1) - f(x_2)| \leqslant K|x_1 - x_2|, \quad (3)$$

with constant $K < 1$. Then f is a contraction mapping, and hence, by

---

[11] A^2x means $A(Ax)$, A^3x means $A(A^2x) = A^2(Ax)$, and so on.

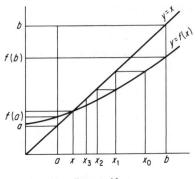

FIGURE 10

Theorem 1, the sequence

$$x_0, \qquad x_1 = f(x_0), \qquad x_2 = f(x_1), \ldots \tag{4}$$

converges to the unique root of the equation $f(x) = x$. In particular, the "contraction condition" (3) holds if f has a continuous derivative f' on $[a, b]$ such that

$$|f'(x)| \leqslant K < 1.$$

The behavior of the successive approximations (4) in the cases $0 < f'(x) < 1$ and $-1 < f'(x) < 0$ is shown in Figures 10 and 11.

Example 2. Consider the mapping A of n-dimensional space into itself given by the system of linear equations

$$y_i = \sum_{j=1}^{n} a_{ij} x_j + b_i \qquad (i = 1, \ldots, n). \tag{5}$$

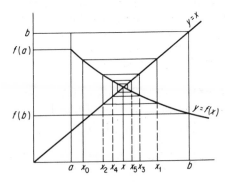

FIGURE 11

If A is a contraction mapping, we can use the method of successive approximations to solve the equation $Ax = x$. The conditions under which A is a contraction mapping depend on the choice of metric. We now examine three cases:

1) The space R_0^n with metric

$$\rho(x, y) = \max_{1 \leqslant i \leqslant n} |x_i - y_i|.$$

In this case,

$$\rho(y, \tilde{y}) = \max_i |y_i - \tilde{y}_i| = \max_i \left| \sum_j a_{ij}(x_j - \tilde{x}_j) \right|$$

$$\leqslant \max_i \sum_j |a_{ij}| \, |x_j - \tilde{x}_j|$$

$$\leqslant \max_i \sum_j |a_{ij}| \max_j |x_j - \tilde{x}_j| = \left(\max_i \sum_j |a_{ij}| \right) \rho(x, \tilde{x}),$$

and the contraction condition

$$\sum_j |a_{ij}| \leqslant \alpha < 1 \qquad (i = 1, \ldots, n). \tag{6}$$

2) The space R_1^n with metric

$$\rho(x, y) = \sum_{i=1}^n |x_i - y_i|.$$

Here

$$\rho(y, \tilde{y}) = \sum_i |y_i - \tilde{y}_i| = \sum_i \left| \sum_j a_{ij}(x_j - \tilde{x}_j) \right|$$

$$\leqslant \sum_i \sum_j |a_{ij}| \, |x_j - \tilde{x}_j| \leqslant \left(\max_j \sum_i |a_{ij}| \right) \rho(x, \tilde{x}),$$

and the contraction condition is now

$$\sum_i |a_{ij}| \leqslant \alpha < 1 \qquad (j = 1, \ldots, n). \tag{7}$$

3) Ordinary Euclidean space R^n with metric

$$\rho(x, y) = \sqrt{\sum_{i=1}^n (x_i - y_i)^2}.$$

Using the Cauchy-Schwarz inequality, we have

$$\rho^2(y, \tilde{y}) = \sum_i \left(\sum_j a_{ij}(x_j - \tilde{x}_j) \right)^2 \leqslant \left(\sum_i \sum_j a_{ij}^2 \right) \rho^2(x, \tilde{x}),$$

and the contraction condition becomes

$$\sum_i \sum_j a_{ij}^2 \leqslant \alpha < 1. \tag{8}$$

Thus, if at least one of the conditions (6)–(8) holds, there exists a unique point $x = (x_1, x_2, \ldots, x_n)$ such that

$$x_i = \sum_{j=1}^{n} a_{ij}x_j + b_i \qquad (i = 1, \ldots, n). \tag{9}$$

The sequence of successive approximations to this solution of the equation $x = Ax$ are of the form

$$x^{(0)} = (x_1^{(0)}, x_2^{(0)}, \ldots, x_n^{(0)}),$$
$$x^{(1)} = (x_1^{(1)}, x_2^{(1)}, \ldots, x_n^{(1)}),$$
$$\cdots \cdots \cdots \cdots \cdots \cdots$$
$$x^{(k)} = (x_1^{(k)}, x_2^{(k)}, \ldots, x_n^{(k)}),$$
$$\cdots \cdots \cdots \cdots \cdots \cdots$$

where

$$x_i^{(k)} = \sum_{j=1}^{n} a_{ij}x_j^{(k-1)} + b_i,$$

and we can choose any point $x^{(0)}$ as the "zeroth approximation."

Each of the conditions (6)–(8) is *sufficient* for applicability of the method of successive approximations, but none of them is *necessary*. In fact, examples can be constructed in which each of the conditions (6)–(8) is satisfied, but not the other two.

Theorem 1 has the following useful generalization, which will be needed later (see Example 2, p. 75):

THEOREM 1′. *Given a continuous mapping of a complete metric space R into itself, suppose A^n is a contraction mapping (n an integer > 1). Then A has a unique fixed point.*

Proof. Choosing any point $x_0 \in R$, let

$$x = \lim_{k \to \infty} A^{kn}x_0.$$

Then, by the continuity of A,

$$Ax = \lim_{k \to \infty} AA^{kn}x_0.$$

But A^n is a contraction mapping, and hence

$$\rho(A^{kn}Ax_0, A^{kn}x_0) \leqslant \alpha\rho(A^{(k-1)n}Ax_0, A^{(k-1)n}x_0) \leqslant \cdots \leqslant \alpha^k\rho(Ax_0, x_0)$$

where $\alpha < 1$. It follows that

$$\rho(Ax, x) = \lim_{k \to \infty} \rho(AA^{kn}x_0, A^{kn}x_0) = 0,$$

i.e., $Ax = x$ so that x is a fixed point of A. To prove the uniqueness of x, we merely note that if A has more than one fixed point, then so does A^n, which is impossible, by Theorem 1, since A^n is a contraction mapping. ∎

8.2. Contraction mappings and differential equations. The most interesting applications of Theorems 1 and 1' arise when the space R is a function space. We can then use these theorems to prove a number of existence and uniqueness theorems for differential and integral equations, as shown in this section and the next.

THEOREM 2 (*Picard*). *Given a function $f(x, y)$ defined and continuous on a plane domain G containing the point (x_0, y_0),[12] suppose f satisfies a Lipschitz condition of the form*

$$|f(x, y) - f(x, \tilde{y})| \leqslant M\,|y - \tilde{y}|$$

in the variable y. Then there is an interval $|x - x_0| \leqslant \delta$ in which the differential equation

$$\frac{dy}{dx} = f(x, y) \tag{10}$$

has a unique solution

$$y = \varphi(x)$$

satisfying the initial condition

$$\varphi(x_0) = y_0. \tag{11}$$

Proof. Together the differential equation (10) and the initial condition (11) are equivalent to the integral equation

$$\varphi(x) = y_0 + \int_{x_0}^{x} f(t, \varphi(t))\,dt. \tag{12}$$

By the continuity of f, we have

$$|f(x, y)| < K \tag{13}$$

in some domain $G' \subset G$ containing the point (x_0, y_0).[13] Choose $\delta > 0$ such that

1) $(x, y) \in G'$ if $|x - x_0| \leqslant \delta$, $|y - y_0| \leqslant K\delta$;
2) $M\delta < 1$,

and let C^* be the space of continuous functions φ defined on the interval

---

[12] By an n-dimensional *domain* we mean an open connected set in Euclidean n-space R^n (connectedness is defined in Problem 12, p. 55).

[13] In fact, f is bounded on $[G']$ if $[G'] \subset G$ (cf. Theorem 2, p. 110).

$|x - x_0| \leqslant \delta$ and such that $|\varphi(x) - y_0| \leqslant K\delta$, equipped with the metric

$$\rho(\varphi, \tilde{\varphi}) = \max_x |\varphi(x) - \tilde{\varphi}(x)|.$$

The space C^* is complete, since it is a closed subspace of the space of all continuous functions on $[x_0 - \delta, x_0 + \delta]$. Consider the mapping $\psi = A\varphi$ defined by the integral equation

$$\psi(x) = y_0 + \int_{x_0}^x f(t, \varphi(t))\, dt \qquad (|x - x_0| \leqslant \delta).$$

Clearly A is a contraction mapping carrying C^* into itself. In fact, if $\varphi \in C^*$, $|x - x_0| \leqslant \delta$ then

$$|\psi(x) - y_0| = \left| \int_{x_0}^x f(t, \varphi(t))\, dt \right| \leqslant \int_{x_0}^x |f(t, \varphi(t))|\, dt \leqslant K\,|x - x_0| \leqslant K\delta$$

by (13), and hence $\psi = A\varphi$ also belongs to C^*. Moreover,

$$|\psi(x) - \tilde{\psi}(x)| \leqslant \int_{x_0}^x |f(t, \varphi(t)) - f(t, \tilde{\varphi}(t))|\, dt \leqslant M\delta \max_x |\varphi(x) - \tilde{\varphi}(x)|,$$

and hence

$$\rho(\psi, \tilde{\psi}) \leqslant M\delta\rho(\varphi, \tilde{\varphi})$$

after maximizing with respect to x. But $M\delta < 1$, so that A is a contraction mapping. It follows from Theorem 1 that the equation $\varphi = A\varphi$, i.e., the integral equation (12), has a unique solution in the space C^*. ∎

Theorem 2 can easily be generalized to the case of *systems* of differential equations:

THEOREM 2′. *Given n functions $f_i(x, y_1, \ldots, y_n)$ defined and continuous on an $(n + 1)$-dimensional domain G containing the point*

$$(x_0, y_{01}, \ldots, y_{0n}),$$

suppose each f_i satisfies a Lipschitz condition of the form

$$|f_i(x, y_1, \ldots, y_n) - f_i(x, \tilde{y}_1, \ldots, {}_n\tilde{y})| \leqslant M \max_{1 \leqslant i \leqslant n} |y_i - \tilde{y}_i|$$

in the variables $y_1, \ldots, y_n$. Then there is an interval $|x - x_0| \leqslant \delta$ in which the system of differential equations

$$\frac{dy_i}{dx} = f_i(x, y_1, \ldots, y_n) \qquad (i = 1, \ldots, n) \tag{14}$$

has a unique solution

$$y_1 = \varphi_1(x), \ldots, y_n = \varphi_n(x)$$

satisfying the initial conditions

$$\varphi_1(x_0) = y_{01}, \ldots, \varphi_n(x_0) = y_{0n}. \tag{15}$$

Proof. Together the differential equations (14) and the initial conditions (15) are equivalent to the system of integral equations

$$\varphi_i(x) = y_{0i} + \int_{x_0}^{x} f_i(t, \varphi_1(t), \ldots, \varphi_n(t))\, dt \qquad (i = 1, \ldots, n). \tag{16}$$

By the continuity of the functions f_i, we have

$$|f_i(x, y_1, \ldots, y_n)| \leqslant K \qquad (i = 1, \ldots, n) \tag{17}$$

in some domain $G' \subset G$ containing the point $(x_0, y_{01}, \ldots, y_{0n})$. Choose $\delta > 0$ such that

1) $(x, y_1, \ldots, y_n) \in G'$ if $|x - x_0| \leqslant \delta$, $|y_i - y_{0i}| \leqslant K\delta$ for all $i = 1, \ldots, n$;

2) $M\delta < 1$.

This time let C^* be the space of ordered n-tuples

$$\varphi = (\varphi_1, \ldots, \varphi_n)$$

of continuous functions $\varphi_1, \ldots, \varphi_n$ defined on the interval $|x - x_0| \leqslant \delta$ such that $|\varphi_i(x) - y_{0i}| \leqslant K\delta$ for all $i = 1, \ldots, n$, equipped with the metric

$$\rho(\varphi, \tilde{\varphi}) = \max_{x,i} |\varphi_i(x) - \tilde{\varphi}_i(x)|.$$

Clearly C^* is complete. Moreover, the mapping $\psi = A\varphi$ defined by the system of integral equations

$$\psi_i(x) = y_{0i} + \int_{x_0}^{x} f_i(t, \varphi_1(t), \ldots, \varphi_n(t))\, dt$$

$$(|x - x_0| \leqslant \delta,\, i = 1, \ldots, n)$$

is a contraction mapping carrying C^* into itself. In fact, if

$$\varphi = (\varphi_1, \ldots, \varphi_n) \in C^*, \qquad |x - x_0| \leqslant \delta,$$

then

$$|\psi_i(x) - y_{0i}| = \left| \int_{x_0}^{x} f_i(t, \varphi_1(t), \ldots, \varphi_n(t))\, dt \right| \leqslant K\delta \qquad (i = 1, \ldots, n)$$

by (17), so that $\psi = (\psi_1, \ldots, \psi_n) = A\varphi$ also belongs to C^*. Moreover,

$$|\psi_i(x) - \tilde{\psi}_i(x)| = \int_{x_0}^{x} |f_i(t, \varphi_1(t), \ldots, \varphi_n(t)) - f_i(t, \tilde{\varphi}_1(t), \ldots, \tilde{\varphi}_n(t))|\, dt$$

$$\leqslant M\delta \max_i |\varphi_i(t) - \tilde{\varphi}_i(t)|,$$

and hence

$$\rho(\psi, \tilde{\psi}) \leqslant M\delta\rho(\varphi, \tilde{\varphi})$$

after maximizing with respect to x and i. But $M\delta < 1$, so that A is a contraction mapping. It follows from Theorem 1 that the equation $\varphi = A\varphi$, i.e., the system of integral equations (16), has a unique solution in the space C^*. ∎

8.3. Contraction mappings and integral equations. We now show how the method of successive approximations can be used to prove the existence and uniqueness of solutions of integral equations.

Example 1. By a *Fredholm equation* (of the second kind) is meant an integral equation of the form

$$f(x) = \lambda \int_a^b K(x, y) f(y) \, dy + \varphi(x), \tag{18}$$

involving two given functions K and φ, an unknown function f and an arbitrary parameter λ. The function K is called the *kernel* of the equation, and the equation is said to be *homogeneous* if $\varphi \equiv 0$ (but otherwise *non-homogeneous*).

Suppose $K(x, y)$ and $\varphi(x)$ are continuous on the square $a \leqslant x \leqslant b$, $a \leqslant y \leqslant b$, so that in particular

$$|K(x, y)| \leqslant M \qquad (a \leqslant x \leqslant b, a \leqslant y \leqslant b).$$

Consider the mapping $g = Af$ of the complete metric space $C_{[a,b]}$ into itself given by

$$g(x) = \lambda \int_a^b K(x, y) f(y) \, dy + \varphi(x).$$

Clearly, if $g_1 = Af_1$, $g_2 = Af_2$, then

$$\rho(g_1, g_2) = \max_x |g_1(x) - g_2(x)| \leqslant |\lambda| \, M(b - a) \max_x |f_1(x) - f_2(x)|$$
$$= |\lambda| \, M(b - a)\rho(f_1, f_2),$$

so that A is a contraction mapping if

$$|\lambda| < \frac{1}{M(b - a)}. \tag{19}$$

It follows from Theorem 1 that the integral equation (18) has a unique solution for any value of λ satisfying (19). The successive approximations $f_0, f_1, \ldots, f_n, \ldots$ to this solution are given by

$$f_n(x) = \lambda \int_a^b K(x, y) f_{n-1}(y) \, dy + \varphi(x) \qquad (n = 1, 2, \ldots),$$

where any function continuous on $[a, b]$ can be chosen as f_0. Note that the method of successive approximations can be applied to the equation (18) only for sufficiently small $|\lambda|$.

Example 2. Next consider the *Volterra equation*

$$f(x) = \lambda \int_a^x K(x, y) f(y) \, dy + \varphi(x), \qquad (20)$$

which differs from the Fredholm equation (18) by having the variable x rather than the fixed number b as the upper limit of integration.[14] It is easy to see that the method of successive approximations can be applied to the Volterra equation (20) for *arbitrary* λ, not just for sufficiently small $|\lambda|$ as in the case of the Fredholm equation (18). In fact, let A be the mapping of $C_{[a,b]}$ into itself defined by

$$Af(x) = \lambda \int_a^x K(x, y) f(y) \, dy + \varphi(x),$$

and let $f_1, f_2 \in C_{[a,b]}$. Then

$$|Af_1(x) - Af_2(x)| = \lambda \int_a^x K(x, y)[f_1(y) - f_2(y)] \, dy$$

$$\leqslant \lambda M(x - a) \max_x |f_1(x) - f_2(x)|,$$

where

$$M = \max_{x,y} |K(x, y)|.$$

It follows that

$$|A^2 f_1(x) - A^2 f_2(x)| < \lambda^2 M^2 \max_x |f_1(x) - f_2(x)| \int_a^x (x - a) \, dx$$

$$= \lambda^2 M^2 \frac{(x - a)^2}{2} \max_x |f_1(x) - f_2(x)|,$$

and in general,

$$|A^n f_1(x) - A^n f_2(x)| < \lambda^n M^n \frac{(x - a)^n}{n!} \max_x |f_1(x) - f_2(x)|$$

$$\leqslant \lambda^n M^n \frac{(b - a)^n}{n!} \max_x |f_1(x) - f_2(x)|,$$

which implies

$$\rho(A^n f_1, A^n f_2) < \lambda^n M^n \frac{(b - a)^n}{n!} \rho(f_1, f_2).$$

----

[14] Equation (20) can be regarded formally as a special case of (18) by extending the definition of the kernel, i.e., by setting

$$K(x, y) = 0 \quad \text{if} \quad y > x.$$

But, given any λ, we can always choose n large enough to make

$$\lambda^n M^n \frac{(b-a)^n}{n!} < 1,$$

i.e., A^n is a contraction mapping for sufficiently large n. It follows from Theorem 1' that the integral equation (20) has a unique solution for arbitrary λ.

Problem 1. Let A be a mapping of a metric space R into itself. Prove that the condition

$$\rho(Ax, Ay) < \rho(x, y) \qquad (x \neq y)$$

is insufficient for the existence of a fixed point of A.

Problem 2. Let $F(x)$ be a continuously differentiable function defined on the interval $[a, b]$ such that $F(a) < 0$, $F(b) > 0$ and

$$0 < K_1 \leqslant F'(x) \leqslant K_2 \qquad (a \leqslant x \leqslant b).$$

Use Theorem 1 to find the unique root of the equation $F(x) = 0$.

Hint. Introduce the auxiliary function $f(x) = x - \lambda F(x)$, and choose λ such that the theorem works for the equivalent equation $f(x) = x$.

Problem 3. Devise a proof of the implicit function theorem based on the use of the fixed point theorem.[15]

Problem 4. Prove that the method of successive approximations can be used to solve the system (9) if $|a_{ij}| < 1/n$ (for all i and j), but not if $|a_{ij}| = 1/n$.

Problem 5. Prove that the condition (6) is necessary for the mapping (5) to be a contraction mapping in the space R_0^n.

Problem 6. Prove that any of the conditions (6)–(8) implies

$$\begin{vmatrix} a_{11} - 1 & a_{12} & \cdots & a_{1n} \\ a_{21} & a_{22} - 1 & \cdots & a_{2n} \\ \cdot & \cdot & \cdots & \cdot \\ a_{n1} & a_{n2} & \cdots & a_{nn} - 1 \end{vmatrix} \neq 0.$$

Comment. Hence the fact that the system (5) has a unique solution (under suitable conditions) follows from Cramer's rule as well as from the fixed point theorem.

---

[15] See e.g., I. G. Petrovski, *Ordinary Differential Equations* (translated by R. A. Silverman), Prentice-Hall, Inc., Englewood Cliffs, N.J. (1966), p. 47.

Problem 7. Consider the nonlinear integral equation

$$f(x) = \lambda \int_a^b K(x, y; f(y)) \, dy + \varphi(x) \tag{21}$$

with continuous K and φ, where K satisfies a Lipschitz condition of the form

$$|K(x, y; z_1) - K(x, y; z_2)| \leqslant M |z_1 - z_2|$$

in its "functional" argument. Prove that (21) has a unique solution for all

$$|\lambda| < \frac{1}{M(b - a)}.$$

Write the successive approximations to this solution.

3

TOPOLOGICAL SPACES

9. Basic Concepts

9.1. Definitions and examples. In our study of metric spaces, we defined a number of key ideas like contact point, limit point, closure of a set, etc. In each case, the definition rests on the notion of a neighborhood, or, what amounts to the same thing, the notion of an open set. These notions (neighborhood and open set) were in turn defined by using the metric (or distance) in the given space. However, instead of introducing a metric in a given set X, we can go about things differently, by specifying a system of open sets in X with suitable properties. This approach leads to the notion of a *topological space*. Metric spaces are topological spaces of a rather special (although very important) kind.

DEFINITION 1. *Given a set X, by a **topology** in X is meant a system τ of subsets $G \subset X$, called **open sets** (relative to τ), with the following two properties*:

1) *The set X itself and the empty set $\varnothing$ belong to τ;*
2) *Arbitrary (finite or infinite) unions $\bigcup_{\alpha} G_\alpha$ and finite intersections $\bigcap_{k=1}^{n} G_k$ of open sets belong to τ.*

DEFINITION 2. *By a **topological space** is meant a pair (X, τ), consisting of a set X and a topology τ defined in X.*

Just as a metric space is a pair consisting of a set X and a metric defined in X, so a topological space is a pair consisting of a set X and a topology defined in X. Thus, to specify a topological space, we must specify both a set X and a topology in X, i.e., we must indicate which subsets of X are to be regarded as "open (in X)." Clearly, we can equip one and the same set with various different topologies, thereby defining various different topological spaces. Nevertheless, we will usually denote a topological space, namely a pair (X, τ), by a single letter like T. Just as in the case of a metric space R, the elements of a topological space T will be called the *points* of T.

By the *closed sets* of a topological space T, we mean the complements $T - G$ of the open sets G of T. It follows from Definition 1 and the "duality principle" (see p. 4) that

    1') *The space T itself and the empty set $\varnothing$ are closed*;
    2') *Arbitrary (finite or infinite) intersections* $\bigcap_\alpha F_\alpha$ *and finite unions* $\bigcup_{k=1}^{n} F_k$ *of closed sets of T are closed.*

The natural way of introducing the concepts of neighborhood, contact point, limit point and closure of a set is now apparent:

    a) By a *neighborhood* of a point x in a topological space T is meant any open set $G \subset T$ containing x;
    b) A point $x \in T$ is called a *contact point* of a set $M \subset T$ if every neighborhood of x contains at least one point of M;
    c) A point $x \in T$ is called a *limit point* of a set $M \subset T$ if every neighborhood of x contains infinitely many points of M;
    d) The set of all contact points of a set $M \subset T$ is called the *closure* of M, denoted by $[M]$.

Example 1. According to Theorem 5, p. 50, the open sets in any metric space satisfy the two properties in Definition 1. Hence every metric space is a topological space as well.

Example 2. Given any set T, suppose we regard *every* subset of T as open. Then T is a topological space (the properties in Definition 1 are obviously satisfied). In particular, every set $M \subset T$ is both open and closed, and every set $M \subset T$ coincides with its own closure. Note that the "discrete metric space" of Example 1, p. 38 has this trivial topology.

Example 3. As another extreme case, consider an arbitrary set T equipped with a topology consisting of just two sets, the whole set T and the empty set $\varnothing$. Then T is a topological space, a kind of "space of coalesced points" (mainly of academic interest). Note that the closure of every nonempty set is the whole space T.

Example 4. Let T be the set $\{a, b\}$, consisting of just two points a and b, and let the open sets in T be T itself, the empty set and the single-element set $\{b\}$. Then the two properties in Definition 1 are satisfied, and T is a topological space. The closed sets in this space are T itself, the empty set and the set $\{a\}$. Note that the closure of $\{b\}$ is the whole space T.

9.2. Comparison of topologies. Let τ_1 and τ_2 be two topologies defined in the same set X.[1] Then we say that the topology τ_1 is *stronger* than the topology τ_2 (or equivalently that τ_2 is *weaker* than τ_1) if $\tau_2 \subset \tau_1$, i.e., if every set of the system τ_2 is a set of the system τ_1.

THEOREM 1. *The intersection* $\tau = \bigcap_\alpha \tau_\alpha$ *of any set of topologies in* X *is itself a topology in* X.

Proof. Clearly $\bigcap_\alpha \tau_\alpha$ contains X and $\varnothing$. Moreover, since every τ_α is closed (algebraically) under the operations of taking arbitrary unions and finite intersections, the same is true of $\bigcap_\alpha \tau_\alpha$. ∎

COROLLARY. *Let $\mathscr{B}$ be any system of subsets of a set* X. *Then there exists a minimal topology in* X *containing* $\mathscr{B}$, *i.e., a topology* $\tau(\mathscr{B})$ *containing* $\mathscr{B}$ *and contained in every topology containing* $\mathscr{B}$.

Proof. A topology containing $\mathscr{B}$ always exists, e.g., the topology in which every subset of X is open. The intersection of all topologies containing $\mathscr{B}$ is the desired minimal topology $\tau(\mathscr{B})$, often called the topology *generated* by the system $\mathscr{B}$. ∎

Let $\mathscr{B}$ be a system of subsets of X and A a fixed subset of X. Then by the *trace* of the system $\mathscr{B}$ on the set A we mean the system $\mathscr{B}_A$ consisting of all subsets of X of the form $A \cap B$, $B \in \mathscr{B}$. It is easy to see that the trace (on A) of a topology τ (defined in X) is a topology τ_A in A. (Such a topology is often called a *relative* topology.) In this sense, every subset A of a given topological space (X, τ) generates a new topological space (A, τ_A), called a *subspace* of the original topological space (X, τ).

9.3. Bases. Axioms of countability. As we have seen, defining a topology in a space T means specifying a system of open sets in T. However, in many concrete problems, it is more convenient to specify, instead of all the open sets, some system of subsets which uniquely determines all the open sets. For example, in the case of a metric space we first introduced the notion of an open sphere (ε-neighborhood) and then defined an open set G as a set such that every point $x \in G$ has a neighborhood $O_\varepsilon(x) \subset G$. In other words, the open sets in a metric space are precisely those which can be represented as finite or infinite unions of open spheres. In particular, the open sets on the real line are precisely those which can be represented as finite or *countable* unions of open intervals (recall Theorem 6, p. 51). These considerations suggest

---

[1] This gives *two* topological spaces $T_1 = (X, \tau_1)$ and $T_2 = (X, \tau_2)$.

DEFINITION 3. *A family $\mathscr{G}$ of open subsets of a topological space T is called a **base** for T if every open set in T can be represented as a union of sets in $\mathscr{G}$.*

Example 1. The set of all open spheres (of all possible radii and with all possible centers) in a metric space R is a base for R. In particular, the set of all open intervals is a base on the real line. The set of all open intervals with rational end points is also a base on the line, since any open interval (and hence any open set on the line) can be represented as a union of such intervals.

It is clear from the foregoing that a topology τ can be defined in a set T by specifying a base $\mathscr{G}$ in T. This topology τ is just the system of sets which can be represented as unions of sets in $\mathscr{G}$. If this way of specifying a topology is to be of practical value, we must find requirements which, when imposed on a system $\mathscr{G}$ of subsets of a given set T, guarantee that the system τ of all possible unions of sets in $\mathscr{G}$ be a topology in T, i.e., that τ have the two properties figuring in Definition 1:

THEOREM 2. *Given a set T, let $\mathscr{G}$ be a system of subsets $G_\alpha \subset T$ with the following two properties:*

1) *Every point $x \in T$ belongs to at least one $G_\alpha \in \mathscr{G}$;*

2) *If $x \in G_\alpha \cap G_\beta$, then there is a $G_\gamma \in \mathscr{G}$ such that $x \in G_\gamma \subset G_\alpha \cap G_\beta$.*

Suppose the empty set $\varnothing$ and all sets representable as unions of sets G_α are designated as open. Then T is a topological space, and $\mathscr{G}$ is a base for T.

Proof. It follows at once from the conditions of the theorem that the whole set T and the empty set $\varnothing$ are open sets, and that the union of any number of open sets is open. We must still show that the intersection of a finite number of open sets is open. It is enough to prove this for just two sets. Thus let

$$A = \bigcup_\alpha G_\alpha, \qquad B = \bigcup_\beta G_\beta.$$

Then

$$A \cap B = \bigcup_{\alpha,\beta} (G_\alpha \cap G_\beta). \qquad (1)$$

By hypothesis, given any point $x \in G_\alpha \cap G_\beta$, there is a $G_\gamma \in \mathscr{G}$ such that $x \in G_\gamma \subset G_\alpha \cap G_\beta$. Hence the set $G_\alpha \cap G_\beta$ is open, being the union of all G_γ contained in $G_\alpha \cap G_\beta$. But then (1) is also open. Therefore T is a topological space. The fact that $\mathscr{G}$ is a base for T is clear from the way open sets in T are defined. $\blacksquare$

The following theorem is a useful tool for deciding whether or not a given system of open sets is a base:

THEOREM 3. *A system $\mathscr{G}$ of open sets G_α in a topological space T is a base for T if and only if, given any open set $G \subset T$ and any point $x \in G$, there is a set $G_\alpha \in \mathscr{G}$ such that $x \in G_\alpha \subset G$.*

Proof. If $\mathscr{G}$ is a base for T, then every open set $G \subset T$ is a union

$$G = \bigcup_\alpha G_\alpha$$

of sets $G_\alpha \in \mathscr{G}$. Therefore every point $x \in G$ is contained in some set $G_\alpha \subset G$. Conversely, given any open set $G \subset T$, suppose that for every point $x \in G$ there is a set $G_\alpha(x) \in \mathscr{G}$ such that $x \in G_\alpha(x) \subset G$. Then

$$G = \bigcup_{x \in G} G_\alpha(x),$$

i.e., G is a union of sets in $\mathscr{G}$. ∎

Example 2. It follows from Theorem 3 that the set of all open spheres with rational radii (and all possible centers) in a metric space R is a base for R (this is obvious anyway). In particular, as already noted in Example 1, the set of all open intervals with rational end points is a base for the real line.

An important class of topological spaces consists of *spaces with a countable base*, i.e., spaces in which there is at least one base containing no more than countably many sets. Such a space is also said to satisfy the *second axiom of countability*.

THEOREM 4. *If a topological space T has a countable base, then T contains a countable everywhere dense subset, i.e., a countable set $M \subset T$ such that $[M] = T$.*

Proof. Let $\mathscr{G} = \{G_1, G_2, \ldots, G_n, \ldots\}$ be a countable base for T, and choose a point x_n in each G_n. Then the set

$$M = \{x_1, x_2, \ldots, x_n, \ldots\}$$

is countable. Moreover, M is everywhere dense in T, since otherwise the nonempty open set $G = T - [M]$ would contain no points of M. But this is impossible, since G is a union of some of the sets G_n in $\mathscr{G}$ and G_n contains the point $x_n \in M$. ∎

For metric spaces, we can say even more:

THEOREM 5. *If a metric space R has a countable everywhere dense subset, then R has a countable base.*

Proof. Suppose R has a countable everywhere dense subset $\{x_1, x_2, \ldots, x_n, \ldots\}$. Then, given any open set $G \subset R$ and any $x \in G$, there is an open sphere $S(x_m, 1/n)$ such that $x \in S(x_m, 1/n) \subset G$ for suitable positive integers m and n (why?). Hence the open spheres $S(x_m, 1/n)$, where m and n range over all positive integers, form a countable base for R. ∎

Combining Theorems 4 and 5, we see that *a metric space R has a countable base if and only if it has a countable everywhere dense subset.*

Example 3. Every separable metric space, i.e., every metric space with a countable everywhere dense subset, is a metric space satisfying the second axiom of countability.

Example 4. The space m of all bounded sequences is not separable (recall Example 7, p. 48) and hence has no countable base.

Remark. In general, Theorem 5 does not hold for arbitrary (nonmetric) topological spaces. In fact, examples can be given of topological spaces which have a countable everywhere dense subset but no countable base. Let us see how this might come about. Given any point x of a metric space R, there is a countable *neighborhood base* (or *local base*) at x, i.e., a countable system $\mathcal{O}$ of neighborhoods of x with the following property: Given any open set G containing x, there is a neighborhood $O \in \mathcal{O}$ such that $O \subset G$ (cf. Theorem 3).[2] Suppose every point x of a topological space T has a countable neighborhood base. Then T is said to satisfy the *first axiom of countability*. However, this axiom need not be satisfied in an arbitrary topological space. Hence the argument used in the case of metric spaces to deduce the existence of a countable base from that of a countable everywhere dense subset does not carry over to the case of an arbitrary topological space.

A system $\mathcal{M}$ of sets M_α is called a *cover* (or *covering*) of a topological space T, and $\mathcal{M}$ is said to *cover* T, if

$$T = \bigcup_\alpha M_\alpha.$$

A cover consisting of open (or closed) sets only is called an *open* (or *closed*) cover. If $\mathcal{M}$ is a cover of a topological space T, then by a *subcover* of $\mathcal{M}$ we mean any subset of $\mathcal{M}$ which also covers T.

THEOREM 6. *If T is a topological space with a countable base $\mathcal{G}$, then every open cover $\mathcal{O}$ has a finite or countable subcover.*

Proof. Since $\mathcal{O}$ covers T, each point $x \in T$ belongs to some open set $O_\alpha \in \mathcal{O}$. Moreover, since $\mathcal{G}$ is a countable base for T, for each $x \in T$ there is a set $G_n(x) \in \mathcal{G}$ such that $x \in G_n(x) \subset O_\alpha$ (recall Theorem 3).

---

[2] For example, the set of open spheres $S(x, 1/n)$ is a countable neighborhood base at any point x of a metric space R.

The collection of all sets $G_n(x)$ selected in this way is finite or countable and covers T. For each $G_n(x)$ we now choose one of the sets O_α containing $G_n(x)$, thereby obtaining a finite or countable subcover of $\mathcal{O}$. ∎

Given any topological space T, the empty set $\varnothing$ and the space T itself are both open and closed, by definition. A topological space T is said to be *connected* if it has no subsets other than $\varnothing$ and T which are both open and closed. For example, the real line R^1 is connected, but not the set $R^1 - \{x\}$ obtained from R^1 by deleting any point x.

9.4. Convergent sequences in a topological space. The concept of a convergent sequence, introduced in Sec. 6.2 for the case of a metric space, generalizes in the natural way to the case of a topological space. Thus a sequence of points $\{x_n\} = x_1, x_2, \ldots, x_n, \ldots$ in a topological space T is said to *converge* to a point $x \in T$ (called the *limit* of the sequence) if every neighborhood $G(x)$ of x contains all points x_n starting from a certain index.[3] However, the concept of a convergent sequence does not play the same basic role for topological spaces as for metric spaces. In fact, in the case of a metric space R, a point x is a contact point of a set $M \subset R$ if and only if M contains a sequence converging to x. On the other hand, in the case of a topological space T, this is in general not true, as shown by Problem 11. In other words, a point x can be a contact point of a set $M \subset T$ (i.e., x can belong to $[M]$) without M containing a sequence converging to x. However, convergent sequences "are given their rights back" if T satisfies the first axiom of countability, i.e., if there is a countable neighborhood base at every point $x \in T$:

THEOREM 7. *If a topological space T satisfies the first axiom of countability, then every contact point x of a set $M \subset T$ is the limit of a convergent sequence of points in M.*

Proof. Let $\mathcal{O}$ be a countable neighborhood base at x, consisting of sets O_n. It can be assumed that $O_{n+1} \subset O_n$ $(n = 1, 2, \ldots)$, since otherwise we need only replace O_n by $\bigcap_{k=1}^{n} O_k$. Let x_n be any point of M contained in O_n. Such a point x_n can always be found, since x is a contact point of M. Then the sequence $\{x_n\}$ obviously converges to x. ∎

Remark. As already noted, every metric space satisfies the first axiom of countability. This, together with Theorem 7, shows why in the case of metric spaces we were able to formulate concepts like contact point, limit

---

[3] More exactly, if, given any $G(x)$, there is an integer N_G such that $G(x)$ contains all points x_n with $n > N_G$.

point, etc. in terms of convergent sequences (recall Theorems 2 and $2'$, p. 48).

9.5. Axioms of separation. Although many basic concepts of the theory of metric spaces carry over easily to the case of topological spaces, an arbitrary topological space is still too general an object for most problems of analysis. In fact, things can happen in an arbitrary topological space which differ in an essential way from what happens in a metric space. Thus, for example, a finite set of points need not be closed in an arbitrary topological space, as shown in Example 4, p. 79. Hence it is desirable to specialize the notion of a topological space somewhat by considering topological spaces more closely resembling metric spaces. This is done by imposing extra conditions on a topological space T, in addition to the two defining properties figuring in Definition 1, p. 78. For example, as we have already seen, the axioms of countability allow us to study topological spaces from the standpoint of the concept of convergence. We now introduce supplementary conditions, called *axioms of separation*, of quite a different type:

DEFINITION 4. *Suppose that for each pair of distinct points x and y in a topological space T, there is a neighborhood O_x of x and a neighborhood O_y of y such that $x \in O_y$, $y \in O_x$. Then T is said to satisfy the **first axiom of separation**, and is called a T_1-space.*

Example 1. The space in Example 2, p. 79 is a T_1-space, but not the space in Example 4.

THEOREM 8. *Every finite subset of a T_1-space is closed.*

Proof. Given any single-element set $\{x\}$, suppose $y \neq x$. Then y has a neighborhood O_y which does not contain x, i.e., $y \notin [\{x\}]$. Therefore $[\{x\}] = \{x\}$, i.e., every "singleton" $\{x\}$ is closed. But every finite union of closed sets is itself closed. Hence every finite subset of the given space is closed. ∎

The next axiom of separation is stronger than the first axiom:

DEFINITION 5. *Suppose that for each pair of distinct points x and y in a topological space T, there is a neighborhood O_x of x and a neighborhood O_y of y such that $O_x \cap O_y = \varnothing$. Then T is said to satisfy the **second** (or **Hausdorff**) **axiom of separation**, and is called a T_2-space or Hausdorff space.*

Thus, roughly speaking, each pair of disjoint points in a Hausdorff space has a pair of disjoint neighborhoods.

Example 2. Every Hausdorff space is a T_1-space, but not conversely (see Problem 10).

Topological spaces more general than Hausdorff spaces are rarely used in analysis. In fact, most of the topological spaces of interest in analysis satisfy a separation condition even stronger than the second axiom of separation:

DEFINITION 6. *A T_1-space T is said to be **normal** if for each pair of disjoint closed sets F_1 and F_2 in T, there is an open set O_1 containing F_1 and an open set O_2 containing F_2 such that $O_1 \cap O_2 = \varnothing$.*

In other words, each pair of disjoint closed sets in a normal space has a pair of disjoint "neighborhoods."

Example 3. Obviously, every normal space is a Hausdorff space.

Example 4. Consider the closed unit interval $[0, 1]$, where neighborhoods of any point $x \neq 0$ are defined in the usual way (i.e., as open sets containing x), but neighborhoods of the point $x = 0$ are all half-open intervals $[0, \alpha)$ with the points

$$1, \frac{1}{2}, \dots, \frac{1}{n}, \dots \tag{2}$$

deleted (and arbitrary unions and finite intersections of these neighborhoods with neighborhoods of nonzero points). This space is Hausdorff, but not normal since the set $\{0\}$ and the set of points (2) are disjoint closed sets without disjoint neighborhoods.

THEOREM 9. *Every metric space is normal.*

Proof. Let X and Y be any two disjoint closed subsets of R. Every point $x \in X$ has a neighborhood O_x disjoint from Y, and hence is at a positive distance ρ_x from Y (recall Problem 9, p. 54). Similarly, every point $y \in Y$ is at a positive distance ρ_y from X. Consider the open sets

$$U = \bigcup_{x \in X} S(x, \tfrac{1}{2}\rho_x), \qquad V = \bigcup_{y \in Y} S(y, \tfrac{1}{2}\rho_y),$$

where, as usual, $S(x, r)$ is the open sphere with center x and radius r. It is clear that $X \subset U$, $Y \subset V$. Moreover, U and V are disjoint. In fact, suppose to the contrary that there is a point $z \in U \cap V$. Then there are points $x_0 \in X$, $y_0 \in Y$ such that

$$\rho(x_0, z) < \tfrac{1}{2}\rho_{x_0}, \qquad \rho(z, y_0) < \tfrac{1}{2}\rho_{y_0}.$$

Assume, to be explicit, that $\rho_{x_0} \leqslant \rho_{y_0}$. Then

$$\rho(x_0, y_0) \leqslant \rho(x_0, z) + \rho(z, y_0) < \tfrac{1}{2}\rho_{x_0} + \tfrac{1}{2}\rho_{y_0} \leqslant \rho_{y_0},$$

i.e., $x_0 \in S(y_0, \rho_{y_0})$. This contradicts the definition of ρ_{y_0}, and shows that there is no point $z \in U \cap V$. ∎

Remark. Every subspace of a metric space is itself a metric space and hence normal. This is not true for normal spaces in general, i.e., a subspace of a normal space need not be normal.[4] A property of a topological space T shared by every subspace of T is said to be *hereditary.* Thus normality of a space is not a hereditary property. These ideas are pursued in Problems 13 and 14.

9.6. Continuous mappings. Homeomorphisms. The concept of a continuous mapping, introduced for metric spaces in Sec. 5.2, generalizes at once to the case of arbitrary topological spaces. Thus, let f be a mapping of one topological space X into another topological space Y, so that f associates an element $y = f(x) \in Y$ with each element $x \in X$. Then f is said to be *continuous at the point* $x_0 \in X$ if, given any neighborhood V_{y_0} of the point $y_0 = f(x_0)$, there is a neighborhood U_{x_0} of the point x_0 such that $f(U_{x_0}) \subset V_{y_0}$. The mapping f is said to be *continuous on* X if it is continuous at every point of X. In particular, a continuous mapping of a topological space X into the real line is called a continuous real function on X.

Remark. These definitions clearly reduce to the corresponding definitions for metric spaces in Sec. 5.2 if X and Y are both metric spaces.

The notion of continuity of a mapping f of one topological space into another[5] is easily stated in terms of open sets, i.e., in terms of the *topologies* of the two spaces:

THEOREM 10. *A mapping f of a topological space X into a topological space Y is continuous if and only if the preimage $\Gamma = f^{-1}(G)$ of every open set $G \subset Y$ is open (in X).*

Proof. Suppose f is continuous on X, and let G be any open subset of Y. Choose any point $x \in \Gamma = f^{-1}(G)$, and let $y = f(x)$. Then G is a neighborhood of the point y. Hence, by the continuity of f, there is a neighborhood U_x of x such that $f(U_x) \subset G$, i.e., $U_x \subset \Gamma$. In other words, every point $x \in \Gamma$ has a neighborhood contained in Γ. But then Γ is open (see Problem 1).

Conversely, suppose $\Gamma = f^{-1}(G)$ is open whenever $G \subset Y$ is open. Given any point $x \in X$, let V_y be any neighborhood of the point $y = f(x)$.

---

[4] See e.g., J. L. Kelley, *General Topology*, D. Van Nostrand Co., Inc., Princeton, N.J. (1955), p. 132.

[5] If desired, the mapping f can always be regarded as "onto," since otherwise we need only replace the space Y by the subspace $f(X) \subset Y$.

Then clearly $x \in f^{-1}(V_y)$, and moreover $f^{-1}(V_y)$ is open, by hypothesis. Therefore $U_x = f^{-1}(V_y)$ is a neighborhood of x such that $f(U_x) \subset V_y$. In other words, f is continuous *at* x and hence *on* X, since x is an arbitrary point of X. ∎

Naturally, Theorem 10 has the following "dual":

THEOREM 10′. *A mapping f of a topological space X into a topological space Y is continuous if and only if the preimage $\Gamma = f^{-1}(F)$ of every closed set $F \subset Y$ is closed (in X).*

Proof. Use the fact that the preimage of a complement is the complement of the preimage. ∎

Remark. Let X and Y be two arbitrary sets, and let f be a mapping of X into Y. Suppose that in Y there is specified a topology τ, i.e., a system of sets containing Y and $\varnothing$, and closed under the operations of taking arbitrary unions and finite intersections. Then since the preimage of a union (or intersection) of sets equals the union (or intersection) of the preimages of the sets, by Theorems 1 and 2, p. 5, the preimage of the topology τ, i.e., the system of all sets $f^{-1}(G)$ where $G \in \tau$, is a topology in X which we denote by $f^{-1}(\tau)$.

Suppose now that X and Y are topological spaces, with topologies τ_X and τ_Y, respectively. Then Theorem 10, giving a necessary and sufficient condition for a mapping f of X into Y to be continuous can be paraphrased as follows: A mapping f of X into Y is continuous if and only if the topology τ_X is *stronger* than the topology $f^{-1}(\tau_Y)$.

Example. It is easy to see that the *image* (as opposed to the preimage) of an open set under a continuous mapping need not be open. Similarly, the image of a closed set under a continuous mapping need not be closed. For example, consider the mapping of the half-open interval $X = [0, 1)$ onto the circle of unit circumference corresponding to "winding" the interval onto the circle. Then the set $[\frac{1}{2}, 1)$, which is closed in $[0, 1)$, goes into a set which is not closed on the circle (see Figure 12).

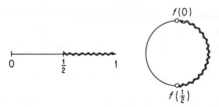

FIGURE 12

The theorem on continuity of composite functions, familiar from elementary calculus, has the following analogue for topological spaces:

THEOREM 11. *Given topological spaces X, Y and Z, suppose f is a continuous mapping of X into Y and φ a continuous mapping of Y into Z. Then the mapping φf, i.e., the mapping carrying x into $\varphi(f(x))$, is continuous.*

Proof. An immediate consequence of Theorem 10. ∎

Given two topological spaces X and Y, let f be a one-to-one mapping of X onto Y, and suppose f and f^{-1} are both continuous. Then f is called a *homeomorphic mapping* or simply a *homeomorphism* (between X and Y). Two spaces X and Y are said to be *homeomorphic* if there exists a homeomorphism between them. Homeomorphic spaces have the same topological properties, and from the topological point of view are merely two "representatives" of one and the same space. In fact, if X and Y have topologies τ_X and τ_Y, respectively, and if f is a homeomorphic mapping of X onto Y, then $\tau_X = f^{-1}(\tau_Y)$ and $\tau_Y = f(\tau_X)$. The relation of being homeomorphic is obviously reflexive, symmetric and transitive, and hence is an equivalence relation. Therefore any given family of topological spaces can be partitioned into disjoint classes of homeomorphic spaces.

Remark. Again these are the natural generalizations of the same notions for metric spaces, introduced in Sec. 2.2. It should be noted that two homeomorphic metric spaces need not have the same "metric properties" (recall Problem 9, p. 66). Note also that the topology of a metric space is uniquely determined by its metric, but not conversely (illustrate this by an example).

9.7. Various ways of specifying topologies. Metrizability. The most direct and in principle the simplest way of specifying a topology in a space T is to indicate which subsets of T are regarded as open. The system of all such subsets must then satisfy properties 1) and 2) of Definition 1. By duality, we could just as well indicate which subsets of X are regarded as *closed*. The system of all such subsets must then satisfy properties 1') and 2') on p. 79. However, this method is of limited practical value. For example, in the case of the plane it is hardly possible to give a direct description of all open sets (as was done in Theorem 6, p. 51 for the case of the line).

A topology is often specified in a space T by giving a *base* for T. In fact, this is precisely what is done in Sec. 6 for the case of a metric space R, where the base for R consists of all open spheres (or even all open spheres with rational radii).

Another way of specifying a topology in a space T is to introduce the notion of convergence in T. As noted in Sec. 9.4, this is not a universal method. It does work, however, in the case of spaces satisfying the first axiom of countability.[6]

Still another way of introducing a topology in a space T is to specify a closure operator in T, i.e., a mapping which assigns to each subset $M \subset T$ a subset $[M] \subset T$ and satisfies the four properties listed in Theorem 1, p. 46. It can be shown that the system of complements of all sets $M \subset T$ such that $[M] = M$ is then a topology in T.[7]

Specifying a metric in a space T is one of the most important ways of introducing a topology in T, but it is again far from being a universal method. As already noted, every metric space is normal and satisfies the first axiom of countability. Hence no metric can be used to introduce a topology in a space which fails to have these two properties. A topological space T is said to be *metrizable* if its topology can be specified by means of some metric (more exactly, if it is homeomorphic to some metric space). As just pointed out, a necessary condition for a topological space T to be metrizable is that it be normal and satisfy the first axiom of countability. However, it can be shown that these conditions are not sufficient for T to be metrizable. On the other hand, in the case of a space with a countable base (i.e., satisfying the second axiom of countability), we have

URYSOHN'S METRIZATION THEOREM. *A necessary and sufficient condition for a topological space with a countable base to be metrizable is that it be normal.*

The necessity follows from Theorem 9. For the sufficiency we refer to the literature.[8]

Problem 1. Given a topological space T, prove that a set $G \subset T$ is open if and only if every point $x \in G$ has a neighborhood contained in G.

Problem 2. Given a topological space T, prove that

a) $[M] = M$ if and only if M is a closed set, i.e., the complement $T - G$ of an open set $G \subset T$;

b) $[M]$ is the smallest closed set containing M;

c) The closure operator, i.e., the mapping of T into T carrying M into $[M]$ satisfies Theorem 1, p. 46.

Problem 3. Consider the set $\mathscr{T}$ of all possible topologies defined in a set X, where $\tau_2 \leqslant \tau_1$ means that τ_2 is weaker than τ_1. Verify that $\leqslant$ is a

---

[6] In fact, by suitably generalizing the notion of convergence (and introducing the concepts of "nets" and "filters"), this method can be made to work quite generally. See e.g., J. L. Kelley, *op. cit.*, p. 83.

[7] J. L. Kelley, *op. cit.*, p. 43.

[8] See e.g., P. S. Alexandroff, *Einführung in die Mengenlehre und die Theorie der Reellen Funktionen*, VEB Deutscher Verlag der Wissenschaften, Berlin (1956), p. 195 ff.

partial ordering of $\mathcal{T}$. Does $\mathcal{T}$ have maximal and minimal elements? If so, what are they?

Problem 4. Can two distinct topologies τ_1 and τ_2 in X generate the same relative topology in a subset $A \subset X$?

Problem 5. Let

$$X = \{a, b, c\}, \quad A = \{a, b\}, \quad B = \{b, c\},$$

and let $\mathcal{G} = \{\varnothing, X, A, B\}$. Is $\mathcal{G}$ a base for a topology in X?

Problem 6. Prove that if M is an uncountable subset of a topological space with a countable base, then some point of M is a limit point of M.

Problem 7. Prove that the topological space T in Example 4, p. 79 is connected.

Comment. T might be called a "connected doubleton."

Problem 8. Prove that a topological space satisfying the second axiom of countability automatically satisfies the first axiom of countability.

Problem 9. Give an example of a topological space satisfying the first axiom of countability but not the second axiom of countability.

Problem 10. Let τ be the system of sets consisting of the empty set and every subset of the closed unit interval $[0, 1]$ obtained by deleting a finite or countable number of points from X. Verify that $T = (X, \tau)$ is a topological space. Prove that T satisfies neither the second nor the first axiom of countability. Prove that T is a T_1-space, but not a Hausdorff space.

Problem 11. Let T be the topological space of the preceding problem. Prove that the only convergent sequences in T are the "stationary sequences," i.e., the sequences all of whose terms are the same starting from some index n. Prove that the set $M = (0, 1]$ has the point 0 as a contact point, but contains no sequence of points converging to 0.

Problem 12. Prove the converse of Theorem 8.

Comment. Hence a topological space T is a T_1-space if and only if every finite subset of T is closed.

Problem 13. Prove the following theorem, known as *Urysohn's lemma*: Given a normal space T and two disjoint closed subsets F_1, $F_2 \in T$, there exists a continuous real function f such that $0 \leqslant f(x) \leqslant 1$ and

$$f(x) = \begin{cases} 0 & \text{if } x \in F_1, \\ 1 & \text{if } x \in F_2. \end{cases}$$

Problem 14. A T_1-space T is said to be *completely regular* if, given any closed set $F \subset T$ and any point $x_0 \in T - F$, there exists a continuous real function f such that $0 \leqslant f(x) \leqslant 1$ and

$$f(x) = \begin{cases} 0 & \text{if } x = x_0, \\ 1 & \text{if } x \in F. \end{cases}$$

(Completely regular spaces are also called *Tychonoff spaces*.) Prove that every normal space is completely regular, but not conversely. Prove that every subspace of a completely regular space (in particular, of a normal space) is completely regular.

Comment. Thus, unlike normality, complete regularity is a hereditary property. It can be shown that a space is completely regular if and only if it is a subspace of a normal space.[9] Completely regular spaces are particularly important in analysis, since they "are able to support sufficiently many continuous functions," i.e., for any two distinct points x and y of a completely regular space T, there is a continuous real function on T taking distinct values at x and y.

10. Compactness

10.1. Compact topological spaces. The reader has presumably already encountered the familiar

HEINE-BOREL THEOREM. *Any cover of a closed interval $[a, b]$ by a system of open intervals (or, more generally, open sets) has a finite subcover.*

Generalizing this property of closed intervals, we are led to a key concept of real analysis:

DEFINITION 1. *A topological space T is said to be **compact** if every open cover of T has a finite subcover. A compact Hausdorff space is called a **compactum**.*

Example. As we will see in Sec. 11.2, any closed bounded subset of Euclidean n-space R^n is compact, for arbitrary n. On the other hand, R^n itself (e.g., the real line or three-dimensional space) is not compact.

DEFINITION 2. *A system of subsets $\{A_\alpha\}$ of a set T is said to be **centered** if every finite intersection $\bigcap_{k=1}^{n} A_k$ is nonempty.*[10]

---

[9] J. L. Kelley, *op. cit.*, p. 145.
[10] A system of sets with typical member A_α will often be denoted by $\{A_\alpha\}$ (this is still another use of curly brackets).

THEOREM 1. *A topological space T is compact if and only if it has the following property*:

$\triangle$) *Every centered system of closed subsets of T has a nonempty intersection.*

Proof. Suppose T is compact, and let $\{F_\alpha\}$ be any centered system of closed subsets of T. Then the sets $G_\alpha = T - F_\alpha$ are open. Hence the fact that no finite intersection $\bigcap_{k=1}^{n} F_k$ is empty implies that no finite system of sets $G_k = T - F_k$ covers T. But then the whole system of sets $\{G_\alpha\}$ cannot cover T, by the compactness, and hence $\bigcap_\alpha F_\alpha \neq \varnothing$. In other words, T has property $\triangle$) if T is compact.

Conversely, suppose T has property $\triangle$), and let $\{G_\alpha\}$ be any open cover of T. Setting $F_\alpha = T - G_\alpha$, we find that $\bigcap_\alpha F_\alpha = \varnothing$, which, by property $\triangle$), implies that the system F_α is not centered, i.e., that there are sets $F_1, \ldots, F_n$ such that $\bigcap_{k=1}^{n} F_k = \varnothing$. But then the corresponding open sets $G_k = T - F_k$ form a finite subcover of the cover $\{G_\alpha\}$. In other words, T is compact if T has property $\triangle$). $\blacksquare$

THEOREM 2. *Every closed subset F of a compact topological space T is itself compact.*

Proof. Let $\{F_\alpha\}$ be any centered system of closed subsets of the subspace $F \subset T$. Then every F_α is closed in T as well, i.e., $\{F_\alpha\}$ is a centered system of closed subsets of T. Therefore $\bigcap_\alpha F_\alpha \neq \varnothing$, by Theorem 1. But then F is compact, by Theorem 1 again. $\blacksquare$

COROLLARY. *Every closed subset of a compactum is itself a compactum.*

Proof. Use Theorem 2 and the fact that every subset of a Hausdorff space is itself a Hausdorff space. $\blacksquare$

THEOREM 3. *Let K be a compactum and T any Hausdorff space containing K. Then K is closed in T.*

Proof. Suppose $y \notin K$, so that $y \in T - K$. Then, given any point $x \in K$, there is a neighborhood U_x of x and a neighborhood V_x of y such that

$$U_x \cap V_x = \varnothing.$$

The neighborhoods $\{U_x\}(x \in K)$ form an open cover of K. Hence, by the compactness of K, $\{U_x\}$ has a finite subcover consisting of sets $U_{x_1}, \ldots, U_{x_n}$. Let

$$V = V_{x_1} \cap \cdots \cap V_{x_n}.$$

Then V is a neighborhood of the point y which does not intersect the set $U_{x_1} \cup \cdots \cup U_{x_n} \supset K$, and hence $y \notin [K]$. It follows that K is closed (in T). ∎

Remark. It is a consequence of Theorems 2 and 3 that compactness is an "intrinsic property," in the sense that a compactum remains a compactum after being "embedded" in any larger Hausdorff space.

THEOREM 4. *Every compactum K is a normal space.*

Proof. Let X and Y be any two disjoint closed subsets of K. Repeating the argument given in the proof of Theorem 3, we easily see that, given any point $y \in Y$, there exists a neighborhood U_y containing y and an open set $O_y \supset X$ such that $U_y \cap O_y = \varnothing$. Since Y is compact, by Theorem 2, the cover $\{U_y\}(y \in Y)$ of the set Y has a finite subcover $U_{y_1}, \ldots, U_{y_n}$. The open sets

$$O^{(1)} = O_{y_1} \cap \cdots \cap O_{y_n}, \qquad O^{(2)} = U_{y_1} \cup \cdots \cup U_{y_n}$$

then satisfy the normality conditions

$$O^{(1)} \supset X, \qquad O^{(2)} \supset Y, \qquad O^{(1)} \cap O^{(2)} = \varnothing. \quad ∎$$

10.2. Continuous mappings of compact spaces. Next we show that the "continuous image" of a compact space is itself a compact space:

THEOREM 5. *Let X be a compact space and f a continuous mapping of X onto a topological space Y. Then $Y = f(X)$ is itself compact.*

Proof. Let $\{V_\alpha\}$ be any open cover of Y, and let $U_\alpha = f^{-1}(V_\alpha)$. Then the sets U_α are open (being preimages of open sets under a continuous mapping) and cover the space X. Since X is compact, $\{U_\alpha\}$ has a finite subcover $U_{x_1}, \ldots, U_{x_n}$. Then the sets $V_{x_1}, \ldots, V_{x_n}$, where $V_k = f(U_k)$, cover Y. It follows that Y is compact. ∎

THEOREM 6. *A one-to-one continuous mapping of a compactum X onto a compactum Y is necessarily a homeomorphism.*

Proof. We must show that the inverse mapping f^{-1} is itself continuous. Let F be a closed set in X and $P = f(F)$ its image in Y. Then P is a compactum, by Theorem 5. Hence, by Theorem 3, P is closed in Y. Therefore the preimage under f^{-1} of any closed set $F \subset X$ is closed. It follows from Theorem 10', p. 88 that f^{-1} is continuous. ∎

10.3. Countable compactness. We begin by proving an important property of compact spaces:

THEOREM 7. *If T is a compact space, then any infinite subset of T has at least one limit point.*

Proof. Suppose T contains an infinite set X with no limit point. Then T contains a countable set

$$X = \{x_1, x_2, \ldots, x_n, \ldots\}$$

with no limit point. But then the sets

$$x_n = \{x_n, x_{n+1}, \ldots\} \qquad (n = 1, 2, \ldots)$$

form a centered system of closed sets in T with an empty intersection, i.e., T is not compact. ∎

These considerations suggest

DEFINITION 3. *A topological space T is said to be **countably compact** if every infinite subset of T has at least one limit point (in T).*

Thus Theorem 7 says that every compact set is countably compact. The converse, however, is not true (see Problem 1). The relation between the concepts of compactness and countable compactness is made clear by

THEOREM 8. *Each of the following two conditions is necessary and sufficient for a topological space T to be countably compact:*

1) *Every countable open cover of T has a finite subcover;*
2) *Every countable centered system of closed subsets of T has a non-empty intersection.*

Proof. The equivalence of conditions 1) and 2) is an immediate consequence of the duality principle. Moreover, if T is not countably compact, then, repeating the argument given in proving Theorem 7, we find that there is a countable centered system of closed subsets of T with an empty intersection. This proves the sufficiency of condition 2). Thus we need only prove the necessity of condition 2). Let T be countably compact, and let $\{F_n\}$ be a countable centered system of closed sets in T. Then, as we now show, $\bigcap\limits_{n} F_n \neq \varnothing$. Let

$$\Phi_n = \bigcap_{k=1}^{n} F_k.$$

Then none of the Φ_n is empty, since $\{F_n\}$ is centered. Moreover,

$$\Phi_1 \supset \Phi_2 \supset \cdots \supset \Phi_n \supset \cdots,$$

and

$$\bigcap_{n} \Phi_n = \bigcap_{n} F_n.$$

There are now just two possibilities:

1) $\Phi_{n_0} = \Phi_{n_0+1} = \cdots$ starting from some index n_0, in which case it is obvious that $\bigcap\limits_{n} \Phi_n = \Phi_{n_0} \neq \varnothing$.

2) There are infinitely many distinct sets Φ_n. In this case, there is clearly no loss of generality in assuming that all the Φ_n are distinct. Let $x_n \in \Phi_n - \Phi_{n+1}$. Then the sequence $\{x_n\}$ consists of infinitely many distinct points of T, and hence, by the countable compactness of T, must have at least one limit point, say x_0. But then x_0 must be a limit point of Φ_n, since Φ_n contains all the points x_n, $x_{n+1}, \ldots$. Moreover $x_0 \in \Phi_n$, since Φ_n is closed. It follows that

$$x_0 \in \bigcap_n \Phi_n, \quad \text{i.e.,} \quad \bigcap_n \Phi_n \neq \varnothing. \quad \blacksquare$$

Thus compact topological spaces are those in which an *arbitrary* open cover has a finite subcover, while countably compact spaces are those in which every *countable* open cover has a finite subcover. Although in general countable compactness does not imply compactness, we have the following important special situation:

THEOREM 9. *The concepts of compactness and countable compactness coincide for a topological space T with a countable base.*

Proof. By Theorem 6, p. 83, every open cover $\mathcal{O}$ of T has a countable subcover. Hence, if T is countably compact, $\mathcal{O}$ has a finite subcover, by Theorem 8. $\blacksquare$

Remark. The concept of a countably compact topological space, unlike that of a compact space, has not turned out to be very natural or fruitful. Its presence in mathematics can be explained in terms of a kind of "historical inertia." The point is that, as will be shown in the next section, the concepts of compactness and countable compactness coincide for metric spaces, as well as for spaces with a countable base. The notion of compactness was originally introduced in connection with metric spaces, with a compact metric space being defined as one in which every infinite subset has at least one limit point (i.e., in terms of what is now called "countable compactness"). The "automatic transcription" of this definition from metric spaces to topological spaces then led to the concept of a countably compact topological space. Sometimes, especially in the older literature, the word "compact" is used in the sense of "countably compact," and a topological space compact in our sense (i.e., such that every open cover has a finite subcover) is said to be "bicompact." In this older language, a compact Hausdorff space (a "compactum" in our terminology) is called a "bicompactum," and the term "compactum" is reserved for a compact metric space. We will adhere to the terminology introduced in Definitions 1 and 3, often using the term "metric compactum" to designate a compact metric space.

10.4. Relatively compact subsets. Among the subsets of a topological space, those whose closures are compact are of special interest:

DEFINITION 4. *A subset M of a topological space T is said to be **relatively compact** (in T) if its closure $\bar{M}$ in T is compact.*

Example 1. According to Theorem 2, every subset of a compact topological space is relatively compact.

Example 2. As we will see in Sec. 11.3, every *bounded* subset of the real line R^1 (or more generally of Euclidean n-space R^n) is relatively compact.

A related concept is given by

DEFINITION 5. *A subset M of a topological space T is said to be **relatively countably compact** (in T) if every infinite subset $A \subset M$ has at least one limit point in T (which may or may not belong to M).*

Relative compactness (unlike compactness) is not an "intrinsic property," i.e., it depends on the space T in which the given set M is "embedded." For example, the set of all rational numbers in the interval $(0, 1)$ is relatively compact if regarded as a subset of the real line, but not if regarded as a subset of the space of all rational numbers. The concept of relative compactness is most important in the case of metric spaces (see Sec. 11.3).

Problem 1. Let X be the set of all ordinal numbers less than the first uncountable ordinal. Let $(\alpha, \beta) \subset X$ denote the set of all ordinal numbers γ such that $\alpha < \gamma < \beta$, and let the open sets in X be all unions of intervals (α, β). Prove that the resulting topological space is countably compact but not compact.

Problem 2. A topological space T is said to be *locally compact* if every point $x \in T$ has at least one relatively compact neighborhood. Show that a compact space is automatically locally compact, but not conversely. Prove that every closed subspace of a locally compact subspace is locally compact.

Problem 3. A point x is said to be a *complete limit point* of a subset A of a topological space if, given any neighborhood U of x, the sets A and $A \cap U$ have the same power (i.e., cardinal number). Prove that every infinite subset of a compact topological space has at least one complete limit point.

Comment. Conversely, it can be shown that if every infinite subset of a topological space T has at least one complete limit point, then T is compact.[11]

11. Compactness in Metric Spaces

11.1. Total boundedness. Since metric spaces are topological spaces of a special kind, the definitions and results of the preceding section apply to

---

[11] P. S. Alexandroff, *op. cit.*, pp. 250–251; J. L. Kelley, *op. cit.*, pp. 163–164.

metric spaces as well. However, in the case of metric spaces, the concept of compactness is intimately connected with another concept, known as *total boundedness*.

DEFINITION 1. *Let R be a metric space and ε any positive number. Then a set $A \subset R$ is said to be an **ε-net** for a set $M \subset R$ if, for every $x \in M$, there is at least one point $a \in A$ such that $\rho(x, a) < \varepsilon$.*

Example 1. The set of all points with integral coordinates is a $(1/\sqrt{2})$-net.

Example 2. Every subset of a totally bounded set is itself totally bounded.

DEFINITION 2. *Given a metric space R and a subset $M \subset R$, suppose M has a **finite** ε-net for every $\varepsilon > 0$. Then M is said to be **totally bounded**.*

If a set M is totally bounded, then obviously so is its closure $[M]$. Every totally bounded set is automatically bounded, being the union of a finite number of bounded sets (recall Problem 5, p. 65). The converse is not true, as shown in Example 4.

Example 3. In Euclidean n-space R^n, total boundedness is equivalent to boundedness. In fact, if $M \subset R$ is bounded, then M is contained in some sufficiently large cube Q. Partitioning Q into smaller cubes of side ε, we find that the vertices of the little cubes form a finite $(\sqrt{n}\varepsilon/2)$-net for Q and hence (a fortiori) for any set contained in Q.

Example 4. The unit sphere Σ in l_2, with equation

$$\sum_{n=1}^{\infty} x_n^2 = 1,$$

is bounded but not totally bounded. In fact, consider the points

$$e_1 = (1, 0, 0, \ldots), \qquad e_2 = (0, 1, 0, \ldots), \ldots,$$

where the nth coordinate of e_n is one and the others are all zero. These points all lie on Σ, and the distance between any two of them is $\sqrt{2}$. Hence Σ cannot have a finite ε-net with $\varepsilon < \sqrt{2}/2$.

Example 5. Let Π be the set of points $x = (x_1, x_2, \ldots, x_n, \ldots)$ in l_2 satisfying the inequalities

$$|x_1| < 1, \qquad |x_2| < \frac{1}{2}, \ldots, \qquad |x_n| < \frac{1}{2^{n-1}}, \ldots$$

The set Π, called the *Hilbert cube* (or *fundamental parallelepiped*)[12] furnishes

---

[12] Another commonly encountered definition of the Hilbert cube is the set of points in l_2 satisfying the inequalities

$$|x_1| < 1, \qquad |x_2| < \frac{1}{2}, \ldots, \qquad |x_n| < \frac{1}{n}, \ldots$$

an example of an infinite-dimensional totally bounded set. The fact that Π is totally bounded can be seen as follows: Given any $\varepsilon > 0$, choose n such that

$$\frac{1}{2^{n-1}} < \frac{\varepsilon}{2},$$

and with each point

$$x = (x_1, x_2, \ldots, x_n, \ldots)$$

in Π associate the point

$$x^* = (x_1, x_2, \ldots, x_n, 0, 0, \ldots) \tag{1}$$

(x^* is also a point in Π). Then

$$\rho(x, x^*) = \sqrt{\sum_{k=n+1}^{\infty} x_k^2} < \sqrt{\sum_{k=n}^{\infty} \frac{1}{4^k}} < \frac{1}{2^{n-1}} < \frac{\varepsilon}{2}.$$

But the set Π^* of all points in Π of the form (1) is totally bounded, being a bounded set in n-space. Let A be a finite $(\varepsilon/2)$-net in Π^*. Then A is a finite ε-net for the whole set Π.

11.2. Compactness and total boundedness. We now show the connection between the concepts of compactness (of both kinds) and total boundedness:

THEOREM 1. *Every countably compact metric space R is totally bounded.*

Proof. Suppose R is not totally bounded. Then there is an $\varepsilon_0 > 0$ such that R has no finite ε_0-net. Choose any point $a_1 \in R$. Then R contains at least one point, say a_2, such that

$$\rho(a_1, a_2) > \varepsilon_0,$$

since otherwise a_1 would be an ε_0-net for R. Moreover, R contains a point a_3 such that

$$\rho(a_1, a_3) > \varepsilon_0, \qquad \rho(a_2, a_3) > \varepsilon_0,$$

since otherwise the pair a_1, a_2 would be an ε_0-net for R. More generally, once having found the points $a_1, a_2, \ldots, a_n$, we choose $a_{n+1} \in R$ such that

$$\rho(a_k, a_{n+1}) > \varepsilon_0 \qquad (k = 1, 2, \ldots, n).$$

This construction gives an infinite sequence of distinct points $a_1, a_2, \ldots,$ $a_n, \ldots$ with no limit points, since $\rho(a_j, a_k) > \varepsilon_0$ if $j \neq k$. But then R cannot be countably compact. ∎

COROLLARY 1. *Every countably compact metric space has a countable everywhere dense subset and a countable base.*

Proof. Since R is totally bounded, by Theorem 1, R has a finite $(1/n)$-net for every $n = 1, 2, \ldots$. The union of all these nets is then a countable everywhere dense subset of R. It follows from Theorem 5, p. 82 that R has a countable base. ∎

COROLLARY 2. *Every countably compact metric space is compact.*

Proof. An immediate consequence of Corollary 1 and Theorem 9, p. 96. ∎

According to Theorem 1, total boundedness is a necessary condition for a metric space to be compact. However, this condition is not sufficient. For example, the set of rational points in the interval $[0, 1]$ with the ordinary definition of distance forms a metric space R which is totally bounded but not compact. In fact, the sequence of points

$$0, \quad 0.4, \quad 0.41, \quad 0.414, \quad 0.4142, \ldots$$

in R, i.e., the sequence of decimal approximations to the irrational number $\sqrt{2} - 1$, has no limit point in R. Necessary and sufficient conditions for compactness of a metric space are given by

THEOREM 2. *A metric space R is compact if and only if it is totally bounded and complete.*

Proof. To see that compactness of R implies completeness of R, we need only note that if R has a Cauchy sequence $\{x_n\}$ with no limit, then $\{x_n\}$ has no limit points in R. This, together with Theorem 1, shows that R is totally bounded and complete if R is compact.

Conversely, suppose R is totally bounded and complete, and let $\{x_n\}$ be any infinite sequence of distinct points in R. Let N_1 be a finite 1-net for R, and construct a closed sphere of radius 1 about every point of N_1. Since these spheres cover R and there are infinitely many of them, at least one of the spheres, say S_1, contains an infinite subsequence

$$x_1^{(1)}, \ldots, x_n^{(1)}, \ldots$$

of the sequence $\{x_n\}$. Let N_2 be a finite $\frac{1}{2}$-net for R, and construct a closed sphere of radius $\frac{1}{2}$ for every point of N_2. Then at least one of these spheres, say S_2, contains an infinite subsequence

$$x_1^{(2)}, \ldots, x_n^{(2)}, \ldots$$

of the sequence $\{x_n^{(1)}\}$. Continue this construction indefinitely, finding a closed sphere S_3 of radius $\frac{1}{4}$ containing an infinite subsequence

$$x_1^{(3)}, \ldots, x_n^{(3)}, \ldots$$

of the sequence $\{x_n^{(2)}\}$, and so on, where S_n has radius $1/2^{n-1}$. Let S_n' be the closed sphere with the same center as S_n but with a radius r_n twice as large (i.e., equal to $1/2^n$). Then clearly

$$S_1' \supset S_2' \supset \cdots \supset S_n' \supset \cdots,$$

and moreover $r_n \to 0$ as $n \to \infty$. Since R is complete, it follows from the nested sphere theorem (Theorem 2, p. 60) that

$$\bigcap_{n=1}^{\infty} S'_n \neq \varnothing.$$

In fact, there is a point $x_0 \in R$ such that

$$\bigcap_{n=1}^{\infty} S'_n = \{x_0\}$$

(recall Problem 3, p. 65). Clearly x_0 is a limit point of the original sequence $\{x_n\}$, since every neighborhood of x_0 contains some sphere S_k and hence some infinite subsequence $\{x_n^{(k)}\}$. Therefore every infinite sequence $\{x_n\}$ of distinct points of R has a limit point in R. It follows that R is countably compact and hence compact, by Corollary 2. $\blacksquare$

Example. As already noted, a subset M of Euclidean n-space R^n is totally bounded if and only if it is bounded. Moreover, M is complete if and only if it is closed (recall Problem 7, p. 66). Hence, by Theorem 2, the set of all compact subsets of R^n coincides with the set of all closed bounded subsets of R^n.

11.3. Relatively compact subsets of a metric space. The concept of relative compactness, introduced in Sec. 10.4 for subsets of an arbitrary topological space, applies in particular to subsets of a metric space. In the case of a metric space, however, there is no longer any distinction between relative compactness and relative countable compactness.

THEOREM 3. *A subset M of a complete metric space R is relatively compact if and only if it is totally bounded.*

Proof. An immediate consequence of Theorem 2 and the fact that a closed subset of a complete metric space is itself complete. $\blacksquare$

Example. Any bounded subset of Euclidean n-space it totally bounded and hence relatively compact (this is our version of the familiar *Bolzano-Weierstrass theorem*).

Remark. The utility of Theorem 3 stems from the fact it is usually easier to prove that a set is totally bounded than to give a direct proof of its relative compactness. On the other hand, compactness is the key property as far as applications are concerned.

11.4. Arzelà's theorem. The problem of proving the compactness of various subsets of a given metric space is encountered quite frequently in analysis. However, the direct application of Theorem 2 is not always easy. This explains the need for special criteria serving as practical tools for proving compactness in particular spaces. For example, as we have seen, the boundedness of a set in Euclidean n-space implies its compactness, but this implication fails in more general metric spaces.

One of the most important metric spaces in analysis is the function space $C_{[a,b]}$, introduced in Example 6, p. 39. For subsets of this space, we have an important and frequently used criterion for relative compactness, called Arzelà's theorem, which will be stated and proved after first introducing two new concepts:

DEFINITION 3. *A family* Φ *of functions* φ *defined on a closed interval* $[a, b]$ *is said to be* **uniformly bounded** *if there exists a number* $K > 0$ *such that*

$$|\varphi(x)| < K$$

for all $x \in [a, b]$ *and all* $\varphi \in \Phi$.

DEFINITION 4. *A family* Φ *of functions* φ *defined on a closed interval* $[a, b]$ *is said to be* **equicontinuous** *if, given any* $\varepsilon > 0$, *there exists a number* $\delta > 0$ *such that* $|x' - x''| < \delta$ *implies*

$$|\varphi(x') - \varphi(x'')| < \varepsilon$$

for all $x', x'' \in [a, b]$ *and all* $\varphi \in \Phi$.

THEOREM 4 (*Arzelà*). *A necessary and sufficient condition for a family* Φ *of continuous functions* φ *defined on a closed interval* $[a, b]$ *to be relatively compact in* $C_{[a,b]}$ *is that* Φ *be uniformly bounded and equicontinuous.*

Proof. We give the proof in two steps:

Step 1 (*Necessity*). Suppose Φ is relatively compact in $C_{[a,b]}$. Then by Theorem 3, given any $\varepsilon > 0$, there is a finite $(\varepsilon/3)$-net $\varphi_1, \ldots, \varphi_n$ in Φ (see Problem 1). Being a continuous function defined on a closed interval, each φ_i is bounded:

$$|\varphi_i(x)| < K_i \qquad (a < x < b).$$

Let

$$K = \max \{K_1, \ldots, K_n\} + \frac{\varepsilon}{3}.$$

By the definition of an $(\varepsilon/3)$-net, given any $\varphi \in \Phi$, there is at least one φ_i such that

$$\rho(\varphi, \varphi_i) = \max_{a \leqslant x \leqslant b} |\varphi(x) - \varphi_i(x)| < \frac{\varepsilon}{3}.$$

Therefore

$$|\varphi(x)| \leqslant |\varphi_i(x)| + \frac{\varepsilon}{3} \leqslant K_i + \frac{\varepsilon}{3} \leqslant K,$$

i.e., Φ is uniformly bounded. Moreover, each function φ_i in the $(\varepsilon/3)$-net is continuous, and hence uniformly continuous, on $[a, b]$. Hence, given any $\varepsilon > 0$, there is a δ_i such that

$$|\varphi_i(x_1) - \varphi_i(x_2)| < \frac{\varepsilon}{3}$$

whenever $|x_1 - x_2| < \delta_i$. Let

$$\delta = \min \{\delta_1, \ldots, \delta_n\}.$$

Then, given any $\varphi \in \Phi$ and choosing φ_i such that $\rho(\varphi, \varphi_i) < \varepsilon/3$, we have

$$|\varphi(x_1) - \varphi(x_2)|$$
$$\leqslant |\varphi(x_1) - \varphi_i(x_1)| + |\varphi_i(x_1) - \varphi_i(x_2)| + |\varphi_i(x_2) - \varphi(x_2)|$$
$$< \frac{\varepsilon}{3} + \frac{\varepsilon}{3} + \frac{\varepsilon}{3} = \varepsilon$$

whenever $|x_1 - x_2| < \delta$. This proves the equicontinuity of Φ.

Step 2 (Sufficiency). Suppose Φ is uniformly bounded and equicontinuous. According to Theorem 3, to prove that Φ is relatively compact in $C_{[a,b]}$, we need only show that Φ is totally bounded, i.e., that given any $\varepsilon > 0$, there exists a finite ε-net for Φ in $C_{[a,b]}$. Suppose $|\varphi(x)| \leqslant K$ for all $\varphi \in \Phi$, and let $\delta > 0$ be such that

$$|\varphi(x_1) - \varphi(x_2)| < \frac{\varepsilon}{5}$$

for all $\varphi \in \Phi$ whenever $|x_1 - x_2| < \delta$. Divide the interval $a \leqslant x \leqslant b$ along the x-axis into subintervals of length less than δ, by introducing points of subdivision $x_0, x_1, x_2, \ldots, x_n$ such that

$$a = x_0 < x_1 < x_2 < \cdots < x_n = b,$$

and then draw a vertical line through each of these points. Similarly, divide the interval $-K \leqslant y \leqslant K$ along the y-axis into subintervals of length less than $\varepsilon/5$, by introducing points of subdivision $y_0, y_1, y_2, \ldots, y_p$ such that

$$-K = y_0 < y_1 < y_2 < \cdots < y_p = K,$$

and then draw a horizontal line through each of these points. In this way, the rectangle $a \leqslant x \leqslant b$, $-M \leqslant y \leqslant M$ is divided into np cells of horizontal side length less than δ and vertical side length less than $\varepsilon/5$.

We now associate with each function $\varphi \in \Phi$ a polygonal line $y = \psi(x)$ which has vertices at points of the form (x_k, y_l) and differs from the function φ by less than $\varepsilon/5$ at every point x_k (the reader should draw a figure and convince himself on the existence of such a function). Since

$$|\varphi(x_k) - \psi(x_k)| < \frac{\varepsilon}{5},$$

$$|\varphi(x_{k+1}) - \psi(x_{k+1})| < \frac{\varepsilon}{5},$$

$$|\varphi(x_k) - \varphi(x_{k+1})| < \frac{\varepsilon}{5},$$

by construction, we have

$$|\psi(x_k) - \psi(x_{k+1})| < \frac{3\varepsilon}{5}.$$

Moreover,

$$|\psi(x_k) - \psi(x)| < \frac{3\varepsilon}{5} \qquad (x_k \leqslant x \leqslant x_{k+1}),$$

since $\psi(x)$ is linear between the points x_k and x_{k+1}. Let x be any point in $[a, b]$ and x_k the point of subdivision nearest to x on the left. Then

$$|\varphi(x) - \psi(x)| \leqslant |\varphi(x) - \varphi(x_k)| + |\varphi(x_k) - \psi(x_k)| + |\psi(x_k) - \psi(x)| \leqslant \varepsilon,$$

i.e., the set of polygonal lines $\psi(x)$ forms an ε-net for Φ. But there are obviously only finitely many such lines. Therefore Φ is totally bounded. ∎

11.5. Peano's theorem. Arzelà's theorem has many applications, among them the following existence theorem for differential equations:

THEOREM 5 (*Peano*). *Let $f(x, y)$ be defined and continuous on a plane domain G. Then at least one integral curve of the differential equation*

$$\frac{dy}{dx} = f(x, y) \tag{2}$$

passes through each point (x_0, y_0) of G.

Proof. By the continuity of f, we have

$$|f(x, y)| \leqslant K$$

in some domain $G' \subset G$ containing the point (x_0, y_0). Draw the lines with slopes K and $-K$ through the point (x_0, y_0). Then draw vertical lines $x = a$ and $x = b$ ($a < x_0 < b$) which together with the first two lines form two isosceles triangles contained in G' with common vertex

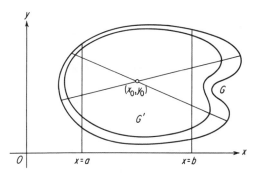

FIGURE 13

(x_0, y_0), as shown in Figure 13. This gives a closed interval $[a, b]$, which will figure in the rest of the proof.

The next step is to construct a family of polygonal lines, called *Euler lines*, associated with the differential equation (2). We begin by drawing the line with slope $f(x_0, y_0)$ through the point (x_0, y_0). Next, choosing a point (x_1, y_1) on the first line, we draw the line with slope $f(x_1, y_1)$ through the point (x_1, y_1). Then, choosing a point (x_2, y_2) on the second line, we draw the line with slope $f(x_2, y_2)$ through the point (x_2, y_2), and so on indefinitely. Suppose we construct a whole sequence $L_1, L_2, \ldots, L_n, \ldots$ of such Euler lines going through the point (x_0, y_0), with the property that the length of the longest line segment making up L_n approaches 0 as $n \to \infty$. Let φ_n be the function with graph L_n. Then this gives a family of functions $\varphi_1, \varphi_2, \ldots, \varphi_n, \ldots$, all defined on the interval $[a, b]$, which is easily seen to be uniformly bounded and equicontinuous (why?). It follows from Arzelà's theorem that the sequence $\{\varphi_n\}$ contains a uniformly convergent subsequence $\varphi^{(1)}, \varphi^{(2)}, \ldots, \varphi^{(n)}, \ldots$ Let

$$\varphi(x) = \lim_{n \to \infty} \varphi^{(n)}(x).$$

Then clearly

$$\varphi(x_0) = y_0,$$

so that the curve $y = \varphi(x)$ passes through the point (x_0, y_0).

We now show that $y = \varphi(x)$ satisfies the differential equation (2) in the open interval (a, b). This means showing that, given any $\varepsilon > 0$ and any points $x', x'' \in (a, b)$, we have

$$\left| \frac{\varphi(x'') - \varphi(x')}{x'' - x'} - f(x', \varphi(x')) \right| < \varepsilon$$

whenever $|x'' - x'|$ is sufficiently small, or equivalently that

$$\left| \frac{\varphi^{(n)}(x'') - \varphi^{(n)}(x')}{x'' - x'} - f(x', \varphi(x')) \right| < \varepsilon \qquad (3)$$

whenever n is sufficiently large and $|x'' - x'|$ is sufficiently small. Let $y' = \varphi(x')$. Then, by the continuity of f, given any $\varepsilon > 0$, there is a number $\eta > 0$ such that

$$f(x', y') - \varepsilon < f(x, y) < f(x', y') + \varepsilon$$

whenever

$$|x - x'| < 2\eta, \qquad |y - y'| < 4K\eta.$$

The set of points (x, y) satisfying these inequalities is a rectangle, which we denote by Q. Let N be so large that for all $n > N$, the length of the longest segment making up L_n is less than η and moreover

$$|\varphi(x) - \varphi^{(n)}(x)| < K\eta.$$

Then all the Euler lines L_n with $n > N$ lie inside the rectangle Q (why?). Suppose L_n has vertices $(a_0, b_0), (a_1, b_1), \ldots, (a_{k+1}, b_{k+1})$, where[13]

$$a_0 \leqslant x' < a_1 < a_2 < \cdots < a_k < x'' \leqslant a_{k+1}.$$

Then

$$\varphi^{(n)}(a_1) - \varphi^{(n)}(x') = f(a_0, b_0)(a_1 - x'),$$

$$\varphi^{(n)}(a_{i+1}) - \varphi^{(n)}(a_i) = f(a_i, b_i)(a_{i+1} - a_i) \qquad (i = 1, 2, \ldots, k - 1),$$

$$\varphi^{(n)}(x'') - \varphi^{(n)}(a_k) = f(a_k, b_k)(x'' - a_k).$$

Hence, if $|x'' - x'| < \eta$,

$$[f(x', y') - \varepsilon](a_1 - x') < \varphi^{(n)}(a_1) - \varphi^{(n)}(x') < [f(x', y') + \varepsilon](a_1 - x'),$$

$$[f(x', y') - \varepsilon](a_{i+1} - a_i) < \varphi^{(n)}(a_{i+1}) - \varphi^{(n)}(a_i)$$
$$< [f(x', y') + \varepsilon](a_{i+1} - a_i) \qquad (i = 1, 2, \ldots, k - 1),$$

$$[f(x', y') - \varepsilon](x'' - a_k) < \varphi^{(n)}(x'') - \varphi^{(n)}(a_k) < [f(x', y') + \varepsilon](x'' - a_k).$$

Adding these inequalities, we get

$$[f(x', y') - \varepsilon](x'' - x') < \varphi^{(n)}(x'') - \varphi^{(n)}(x') < [f(x', y') + \varepsilon](x'' - x')$$

if $|x'' - x'| < \eta$, which is equivalent to (3). ∎

Remark. Different subsequences of a sequence of Euler lines may converge to different solutions of the differential equation (2). Hence the solution φ found in the proof of Theorem 5 may not be the unique solution of (2) passing through the point (x_0, y_0).

Problem 1. Let M be a totally bounded subset of a metric space R. Prove that the ε-nets figuring in the definition of total boundedness of M can always be chosen to consist of points of M rather than of R.

---

[13] To be explicit, we assume that $x'' > x'$. The case $x'' < x'$ is treated similarly.

Hint. Given an ε-net for M consisting of points $a_1, a_2, \ldots, a_n \in R$, all within ε of some point of M, replace each point a_k by a point $b_k \in M$ such that $\rho(a_k, b_k) \leqslant \varepsilon$.

Problem 2. Prove that every totally bounded metric space is separable.

Hint. Construct a finite $(1/n)$-net for every $n = 1, 2, \ldots$ Then take the union of these nets.

Problem 3. Let M be a bounded subset of the space $C_{[a,b]}$. Prove that the set of all functions

$$F(x) = \int_a^x f(t)\, dt$$

with $f \in M$ compact.

Problem 4. Given two metric compacta X and Y, let C_{XY} be the set of all continuous mappings of X into Y. Let distance be defined in C_{XY} by the formula

$$\rho(f, g) = \sup_{x \in X} \rho(f(x), g(x)). \tag{4}$$

Prove that C_{XY} is a metric space. Let M_{XY} be the set of all mappings of X into Y, with the same metric (4). Prove that C_{XY} is closed in M_{XY}.

Hint. Use the method of Problem 1, p. 65 to prove that the limit of a uniformly convergent sequence of continuous mappings is itself a continuous mapping.

Problem 5. Let X, Y and C_{XY} be the same as in the preceding problem. Prove the following generalization of Arzelà's theorem: A necessary and sufficient condition for a set $D \subset C_{XY}$ to be relatively compact is that D be an *equicontinuous* family of functions, in the sense that given any $\varepsilon > 0$, there exists a number $\delta > 0$ such that $\rho(x', y') < \delta$ implies $\rho(f(x'), f(x'')) < \varepsilon$ for all $x', x'' \in X$ and all $f \in D$.

Hint. To prove the sufficiency, show that D is relatively compact in M_{XY} (defined in the preceding problem) and hence in C_{XY}, since C_{XY} is closed in M_{XY}. To prove the relative compactness of D in M_{XY}, first represent X as a union of finitely many pairwise disjoint sets E_i such that $x', x'' \in E_i$ implies $\rho(x', x'') < \delta$. For example, let $x_1, \ldots, x_n$ be a $(\delta/2)$-net for X, and let

$$E_i = S[x_i, \delta] - \bigcup_{j < i} S[x_j, \delta].$$

Then let $y_1, \ldots, y_p$ be an ε-net in Y, and let L be the set of all functions taking the values y_j on the sets E_i. Given any $f \in D$ and any $x_i \in \{x_1, \ldots, x_n\}$, let $y_j \in \{y_1, \ldots, y_n\}$ be such that $\rho(f(x_i), y_j) < \varepsilon$ and let $g \in L$ be such that $g(x_i) = y_i$. Show that $\rho(f(x), g(x)) < 2\varepsilon$, thereby proving that L is a finite 2ε-net for D in M_{XY}.

12. Real Functions on Metric and Topological Spaces

12.1. Continuous and uniformly continuous functions and functionals. Let T be a topological space, in particular a metric space. Then by a *real function* on T we mean a mapping of T into the space R^1 (the real line). For example, a real function on Euclidean n-space R^n is just the usual "function of n variables." Suppose T is a *function space*, i.e., a space whose elements are functions. Then a real function on T is called a *functional*.

Example 1. Let $x(t)$ be a function defined on the interval $[0, 1]$, let $\varphi(s_0, s_1, \ldots, s_n)$ be a function of $n + 1$ variables defined for all real values of its arguments, and let $\psi(t, u)$ be a function of two variables defined for all $t \in [0, 1]$ and all real u. Then the following are all functionals:

$$F_1(x) = \sup_{0 \leqslant t \leqslant 1} x(t),$$

$$F_2(x) = \inf_{0 \leqslant t \leqslant 1} x(t),$$

$$F_3(x) = x(t_0) \qquad \text{where} \quad t_0 \in [0, 1],$$

$$F_4(x) = \varphi[x(t_0), x(t_1), \ldots, x(t_n)],$$

$$F_5(x) = \int_0^1 \psi[t, x(t)]\, dt$$

$$F_6(x) = x'(t_0) \qquad \text{where} \quad t_0 \in [0, 1],$$

$$F_7(x) = \int_0^1 \sqrt{1 + x'^2(t)}\, dt,$$

$$F_8(x) = \int_0^1 |x'(t)|\, dt.$$

The functionals F_1, F_2, F_3, F_4 and F_5 are defined on the space C of all functions continuous on the interval $[0, 1]$. On the other hand, F_6 is defined only for functions differentiable at the point t_0, F_7 is defined only for functions such that the expression $\sqrt{1 + x'^2(t)}$ is integrable, and F_8 is defined only for functions with integrable $|x'(t)|$.

Example 2. The functional F_1 is continuous on C, since

$$\rho(x, y) = \sup |x - y|, \qquad |\sup x - \sup y| \leqslant \sup |x - y|.$$

Example 3. The functional F_6 is discontinuous on C at any point x_0 where it is defined. In fact, let $x(t)$ be such that $x'(t_0) = 1$ and $|x(t)| < \varepsilon$, and let $y = x_0 + x$. Then $y'(t_0) = x_0'(t_0) + 1$ even though $\rho(x_0, y) < \varepsilon$. However, F_6 is continuous if it is defined on the space $C^{(1)}$ of all functions *continuously differentiable* on the interval $[0, 1]$, with metric

$$\rho(x, y) = \sup_{0 \leqslant t \leqslant 1} [|x(t) - y(t)| + |x'(t) - y'(t)|]$$

(why?).

Example 4. The function F_7 is also discontinuous on C. In fact, let

$$x_0(t) \equiv 0, \qquad x_n(t) = \frac{1}{n} \sin 2\pi nt.$$

Then

$$\rho(x_n, x_0) = \frac{1}{n} \to 0,$$

but $F_7(x_n) > 4$ for all n while $F_7(x_0) = 1$. Hence $F_7(x_n)$ fails to approach $F_7(x_0)$ even though $x_n \to x_0$.

The ordinary concept of *uniform continuity* generalizes at once to the case of arbitrary metric spaces:

DEFINITION 1. *A real function $f(x)$ defined on a metric space R is said to be **uniformly continuous** on R if, given any $\varepsilon > 0$, there is a $\delta > 0$ such that $\rho(x_1, x_2) < \delta$ implies $|f(x_1) - f(x_2)| < \varepsilon$ for all $x_1, x_2 \in R$.*

The reader will recall from calculus that a real function continuous on a closed interval $[a, b]$ is uniformly continuous on $[a, b]$. This fact is a special case of

THEOREM 1. *A real function f continuous on a compact metric space R is uniformly continuous on R.*

Proof. Suppose f is continuous but not uniformly continuous on R. Then for some positive ε and every n there are points x_n and x'_n in R such that

$$\rho(x_n, x'_n) < \frac{1}{n} \tag{1}$$

but

$$|f(x_n) - f(x'_n)| \geq \varepsilon. \tag{2}$$

Since R is compact, the sequence $\{x_n\}$ has a subsequence $\{x_{n_k}\}$ converging to a point $x \in R$. Hence $\{x'_{n_k}\}$ also converges to x, because of (1). But then at least one of the inequalities

$$|f(x) - f(x_{n_k})| > \frac{\varepsilon}{2}, \qquad |f(x) - f(x'_{n_k})| > \frac{\varepsilon}{2}$$

must hold for arbitrary k, because of (2). This contradicts the assumed continuity of f at x. ∎

12.2. Continuous and semicontinuous functions on compact spaces. As just shown, the theorem on uniform continuity of a function continuous on a closed interval generalizes to functions continuous on arbitrary metric compacta. There are other properties of functions continuous on a closed interval which generalize to arbitrary compact spaces (not necessarily metric spaces):

THEOREM 2. *A real function f continuous on a compact topological space T is bounded on T.*[14] *Moreover f achieves its least upper bound and greatest lower bound on T.*

Proof. A continuous real function on T is a continuous mapping of T into the real line R^1. The image of T in R^1 is compact, by Theorem 5, p. 94. But every compact subset of R^1 is bounded and closed (see p. 101). Hence f is bounded on T. Moreover, f not only has a least upper bound and greatest lower bound on T, but actually achieves these bounds at points of T. ∎

Theorem 2 can be generalized to a larger class of functions, which we now introduce:

DEFINITION 2. *A (real) function f defined on a topological space T is said to be **upper semicontinuous** at a point $x_0 \in T$ if, given any $\varepsilon > 0$, there exists a neighborhood of x_0 in which $f(x) < f(x_0) + \varepsilon$. Similarly, f is said to be **lower semicontinuous** at x_0 if, given any $\varepsilon > 0$, there exists a neighborhood of x_0 in which $f(x) > f(x_0) - \varepsilon$.*

Example 1. Let $[x]$ be the integral part of x, i.e., the largest integer $\leqslant x$. Then $f(x) = [x]$ is upper semicontinuous for all x.

Example 2. Given a continuous function f, suppose we increase the value $f(x_0)$ taken by f at the point x_0. Then f becomes upper semicontinuous at x_0. Similarly, f becomes lower semicontinuous at x_0 if we decrease $f(x_0)$. Moreover, f is upper semicontinuous if and only if $-f$ is lower semicontinuous. These facts can be used to construct many more examples of semicontinuous functions.

In studying the properties of semicontinuous functions, it is convenient to allow them to take infinite values. If $f(x_0) = +\infty$, we regard f as upper semicontinuous at x_0. The function f is also regarded as lower semicontinuous at x_0 if, given any $h > 0$, there is a neighborhood of x_0 in which $f(x) > h$. Similarly, if $f(x_0) = -\infty$, we regard f as lower semicontinuous at x_0, and at the same time upper semicontinuous at x_0 if, given any $h > 0$, there is a neighborhood of x_0 in which $f(x) < -h$.

We now prove the promised generalization of Theorem 2:

---

[14] A real function (or functional) f is said to be *bounded* on a set E if $f(E)$ is contained in some interval $[-C, C]$.

THEOREM 2′. *A finite lower semicontinuous function f defined on a compact topological space T is bounded from below.*

Proof. Suppose to the contrary that $\inf f(x) = -\infty$. Then there exists a sequence $\{x_n\}$ such that $f(x_n) < -n$. Since T is compact, the infinite set $E = \{x_1, x_2, \ldots, x_n, \ldots\}$ has at least one limit point x_0. Since f is finite and lower semicontinuous at x_0, there is a neighborhood U of x_0 in which $f(x) > f(x_0) - 1$. But then U can only contain finitely many points of E, so that x_0 cannot be a limit point of E. ∎

THEOREM 2″. *A finite lower semicontinuous function f defined on a compact topological space T achieves its greatest lower bound on T.*

Proof. By Theorem 2′, $\inf f(x)$ is finite. Clearly, there exists a sequence $\{x_n\}$ such that

$$f(x_n) \leqslant \inf f(x) + \frac{1}{n}.$$

By the compactness of T, the set $E = \{x_1, x_2, \ldots, x_n, \ldots\}$ has at least one limit point x_0. If $f(x_0) > \inf f$, then, by the semicontinuity of f at x_0, there is a neighborhood U of the point x_0 and a $\delta > 0$ such that $f(x) > \inf f + \delta$ for all $x \in U$. But then U cannot contain an infinite subset of E, i.e., x_0 cannot be a limit point of x_0. It follows that $f(x_0) = \inf f$. ∎

Remark. Theorems 2′ and 2″ remain true if the words "lower," "below," and "greatest" are replaced by "upper," "above," and "least." The details are left as an exercise.

We conclude this section with some useful terminology:

DEFINITION 3. *Given a real function f defined on a metric space R, the (finite or infinite) quantity*

$$\bar{f}(x_0) = \lim_{\varepsilon \to 0} \left\{ \sup_{x \in S(x_0, \varepsilon)} f(x) \right\}$$

*is called the **upper limit** of f at x_0, while the (finite or infinite) quantity*

$$\underline{f}(x_0) = \lim_{\varepsilon \to 0} \left\{ \inf_{x \in S(x_0, \varepsilon)} f(x) \right\}$$

*is called the **lower limit** of f at x_0. The difference*

$$\omega f(x_0) = \bar{f}(x_0) - \underline{f}(x_0),$$

provided it exists,[15] *is called the **oscillation** of f at x_0.*

---

[15] I.e., provided at least one of the numbers $\bar{f}(x_0), \underline{f}(x_0)$ is finite.

(a)   (b)   (c)

FIGURE 14

12.3. Continuous curves in metric spaces. Instead of mappings of a metric space into the real line, we now consider mappings of a subset of the real line into a metric space. More exactly, let $P = f(t)$ be a continuous mapping of the interval $a \leqslant t \leqslant b$ into a metric space R. As t "traverses" the interval from a to b, the point $P = f(t)$ "traverses a continuous curve" in the space R. Before giving a formal definition corresponding to this rough idea of a "curve," we make two key observations:

1) The order in which points are traversed will be regarded as an essential property of a curve. For example, the set of points shown in Figure 14(a) gives rise to two distinct curves when traversed in the two distinct ways shown in Figures 14(b) and 14(c). Similarly, the function shown in Figure 15(a), defined in the interval $0 \leqslant t \leqslant 1$, determines a "curve" filling up the segment $0 \leqslant y \leqslant 1$ of the y-axis, but this curve is traversed three times (twice upward and once downward) and hence is distinct from the segment $0 \leqslant y \leqslant 1$ traversed just once from the point $y = 0$ to the point $y = 1$.

2) The choice of the parameter t will be regarded as unimportant, provided a change in parameter does not change the order in which the points of the curve are traversed. Thus the functions shown in Figures 15(a) and 15(b) represent the same curve, even though a given point of the curve corresponds to different parameter values in the two cases. For example, the point A in Figure 15(a) corresponds to

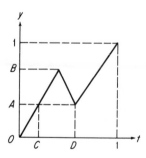 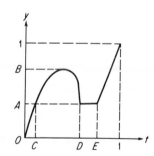

FIGURE 15

two isolated points C and D on the t-axis, while in Figure 15(b) the same point A corresponds to an isolated point C and a whole line segment DE (note that the point on the curve does not move at all as t traverses the segment DE).

We now give a formal definition of a curve, embodying these qualitative ideas. Two continuous functions

$$P = f(t'), \qquad P = g(t''),$$

defined on intervals

$$a' \leqslant t' \leqslant b', \qquad a'' \leqslant t'' \leqslant b''$$

and taking values in a metric space R, are said to be *equivalent* if there exist two continuous nondecreasing functions

$$t' = \varphi(t), \qquad t'' = \psi(t),$$

defined on the same interval

$$a \leqslant t \leqslant b,$$

such that

$$\varphi(a) = a', \qquad \varphi(b) = b',$$

$$\psi(a) = a'', \qquad \psi(b) = b''$$

and

$$f(\varphi(t)) = g(\psi(t)) \quad \text{for all} \quad t \in [a, b].$$

It is easy to see that this relation of equivalence is reflexive (f is equivalent to f), symmetric (if f is equivalent to g, then g is equivalent to f) and transitive (if f is equivalent to g and g is equivalent to h, then f is equivalent to h). Hence the set of all continuous functions of the given type can be partitioned into classes of equivalent functions (cf. Sec. 1.4), and each such class is said to define a (*continuous*) *curve* in the space R.

For each function $P = f(t')$ defined on an interval $[a', b']$, there is an equivalent function defined on the interval $[a'', b''] = [0, 1]$. In fact, we need only make the choice

$$t' = \varphi(t) = (b' - a')t + a', \qquad t'' = \psi(t) = t.$$

Thus every curve can be regarded as specified parametrically in terms of a function defined on the *unit* interval $I = [0, 1]$. By the same token, it is often convenient[16] to introduce the space $C(I, R)$ of continuous mappings f of the interval I into the space R, equipped with the metric

$$\tilde{\rho}(f, g) = \sup_{0 \leqslant t \leqslant 1} \rho(f(t), g(t)), \tag{3}$$

where ρ is the metric in the space R.

---

[16] Cf. Problems 7–12.

Problem 1. Let the functionals F_1, F_2, F_3, F_4, F_5 and the space C be the same as on p. 108. Prove that

a) F_2, F_3 and F_5 are continuous on C;
b) F_4 is continuous on C if the function φ is continuous in all its arguments;
c) F_1 is uniformly continuous on C.

Define F_1, F_2, F_3 and F_4 on a space larger than C.

Problem 2. Let the functionals F_7, F_8 and the spaces C, $C^{(1)}$ be the same as on p. 108. Prove that

a) F_8 is discontinuous on C;
b) F_7 and F_8 are continuous on $C^{(1)}$.

Problem 3. Let M be the space of all bounded real functions defined on the interval $[a, b]$, with metric $\rho(f, g) = \sup |f - g|$. By the *length* of the curve

$$y = f(x) \qquad (a \leqslant x \leqslant b)$$

is meant the functional

$$L(f) = \sup \sum_{i=1}^{n} \sqrt{(x_i - x_{i-1})^2 + (f(x_i) - f(x_{i-1}))^2},$$

where the least upper bound (which may equal $+\infty$) is taken over all possible partitions of $[a, b]$ obtained by introducing points of subdivision x_0, x_1, x_2, ..., x_n such that

$$a = x_0 < x_1 < x_2 < \cdots < x_n = b.$$

Prove that

a) For continuous functions

$$L(f) = \lim_{\max |x_i - x_{i-1}| \to 0} \sqrt{\sum_{i=1}^{n} (x_i - x_{i-1})^2 + (f(x_i) - f(x_{i-1}))^2};$$

b) For continuously differentiable functions

$$L(f) = \int_a^b \sqrt{1 + f'^2(x)}\, dx;$$

c) The functional $L(f)$ is lower semicontinuous on M.

Problem 4. Let $\bar{f}, \underline{f}$ and ω be the same as in Definition 3. Prove that

a) $\bar{f}$ is upper semicontinuous;
b) $\underline{f}$ is lower semicontinuous;
c) f is continuous at x_0 if and only if $-\infty < \underline{f}(x_0) = \bar{f}(x_0) < \infty$, i.e., if and only if $\omega f(x_0) = 0$.

Problem 5. Let K be a metric compactum and A a mapping of K into itself such that $\rho(Ax, Ay) < \rho(x, y)$ if $x \neq y$. Prove that A has a unique fixed point in K. Reconcile this with Problem 1, p. 76.

Problem 6. Let K be a metric compactum and $\{f_n(x)\}$ a sequence of continuous functions on K, increasing in the sense that

$$f_1(x) \leqslant f_2(x) \leqslant \cdots \leqslant f_n(x) \leqslant \cdots$$

Prove that if $\{f_n(x)\}$ converges to a continuous function on K, then the covergence is uniform (*Dini's theorem*).

Problem 7. A sequence of curves $\{\Gamma_n\}$ in a metric space R is said to *converge* to a curve Γ in R if the curves Γ_n and Γ have parametric representations

$$P = f_n(t) \qquad (0 \leqslant t \leqslant 1)$$

and

$$P = f(t) \qquad (0 \leqslant t \leqslant 1),$$

respectively, such that

$$\lim_{n \to \infty} \tilde{\rho}(f, f_n) = 0,$$

where $\tilde{\rho}$ is the metric (3) of the space $C(I, R)$ introduced on p. 113. Prove that if a sequence of curves in a compact metric space R can be represented parametrically by an equicontinuous family of functions on $[0, 1]$, then the sequence contains a convergent subsequence.

Hint. Use Problem 5, p. 107.

Problem 8. Let Γ be a curve in a metric space R, with parametric representation

$$P = f(t) \qquad (a \leqslant t \leqslant b).$$

By the *length* of Γ is meant the functional

$$L(\Gamma) = L(f) = \sup \sum_{i=1}^{n} \rho(f(t_{i-1}), f(t_i)),$$

where ρ is the metric in R and the least upper bound (which may equal $+\infty$) is taken over all possible partitions of $[a, b]$ obtained by introducing points of subdivision $t_0, t_1, t_2, \ldots, t_n, \ldots$ such that

$$a = t_0 < t_1 < t_2 < \cdots < t_n = b.$$

Prove that $L(\Gamma)$ is independent of the parametric representation of Γ. Suppose we choose $a = 0$, $b = 1$, thereby confining ourselves to parametric representations of the form

$$P = f(t) \qquad (0 \leqslant t \leqslant 1).$$

Prove that $L(f)$ is then a lower semicontinuous functional on the space $C(I, R)$ introduced on p. 113. Equivalently, prove that if a sequence of curves $\{\Gamma_n\}$ converges to a curve Γ, in the sense of Problem 7, then $L(\Gamma)$ does not exceed the smallest limit point (i.e., the lower limit) of the sequence $\{L(\Gamma_n)\}$.

Problem 9. Given a metric space R with metric ρ, let Γ be a curve in R of finite length S with parametric representation

$$P = f(t) \qquad (a < t < b).$$

Let $s = \varphi(T)$ be the length of the arc

$$P = f(t) \qquad (a \leqslant t \leqslant T)$$

(where $T < b$), i.e., the arc of Γ going from the "initial point" $P_a = f(a)$ to the "final point" $P_T = f(T)$. Then Γ has a parametric representation of the form

$$P = g(s) \qquad (0 \leqslant s < S),$$

where $g(s) \equiv f(\varphi^{-1}(s))$ if φ is one-to-one. Prove that

$$\rho(g(s_1), g(s_2)) \leqslant |s_1 - s_2|.$$

Hint. The length of an arc is no less than the length of the inscribed chord.

Problem 10. In the preceding problem, let $\tau = s/S$. Then Γ has a parametric representation

$$P = F(\tau) \equiv g(S\tau) \qquad (0 \leqslant t < 1)$$

in terms of a function F defined on the unit interval $[0, 1]$. Prove that F satisfies a Lipschitz condition of the form

$$\rho(F(\tau_1), F(\tau_2)) \leqslant S |\tau_1 - \tau_2|.$$

Suppose R is compact and let $\{\Gamma_n\}$ be a sequence of curves, all of length less than some finite number M. Prove that $\{\Gamma_n\}$ contains a convergent subsequence, where convergence of curves is defined as in Problem 7.

Problem 11. Given a compact metric space R, suppose two points A and B in R can be joined by a continuous curve of finite length. Prove that among all such curves, there is a curve of least length.

Comment. Even in the case where R is a "smooth" (i.e., sufficiently differentiable) closed surface in Euclidean 3-space, this result is not amenable to the methods of elementary differential geometry, which ordinarily deals only with the case of "neighboring" points A and B.

Problem 12. Let $\mathscr{C}$ be the set of all curves in a given metric space R.

Define the distance between two curves Γ_1, $\Gamma_2 \in \mathscr{C}$ by the formula

$$\hat{\rho}(\Gamma_1, \Gamma_2) = \inf \tilde{\rho}(f_1, f_2), \tag{4}$$

where $\tilde{\rho}$ is the metric (3) in the space $C(I, R)$, and the greatest lower bound is taken over all possible representations

$$P = f_1(t) \qquad (0 \leqslant t \leqslant 1) \tag{5}$$

of Γ_1 and

$$P = f_2(t) \qquad (0 \leqslant t \leqslant 1) \tag{6}$$

of Γ_2. Prove that the metric $\hat{\rho}$ makes $\mathscr{C}$ into a metric space.

Comment. The fact that $\hat{\rho}(\Gamma_1, \Gamma_2) = 0$ implies the identity of Γ_1 and Γ_2 follows from the (not very easily proved) fact that the greatest lower bound in (4) is achieved for a suitable choice of the parametric representations (5) and (6).

4

LINEAR SPACES

---

13. Basic Concepts

13.1. Definitions and examples. One of the most important concepts in mathematics is that of a *linear space*, which will play a key role in the rest of this book:

DEFINITION 1. *A nonempty set L of elements* $x, y, z, \ldots$ *is said to be a* **linear space** *(or* **vector space***) if it satisfies the following three axioms:*

1) *Any two elements* $x, y \in L$ *uniquely determine a third element* $x + y \in L$, *called the* **sum** *of x and y, such that*
   a) $x + y = y + x$ *(commutativity)*;
   b) $(x + y) + z = x + (y + z)$ *(associativity)*;
   c) *There exists an element* $0 \in L$, *called the* **zero element,** *with the property that* $x + 0 = x$ *for every* $x \in L$;
   d) *For every* $x \in L$ *there exists an element* $-x$, *called the* **negative** *of x, with the property that* $x + (-x) = 0$;

2) *Any number* α *and any element* $x \in L$ *uniquely determine an element* $\alpha x \in L$, *called the* **product** *of* α *and x, such that*
   a) $\alpha(\beta x) = (\alpha\beta)x$;
   b) $1x = x$;

3) *The operations of addition and multiplication obey two distributive laws:*
   a) $(\alpha + \beta)x = \alpha a + \beta x$;
   b) $\alpha(x + y) = \alpha x + \alpha y$.

Remark. The elements of L are called "points" or "vectors," while the numbers $\alpha, \beta, \ldots$ are often called "scalars." If α is an arbitrary real number, L is called a *real* linear space, while if α is an arbitrary complex number, L is called a *complex* linear space.[1] Unless the contrary is explicitly stated, the considerations that follow will be valid for both real and complex spaces. Clearly, any complex linear space reduces to a real linear space if we allow vectors to be multiplied by real numbers only.

We now give some examples of linear spaces, leaving it to the reader to verify in detail that the conditions in Definition 1 are satisfied in each case.[2]

Example 1. The real line (the set of all real numbers) with the usual arithmetic operations of addition and multiplication is a linear space.

Example 2. The set of all ordered n-tuples

$$x = (x_1, x_2, \ldots, x_n)$$

of real or complex numbers $x_1, x_2, \ldots, x_n$, with sums and "scalar multiples" defined by the formulas

$$(x_1, x_2, \ldots, x_n) + (y_1, y_2, \ldots, y_n) = (x_1 + y_1, x_2 + y_2, \ldots, x_n + y_n),$$
$$\alpha(x_1, x_2, \ldots, x_n) = (\alpha x_1, \alpha x_2, \ldots, \alpha x_n),$$

is also a linear space. This space is called *n-dimensional (vector) space*, or simply *n-space*, denoted by R^n in the real case and C^n in the complex case. (Concerning the precise meaning of the term "n-dimensional," see Sec. 13.2.)

Example 3. The set of all (real or complex) functions continuous on an interval $[a, b]$, with the usual operations of addition of functions and multiplication of functions by numbers, forms a linear space $C_{[a,b]}$, one of the most important spaces in analysis.

Example 4. The set l_2 of all infinite sequences

$$x = (x_1, x_2, \ldots, x_k, \ldots) \tag{1}$$

of real or complex numbers $x_1, x_2, \ldots, x_k, \ldots$ satisfying the convergence condition

$$\sum_{k=1}^{\infty} |x_k|^2 < \infty,$$

---

[1] More generally, one can consider linear spaces over an arbitrary field.

[2] It will be noted that certain symbols like R^n, $C_{[a,b]}$, l_2 and m are used here with somewhat different meanings than in Sec. 5.1. The point is that there is no metric here, at least for the time being, while on the other hand, sums and scalar multiples of vectors were not defined in Chaps. 2 and 3.

equipped with operations

$$(x_1, x_2, \ldots, x_k, \ldots) + (y_1, y_2, \ldots, y_k, \ldots)$$
$$= (x_1 + y_1, x_2 + y_2, \ldots, x_k + y_k, \ldots),$$
$$\alpha(x_1, x_2, \ldots, x_k, \ldots) = (\alpha x_1, \alpha x_2, \ldots, \alpha x_k, \ldots), \qquad (2)$$

is a linear space. The fact that

$$\sum_{k=1}^{\infty} |x_k|^2 < \infty, \qquad \sum_{k=1}^{\infty} |y_k|^2 < \infty$$

implies

$$\sum_{k=1}^{\infty} |x_k + y_k|^2 < \infty$$

is an immediate consequence of the elementary inequality

$$(x_k + y_k)^2 \leqslant 2(x_k^2 + y_k^2).$$

Example 5. Let c be the set of all *convergent* sequences (1), c_0 the set of all sequences (1) *converging to zero*, m the set of all *bounded* sequences, and R^{∞} the set of *all* sequences (1). Then c, c_0, m and R^{∞} are all linear spaces, provided that in each case addition of sequences and multiplication of sequences by numbers are defined by (2).

Since linear spaces are defined in terms of two operations, addition of elements and multiplication of elements by numbers, it is natural to introduce

DEFINITION 2. *Two linear spaces L and L^* are said to be* **isomorphic** *if there is a one-to-one correspondence $x \leftrightarrow x^*$ between L and L^* which preserves operations, in the sense that*

$$x \leftrightarrow x^*, \qquad y \leftrightarrow y^*$$

(where $x, y \in L$, $x^*, y^* \in L^*$) implies

$$x + y \leftrightarrow x^* + y^*$$

and

$$\alpha x \leftrightarrow \alpha x^*$$

(α an arbitrary number).

Remark. It is sometimes convenient to regard isomorphic linear spaces as different "realizations" of one and the same linear space.

13.2. Linear dependence. We say that the elements $x, y, \ldots, w$ of a linear space L are *linearly dependent* if there exist numbers $\alpha, \beta, \ldots, \lambda$, *not all zero*, such that[3]

$$\alpha x + \beta y + \cdots + \lambda w = 0. \qquad (3)$$

_____

[3] The left-hand side of (3) is called a *linear combination* of the elements $x, y, \ldots, w$.

If no such numbers exist, the elements $x, y, \ldots, w$ are said to be *linearly independent*. In other words, the elements $x, y, \ldots, w$ are linearly independent if and only if (3) implies

$$\alpha = \beta = \cdots = \lambda = 0.$$

More generally, the elements $x, y, \ldots$ belonging to some infinite set $E \subset L$ are said to be linearly independent if the elements belonging to every finite subset of E are linearly independent.

A linear space L is said to be *n-dimensional* (or of *dimension n*) if n linearly independent elements can be found in L, but any $n + 1$ elements of L are linearly dependent. Suppose n linearly independent elements can be found in L *for every n*. Then L is said to be *infinite-dimensional*, but otherwise L is said to be *finite-dimensional*. Any set of n linearly independent elements of an n-dimensional space L is called a *basis* in L.

Remark. The typical course on linear algebra deals with finite-dimensional linear spaces. Here, however, we will be primarily concerned with infinite-dimensional spaces, the case of greater interest from the standpoint of mathematical analysis.

13.3. Subspaces. Given a nonempty subset L' of a linear space L, suppose L is itself a linear space with respect to the operations of addition and multiplication defined in L. Then L' is said to be a *subspace* (of L). In other words, we say that $L' \subset L$ is a subspace if $x \in L'$, $y \in L'$ implies $\alpha x + \beta y \in L'$ for arbitrary α and β. The "trivial space" consisting of the zero element alone is a subspace of every linear space L. At the opposite extreme, L can always be regarded as a subset of itself. By a *proper subspace* of a linear space L, we mean a subspace which is distinct from L itself and contains at least one nonzero element.

Example 1. Let L be any linear space, and x any nonzero element of L. Then the set $\{\lambda x\}$ of all scalar multiples of x, where λ ranges over all (real or complex) numbers is obviously a one-dimensional subspace of L, in fact a proper subspace if the dimension of L exceeds 1.

Example 2. The set $P_{[a,b]}$ of all polynomials on $[a, b]$ is a proper subspace of the set $C_{[a,b]}$ of all continuous functions on $[a, b]$. Like $C_{[a,b]}$ itself, $P_{[a,b]}$ is infinite-dimensional. At the same time, $C_{[a,b]}$ is itself a proper subspace of the set of all functions on $[a, b]$, both continuous and discontinuous.

Example 3. Each of the linear spaces l_2, c_0, c, m and R^∞ (in that order) is a proper subspace of the next one.

Given a linear space L, let $\{x_\alpha\}$ be any nonempty set of elements $x_\alpha \in L$. Then L has a smallest subspace (possibly L itself) containing $\{x_\alpha\}$.[4] In fact,

---

[4] Here we use curly brackets in the same way as in footnote 10, p. 92.

there is at least one such subspace, namely L itself. Moreover, it is clear that the intersection of any system of subspaces $\{L_\gamma\}$ is itself a subspace, since if $L^* = \bigcap_\gamma L_\gamma$ and $x, y \in L^*$, then $\alpha x + \beta y \in L^*$ for all α and β (why?). The smallest subspace of L containing the set $\{x_\alpha\}$ is then just the intersection of all subspaces containing $\{x_\alpha\}$. This minimal subspace, denoted by $L(\{x_\alpha\})$, is called the *(linear) subspace generated by* $\{x_\alpha\}$ or the *linear hull of* $\{x_\alpha\}$.

13.4. Factor spaces. Let L be a linear space and L' a subspace of L. Then two elements $x, y \in L$ are said to belong to the same *(residue) class generated by* L' if the difference $x - y$ belongs to L'. The set of all such classes is called the *factor space* (or *quotient space*) of L relative to L', denoted by L/L'. The operations of addition of elements and multiplication of elements by numbers can be introduced in a factor space L/L' in the following natural way: Given two elements of L/L', i.e., two classes ξ and η, we choose a "representative" from each class, say x from ξ and y from η. We then define the sum $\xi + \eta$ of the classes ξ and η to be the class containing the element $x + y$, while the product $\alpha \xi$ of the number α and the class ξ is defined to be the class containing the element αx. Here we rely on the fact that the classes $\xi + \eta$ and $\alpha \xi$ are independent of the choice of the "representatives" x and y (why?).

THEOREM 1. *Every factor space* L/L', *with operations defined in the way just described, is a linear space.*

Proof. We need only verify that L/L' satisfies the three axioms in Definition 1. This is almost trivial (give the details). ∎

Let L be a linear space and L' a subspace of L. Then the dimension of the factor space L/L' is called the *codimension* of L' in L.

THEOREM 2. *Let* L' *be a subspace of a linear space* L. *Then* L' *has finite codimension* n *if and only if there are linearly independent elements* $x_1, \ldots, x_n$ *in* L *such that every element* $x \in L$ *has a unique representation of the form*

$$x = \alpha_1 x_1 + \cdots + \alpha_n x_n + y, \tag{4}$$

where $\alpha_1, \ldots, \alpha_n$ *are numbers and* $y \in L'$.

Proof. Suppose every element $x \in L$ has a unique representation of the form (4). Given any class $\xi \in L/L'$, let x be any element of ξ, and let ξ_k be the class containing x_k $(k = 1, \ldots, n)$. Then (4) clearly implies

$$\xi = \alpha_1 \xi_1 + \cdots + \alpha_n \xi_n.$$

Hence $\xi_1, \ldots, \xi_n$ is a basis for L/L' (the linear independence of $\xi_1, \ldots, \xi_n$ follows from that of $x_1, \ldots, x_n$). In other words, L/L' has dimension n, or equivalently L' has codimension n.

Conversely, suppose L' has codimension n, so that L/L' has dimension n. Then L/L' has a basis $\xi_1, \ldots, \xi_n$. Given any $x \in L$, let ξ be the class in L/L' containing x. Then

$$\xi = \alpha_1\xi_1 + \cdots + \alpha_n\xi_n$$

for suitable numbers $\alpha_1, \ldots, \alpha_n$. But this means that every element in ξ, in particular x, differs only by an element $y \in L'$ from a linear combination of elements $x_1, \ldots, x_n$ where x_k is any fixed element of ξ_k ($k = 1, \ldots, n$), i.e.,

$$x = \alpha_1x_1 + \cdots + \alpha_nx_n + y \qquad (y \in L') \tag{5}$$

(the linear independence of $x_1, \ldots, x_n$ follows from that of $\xi_1, \ldots, \xi_n$). Suppose there is another such representation

$$x = \alpha'_1x_1 + \cdots + \alpha'_nx_n + y' \qquad (y' \in L'). \tag{5'}$$

Then, subtracting (5') from (5), we get

$$0 = (\alpha_1 - \alpha'_1)x_1 + \cdots + (\alpha_n - \alpha'_n) + y'' \qquad (y'' \in L'),$$

and hence

$$0 = (\alpha_1 - \alpha'_1)\xi_1 + \cdots + (\alpha_n - \alpha'_n)\xi_n,$$

where in the last equation 0 means the class containing the zero element of L, i.e., the space L' itself. But $\xi_1, \ldots, \xi_n$ are linearly independent, and hence $\alpha_1 = \alpha'_1, \ldots, \alpha_n = \alpha'_n$. ∎

13.5. Linear functionals. A numerical function f defined on a linear space L is called a *functional* (on L).[5] A functional f is said to be *additive* if

$$f(x + y) = f(x) + f(y)$$

for all $x, y \in L$ and *homogeneous* if

$$f(\alpha x) = \alpha f(x)$$

for every number α. A functional defined on a *complex* linear space is called *conjugate-homogeneous* if

$$f(\alpha x) = \bar{\alpha}f(x)$$

for every number α, where $\bar{\alpha}$ is the complex conjugate of α. An additive

---

[5] The word "functional" has already been used in a somewhat different sense in Sec. 12.1, where a functional means a real function defined on a function space (topological or metric). Later on, we will deal with linear spaces which are also metric spaces and have functions as their elements. The two uses of the word "functional" will then coincide (if we allow complex-valued functionals).

homogeneous functional is called a *linear functional*, while an additive conjugate-homogeneous functional is called a *conjugate-linear functional*.

Example 1. Let R^n be real n-space, with elements $x = (x_1, \ldots, x_n)$, and let $a = (a_1, \ldots, a_n)$ be a fixed element of R^n. Then

$$f(x) = \sum_{k=1}^{n} a_k x_k$$

is a linear functional on R^n. Similarly,

$$f(x) = \sum_{k=1}^{n} a_k \bar{x}_k$$

is a conjugate-linear functional on complex n-space C^n.

Example 2. Consider the integral

$$I(x) = \int_a^b x(t)\, dt,$$

or more generally

$$I(x) = \int_a^b x(t)\varphi(t)\, dt,$$

where $\varphi(t)$ is a fixed continuous function on $[a, b]$. It follows at once from elementary properties of integrals that $I(x)$ is a linear functional. Similarly, the integral

$$\bar{I}(x) = \int_a^b \overline{x(t)}\, dt,$$

or more generally

$$\bar{I}(x) = \int_a^b \overline{x(t)}\varphi(t)\, dt,$$

is a conjugate-linear functional on $C_{[a,b]}$.

Example 3. Another kind of linear functional on the space $C_{[a,b]}$ is the functional

$$\delta_{t_0}(x) = x(t_0),$$

which assigns to each function $x(t) \in C_{[a,b]}$ its value at some fixed point $t_0 \in [a, b]$. In mathematical physics, particularly in quantum mechanics, this functional is often written in the form

$$\delta_{t_0}(x) = \int_a^b x(t)\delta(t - t_0)\, dt,$$

where $\delta(t - t_0)$ is a "fictitious" or "generalized" function, called the (*Dirac*) *delta function*, which equals zero everywhere except at $t = 0$ and has an integral equal to 1.[6] As we will see in Sec. 20.3, the delta function can be

---

[6] Clearly, no "true" function can have these properties!

represented as the limit, in a suitable sense, of a sequence of "true" functions φ_n, each vanishing outside of some ε_n-neighborhood of the point $t = 0$ and satisfying the condition

$$\int_a^b \varphi_n(t) \, dt = 1$$

$(\varepsilon_n \to 0$ as $n \to \infty)$.

Example 4. Let n be a fixed positive integer, and let

$$x = (x_1, x_2, \ldots, x_k, \ldots)$$

be an arbitrary element of l_2. Then

$$f_n(x) = x_n$$

is obviously a linear functional on l_2. The same functional can be defined on other spaces whose elements are sequences, e.g., on the spaces c_0, c, m and R^∞ considered in Example 5, p. 120.

13.6. The null space of a functional. Hyperplanes. Let f be a linear functional defined on a linear space L. Then the set L_f of all elements $x \in L$ such that

$$f(x) = 0$$

is called the *null space* of f. It will be assumed that f is *nontrivial*, i.e., that $f(x) \neq 0$ for at least one (and hence infinitely many) $x \in L$, so that the set $L - L_f$ is nonempty. Obviously L_f is a subspace of L, since $x, y \in L_f$ implies

$$f(\alpha x + \beta y) = \alpha f(x) + \beta f(y) = 0.$$

THEOREM 3. *Let x_0 be any fixed element of $L - L_f$. Then every element $x \in L$ has a unique representation of the form*

$$x = \alpha x_0 + y,$$

where $y \in L_f$.

Proof. Clearly $f(x_0) \neq 0$, and in particular $x_0 \neq 0$. There is no loss of generality in assuming that $f(x_0) = 1$, since otherwise we need only replace x_0 by $x_0/f(x_0)$, noting that

$$f\left(\frac{x_0}{f(x_0)}\right) = \frac{f(x_0)}{f(x_0)} = 1.$$

Given any $x \in L$, let

$$y = x - \alpha x_0,$$

where

$$\alpha = f(x).$$

Then $y \in L_f$, since

$$f(y) = f(x - \alpha x_0) = f(x) - \alpha f(x_0) = f(x) - \alpha = 0.$$

Thus

$$x = \alpha x_0 + y \qquad (y \in L_f). \tag{6}$$

Moreover, the representation (6) is unique. In fact, if there is another such representation

$$x = \alpha' x_0 + y' \qquad (y \in L_f), \tag{6'}$$

then, subtracting (6') from (6), we get

$$(\alpha - \alpha')x_0 = y' - y.$$

If $\alpha = \alpha'$, then obviously $y' = y$. On the other hand, if $\alpha \neq \alpha'$, then

$$x_0 = \frac{y' - y}{\alpha - \alpha'} \in L_f,$$

contrary to the choice of x_0. ∎

COROLLARY 1. *Two elements x_1 and x_2 belong to the same class generated by L_f if and only if $f(x_1) = f(x_2)$.*

Proof. It follows from

$$x_1 = f(x_1)x_0 + y_1,$$
$$x_2 = f(x_2)x_0 + y_2$$

that

$$x_1 - x_2 = (f(x_1) - f(x_2))x_0 + (y_1 - y_2).$$

Hence $x_1 - x_2 \in L_f$ if and only if the coefficient of x_0 vanishes. ∎

COROLLARY 2. *L_f has codimension 1.*

Proof. Given any class ξ generated by L_f, let x be any element of ξ and choose $f(x)x_0 = \alpha x_0$ as the "representative" of ξ. By Corollary 1, this representative is unique, and there is obviously a nonzero class since $x_0 \neq 0$ and $f(x) \neq 0$ for some $x \in L$. Moreover, given any two distinct classes ξ and η with representatives αx_0 and βx_0, respectively, we have

$$\beta(\alpha x_0) - \alpha(\beta x_0) = 0$$

and hence

$$\beta\xi - \alpha\eta = 0,$$

where at least one of the numbers α, β is nonzero (why?). Therefore any two distinct elements of L/L_f are linearly dependent. It follows that L/L_f is one-dimensional, i.e., L_f has codimension 1. ∎

COROLLARY 3. *Two nontrivial linear functionals f and g with the same null space are proportional.*

Proof. Again let x_0 be such that $f(x_0) = 1$. Then $g(x_0) \neq 0$. In fact,

$$x = f(x)x_0 + y \qquad (y \in L_f),$$

and hence

$$g(x) = f(x)g(x_0) + g(y) = f(x)g(x_0),$$

since $L_g = L_f$. But then $g(x_0) = 0$ would imply that g is trivial, contrary to hypothesis. It follows that

$$g(x) = g(x_0)f(x),$$

i.e., $g(x)$ is proportional to $f(x)$ with constant of proportionality $g(x_0)$. ∎

Given a linear space L, let $L' \subset L$ be any subspace of codimension 1. Then every class in L generated by L' is called a *hyperplane* "parallel to L" (in particular, L' itself is a hyperplane containing 0, i.e., "going through the origin"). In other words, a hyperplane M' parallel to a subspace L' is the set obtained by subjecting L' to the parallel displacement (or shift) determined by the vector $x_0 \in L$, so that[7]

$$M' = L' + x_0 = \{x : x = x_0 + y, y \in L'\}.$$

It is clear that $M' = L'$ if and only if $x_0 \in L'$. We can now give a simple geometric interpretation of linear functionals:

THEOREM 4. *Given a linear space L, let f be a nontrivial linear functional on L. Then the set $M_f = \{x : f(x) = 1\}$ is a hyperplane parallel to the null space L_f of the functional. Conversely, let $M' = L' + x_0$ $(x_0 \notin L')$ be any hyperplane parallel to a subspace $L' \subset L$ of codimension 1 and not passing through the origin. Then there exists a unique linear functional f on L such that $M' = \{x : f(x) = 1\}$.*

Proof. Given f, let x_0 be such that $f(x_0) = 1$ (such an x_0 can always be found). Then, by Theorem 3, every vector $x \in M_f$ can be represented in the form $x = x_0 + y$, where $y \in L_f$.

Conversely, given $M' = L' + x_0$ $(x_0 \notin L')$, it follows from Theorem 2 and its proof that every element $x \in L$ can be uniquely represented in the form $x = \alpha x_0 + y$, where $y \in L'$. Setting $f(x) = \alpha$, we get the desired linear functional. The uniqueness of f follows from the fact that if $g(x) \equiv 1$ for $x \in M'$, then $g(y) \equiv 0$ for $y \in L'$ (why?), so that

$$g(\alpha x_0 + y) = \alpha = f(\alpha x_0 + y).$$ ∎

Remark. Thus we have established a one-to-one correspondence between the set of all nontrivial linear functionals on L and the set of all hyperplanes in L which do not pass through the origin.

---

[7] The expression on the right is shorthand for the set of all x such that $x = x_0 + y$, $y \in L'$ (the colon is read "such that"). Similarly, $\{x : f(x) = 1\}$ is the set of all x such that $f(x) = 1$, and so on.

Problem 1. Prove that the set of all polynomials of degree $n - 1$ with real (or complex) coefficients is a linear space, isomorphic to the n-dimensional vector space R^n (or C^n).

Problem 2. Verify that R^n and C^n are n-dimensional, as anticipated by the terminology in Example 2, p. 119.

Problem 3. Verify that the spaces $C_{[a,b]}$, l_2, c, c_0, m and R^∞ are all infinite-dimensional.

Problem 4. Given a linear space L, a set $\{x_\alpha\}$ of linearly independent elements of L is said to be a *Hamel basis* (in L) if the linear subspace generated by $\{x_\alpha\}$ coincides with L. Prove that
   a) Every linear space has a Hamel basis;
   b) If $\{x_\alpha\}$ is a Hamel basis in L, then every vector $x \in L$ has a unique representation as a finite linear combination of vectors from the set $\{x_\alpha\}$;
   c) Any two Hamel bases in a linear space L have the same power (cardinal number), called the *algebraic dimension* of L;
   d) Two linear spaces are isomorphic if and only if they have the same algebraic dimension.

Problem 5. Let L' be a k-dimensional subspace of an n-dimensional linear space L. Prove that the factor space L/L' has dimension $n - k$.

Problem 6. Let $f, f_1, \ldots, f_n$ be linear functionals on a linear space L such that $f_1(x) = \cdots = f_n(x) = 0$ implies $f(x) = 0$. Prove that there exist constants $a_1, \ldots, a_n$ such that

$$f(x) = \sum_{k=1}^{n} a_k f_k(x)$$

for every $x \in L$.

14. Convex Sets and Functionals. The Hahn-Banach Theorem

14.1. Convex sets and bodies. Many important topics in the theory of linear spaces rely on the notion of *convexity*. This notion, stemming from intuitive geometric ideas, can be formulated purely analytically. Given a *real* linear space L, let x and y be any two points of L. Then by the (*closed*) *segment* in L joining x and y we mean the set of all points in L of the form $\alpha x + \beta y$ where α, $\beta \geqslant 0$ and $\alpha + \beta = 1$. Such a segment minus its end points x and y is called an *open segment*. By the *interior* of a set $M \subset L$, denoted by $I(M)$, we mean the set of all points $x \in M$ with the following property: Given any $y \in L$, there exists a number $\varepsilon = \varepsilon(y) > 0$ such that $x + ty \in M$ if $|t| < \varepsilon$.

DEFINITION 1. *A set $M \subset L$ is said to be **convex** if whenever it contains two points x and y, it also contains the segment joining x and y.*

DEFINITION 2. *A convex set is called a **convex body** if its interior is nonempty.*

Example 1. The cube, ball, tetrahedron and half-space are all convex bodies in three-dimensional Euclidean space R^3. On the other hand, the line segment, plane and triangle are convex sets in R^3, but not convex bodies.

Example 2. As usual, let $C_{[a,b]}$ be the space of all functions continuous on the interval $[a, b]$, and let M be the subset of $C_{[a,b]}$ consisting of all functions satisfying the extra condition

$$|f(t)| \leqslant 1.$$

Then M is convex, since

$$|f(t)| \leqslant 1, \qquad |g(t)| \leqslant 1$$

together with $\alpha, \beta \geqslant 0$, $\alpha + \beta = 1$ implies

$$|\alpha f(t) + \beta g(t)| \leqslant \alpha + \beta = 1.$$

Example 3. The closed unit sphere in l_2, i.e., the set of all points $x = (x_1, x_2, \ldots, x_n, \ldots)$ such that

$$\sum_{n=1}^{\infty} x_n^2 \leqslant 1,$$

is a convex body. Its interior consists of all points $x = (x_1, x_2, \ldots, x_n, \ldots)$ satisfying the condition

$$\sum_{n=1}^{\infty} x_n^2 < 1.$$

Example 4. The Hilbert cube Π (see Example 5, p. 98) is a convex set in l_2, but not a convex body. In fact,

$$|x_n| \leqslant \frac{1}{2^{n-1}} \qquad (n = 1, 2, \ldots)$$

if $x \in \Pi$. Let

$$y_0 = \left(1, \frac{1}{2}, \ldots, \frac{1}{n}, \ldots\right),$$

and suppose $x + ty_0 \in \Pi$, i.e.,

$$\left| x_0 + \frac{t}{n} \right| \leqslant \frac{1}{2^{n-1}}.$$

Then

$$\left| \frac{t}{n} \right| \leqslant \left| x_n + \frac{t}{n} \right| + |x_n| \leqslant \frac{1}{2^{n-1}} + \frac{1}{2^{n-1}} = \frac{1}{2^{n-2}}$$

for all $n = 1, 2, \ldots$, which implies $t = 0$. Therefore the interior of Π is empty.

THEOREM 1. *If M is a convex set, then so is its interior $I(M)$.*

Proof. Suppose x, $y \in I(M)$, and let $z = \alpha x + \beta y$, α, $\beta \geqslant 0$, $\alpha + \beta = 1$. Then, given any $a \in L$, there are numbers $\varepsilon_1 > 0$, $\varepsilon_2 > 0$ such that the points $x + t_1 a$, $y + t_2 a$ belong to M if $|t_1| < \varepsilon_1$, $|t_2| < \varepsilon_2$. Therefore

$$\alpha(x + ta) + \beta(y + ta) = z + ta$$

belongs to M if $|t| < \varepsilon = \min \{\varepsilon_1, \varepsilon_2\}$, i.e., $z \in I(M)$. ∎

THEOREM 2. *The intersection*

$$M = \bigcap_\alpha M_\alpha$$

of any number of convex sets M_α is itself a convex set.

Proof. Let x and y be any two points of M. Then x and y belong to every M_α, and hence so does the segment joining x and y. But then the segment joining x and y belongs to M. ∎

Given any subset A of a linear space L, there is a smallest convex set containing A, i.e., the intersection of all convex sets containing A (there is at least one convex set containing A, namely L itself). This minimal convex set containing A is called the *convex hull* of A. For example, the convex hull of three noncollinear points is the triangle with these points as vertices.

14.2. Convex functionals. Next we introduce the important concept of a *convex functional*:

DEFINITION 3. *A functional p defined on a real linear space L is said to be* **convex** *if*

1) $p(x) \geqslant 0$ *for all $x \in L$ (nonnegativity)*;
2) $p(\alpha x) = \alpha p(x)$ *for all $x \in L$ and all $\alpha \geqslant 0$*;
3) $p(x + y) \leqslant p(x) + p(y)$ *for all $x, y \in L$.*

Remark. Here, unlike the case of linear functionals, we do not assume that $p(x)$ is finite for all $x \in L$, i.e., we allow the case where $p(x) = +\infty$ for some $x \in L$.

Example 1. The length of a vector in Euclidean n-space R^n is a convex functional. The first and second conditions are immediate consequences of the definition of length in R^n (length is inherently nonnegative), while the third condition means that the length of the sum of two vectors does not exceed the sum of their lengths (the triangle inequality).

Example 2. Let M be the space of bounded functions of x defined on some set S, and let s_0 be a fixed point of S. Then

$$p_{s_0}(x) = |x(s_0)|$$

is a convex functional.

Example 3. Let m be the space of bounded numerical sequences $x = (x_1, x_2, \ldots, x_k, \ldots)$. Then the functional

$$p(x) = \sup_k |x_k|$$

is convex.

14.3. The Minkowski functional. Next we consider the connection between convex functionals and convex sets:

THEOREM 3. *If p is a convex functional on a linear space L and k is any positive number, then the set*

$$E = \{x : p(x) \leqslant k\}$$

is convex. If p is finite, then E is a convex body with interior

$$I(E) = \{x : p(x) < k\}$$

(so that in particular $0 \in I(E)$).

Proof. If $x, y \in E$, $\alpha, \beta \geqslant 0$, $\alpha + \beta = 1$, then

$$p(\alpha x + \beta y) \leqslant \alpha p(x) + \beta p(y) \leqslant k,$$

i.e., E is a convex set. Now suppose p is finite, and let $p(x) < k, t > 0$, $y \in L$. Then

$$p(x \pm ty) \leqslant p(x) + tp(\pm y).$$

If $p(-y) = p(y) = 0$, then $x \pm ty \in E$ for all t. On the other hand, if at least one of the numbers $p(y), p(-y)$ is nonzero, then $x \pm ty \in E$ if

$$t < \frac{k - p(x)}{\max \{p(y), p(-y)\}}. \quad \blacksquare$$

Suppose we choose a definite value of k, say $k = 1$. Then every finite convex functional p uniquely determines a convex body E in L, such that $0 \in I(E)$. Conversely, suppose E is a convex body whose interior contains the point 0, and consider the functional

$$p_E(x) = \inf \left\{ r : \frac{x}{r} \in E, r > 0 \right\}, \tag{1}$$

called the *Minkowski functional* of the convex body E. Then we have

THEOREM 4. *The Minkowski functional* (1) *is finite and convex.*

Proof. Given any $x \in L$, the element x/r belongs to E if r is sufficiently large (why?), and hence $p_E(x)$ is nonnegative and finite. Clearly $p_E(0) = 0$. If $\alpha > 0$, then

$$p_E(\alpha x) = \inf \left\{ r > 0 : \frac{\alpha x}{r} \in E \right\} = \inf \left\{ \alpha r' > 0 : \frac{x}{r'} \in E \right\}$$

$$= \alpha \inf \left\{ r' > 0 : \frac{x}{r'} \in E \right\} = \alpha p_E(x). \tag{2}$$

Next, given any $\varepsilon > 0$ and any $x_1, x_2 \in L$, choose numbers r_i ($i = 1, 2$) such that

$$p_E(x_i) < r_i < p_E(x_i) + \varepsilon.$$

Then $x_i/r_i \in E$. If $r = r_1 + r_2$, then

$$\frac{x_1 + x_2}{r} = \frac{r_1 x_1}{r r_1} + \frac{r_2 x_2}{r r_2}$$

belongs to the segment with end points x_1/r_1 and x_2/r_2. Since E is convex, this segment and hence the point $(x_1 + x_2)/r$ belongs to E. It follows that

$$p_E(x_1 + x_2) \leqslant r = r_1 + r_2 < p_E(x_1) + p_E(x_2) + 2\varepsilon$$

or

$$p_E(x_1 + x_2) \leqslant p_E(x_1) + p_E(x_2), \tag{3}$$

since ε is arbitrary. Together (2) and (3) imply that $p_E(x)$ is convex. ∎

13.4. The Hahn-Banach theorem. Given a real linear space L and any subspace $L_0 \subset L$, let f_0 be a linear functional defined on L_0. Then a linear functional f defined on the whole space L is said to be an *extension* of the functional f_0 if

$$f(x) = f_0(x) \quad \text{for all} \quad x \in L_0.$$

A problem frequently encountered in analysis is that of extending an arbitrary linear functional, originally defined on some subspace, onto a larger space. A central role in problems of this kind is played by

THEOREM 5 (*Hahn-Banach*). *Let p be a finite convex functional defined on a real linear space L, and let L_0 be a subspace of L. Suppose f_0 is a linear functional on L_0 satisfying the condition*

$$f_0(x) \leqslant p(x) \tag{4}$$

on L_0. Then f_0 can be extended to a linear functional on L satisfying (4) on the whole space L. More exactly, there is a linear functional f defined on L and equal to f_0 at every point of L_0, such that $f(x) \leqslant p(x)$ on L.

Proof. Suppose $L_0 \neq L$, since otherwise the theorem is trivial. We begin by showing that f_0 can be extended onto a larger space L without violating the condition (4). Let z be any element of $L - L_0$, and let $\tilde{L}$ be the subspace generated by L_0 and the element z, i.e., the set of all linear combinations

$$x + tz \qquad (x \in L_0).$$

If $\tilde{f}$ is to be an extension of f_0 onto $\tilde{L}$, we must have

$$\tilde{f}(x + tz) = f_0(x) + t\tilde{f}(z)$$

or

$$\tilde{f}(x + tz) = f_0(x) + tc \tag{5}$$

after setting $\tilde{f}(z) = c$. We now choose c such that the "majorization" condition $\tilde{f}(x + tz) \leqslant p(x + tz)$ is satisfied, i.e., such that

$$f_0(x) + tc \leqslant p(x + tz).$$

We can write this condition as

$$f_0\left(\frac{x}{t}\right) + c \leqslant p\left(\frac{x}{t} + c\right)$$

or

$$c \leqslant p\left(\frac{x}{t} + c\right) - f_0\left(\frac{x}{t}\right) \tag{6}$$

if $t > 0$, and as

$$f_0\left(\frac{x}{t}\right) + c \geqslant -p\left(-\frac{x}{t} - z\right)$$

or

$$c \geqslant -p\left(-\frac{x}{t} - z\right) - f_0\left(\frac{x}{t}\right) \tag{7}$$

if $t < 0$. Hence we want to show that there is always a value of c satisfying (6) and (7). Let y' and y'' be arbitrary elements of L_0. Then it follows from the inequality

$$f_0(y'') - f_0(y') \leqslant p(y'' - y') = p((y'' + z) - (y' + z))$$
$$\leqslant p(y'' + z) + p(-y' - z)$$

that

$$-f_0(y'') + p(y'' + z) \geqslant -f_0(y') - p(-y' - z). \tag{8}$$

Let

$$c' = \sup_{y'} [-f_0(y') - p(-y' - z)],$$

$$c'' = \inf_{y''} [-f_0(y'') + p(y'' + z)].$$

Then

$$c'' \geqslant c',$$

by (8) and the fact that y' and y'' are arbitrary. Hence, choosing c such that

$$c'' \geqslant c \geqslant c',$$

we find that the functional $\tilde{f}$ defined on $\tilde{L}$ by the formula (5) satisfies the condition $\tilde{f}(x) \leqslant p(x)$. Thus we have succeeded in showing that if f_0 is defined on a subspace $L_0 \subset L$ and satisfies (4) on L_0, then f_0 can be extended onto a larger subspace $\tilde{L}$ with the condition (4) being preserved.

To complete the proof, suppose first that L is generated by a countable set of elements $x_1, x_2, \ldots, x_n, \ldots$ in L. Then we construct a functional on L by induction, i.e., by constructing a sequence of subspaces

$$L^{(1)} = \{L, x_1\}, \qquad L^{(2)} = \{L^{(1)}, x_2\}, \ldots,$$

each contained in the next. Here $\{L^{(k)}, x_{k+1}\}$ denotes the minimal linear subspace of L containing $L^{(k)}$ and x_{k+1}. This process extends the functional onto the whole space L, since every element $x \in L$ belongs to some subspace $L^{(k)}$.

More generally, i.e., in the case where there is no countable set generating L, the theorem is proved by applying Zorn's lemma (see p. 28). The set $\mathscr{F}$ of all possible extensions of the functional f_0 satisfying the majorization condition (5) is partially ordered, and each *linearly* ordered subset $\mathscr{F}_0 \subset \mathscr{F}$ has an upper bound. This upper bound is the functional which is defined on the union of the domains of all functionals $\tilde{f} \in \mathscr{F}_0$ and coincides with every such functional $\tilde{f}$ on the domain of $\tilde{f}$. Hence, by Zorn's lemma, $\mathscr{F}$ has a maximal element f. Clearly f must be the desired functional extending f_0 onto L and satisfying the condition $p(x) \leqslant f(x)$, since otherwise we could extend f in turn, by the method described above, from the proper subspace on which it is defined onto a large subspace, thereby contradicting the maximality of f. ∎

Next we turn to the case of *complex* linear spaces:

DEFINITION 3′. *A functional p defined on a complex linear space L is said to be **convex** if*

1) $p(x) \geqslant 0$ *for all $x \in L$ (nonnegativity)*;
2) $p(\alpha x) = |\alpha|\, p(x)$ *for all $x \in L$ and all complex α*;
3) $p(x + y) \leqslant p(x) + p(y)$ *for all $x, y \in L$.*

The corresponding complex version of the Hahn-Banach theorem is given by

THEOREM 5′. *Let p be a finite convex functional, defined on a complex linear space L, and let L_0 be a subspace of L. Suppose f_0 is a linear functional on L_0 satisfying the condition*

$$|f_0(x)| \leqslant p(x) \tag{4′}$$

on L_0. *Then f_0 can be extended to a linear functional on L satisfying* (4′) *on the whole space L.*

Proof. Let L_R and L_{0R} denote the spaces L and L_0, regarded as *real* linear spaces. Clearly p is a finite convex functional on L_R, while

$$f_{0R}(x) = \operatorname{Re} f_0(x)$$

is a real linear functional on L_{0R} satisfying the condition

$$|f_{0R}(x)| \leqslant p(x)$$

and hence (a fortiori) the condition

$$f_{0R}(x) \leqslant p(x).$$

By Theorem 5, there exists a real linear functional f_R defined on all of L_R, satisfying the conditions

$$f_R(x) \leqslant p(x) \qquad \text{if} \quad x \in L_R \ (= L),$$
$$f_R(x) = f_{0R}(x) \qquad \text{if} \quad x \in L_{0R} \ (= L_0).$$

Clearly

$$-f_R(x) = f_R(-x) \leqslant p(-x) = p(x),$$

and hence

$$|f_R(x)| \leqslant p(x) \qquad \text{if} \quad x \in L_R \ (= L). \tag{9}$$

We now define the functional

$$f(x) = f_R(x) - if_R(ix)$$

on L, using the fact that L is a *complex* linear space in which multiplication by complex numbers is defined. It is easily verified that f is a complex linear functional on L such that

$$f(x) = f_0(x) \qquad \text{if} \quad x \in L_0,$$
$$\operatorname{Re} f(x) = f_R(x) \qquad \text{if} \quad x \in L.$$

Finally, to show that $|f(x)| \leqslant p(x)$ for all $x \in L$, suppose to the contrary that $|f(x_0)| > p(x_0)$ for some $x_0 \in L$. Writing $f(x_0) = \rho e^{i\varphi}$ where $\rho > 0$, we set $y_0 = e^{-i\varphi} x_0$. Then

$$f_R(y_0) = \operatorname{Re} f(y_0) = \operatorname{Re} \left[e^{-i\varphi} f(x_0) \right] = \rho > p(x_0) = p(y_0)$$

which contradicts (9). ∎

14.5. Separation of convex sets in a linear space. Given a real linear space L, let M and N be two subsets of L. Then a linear functional f defined on

L is said to *separate* M and N if there exists a number C such that

$$f(x) \geqslant C \quad \text{if} \quad x \in M,$$
$$f(x) \leqslant C \quad \text{if} \quad x \in N.$$

It follows at once from this definition that

1) A linear functional f separates two sets M and N if and only if it separates $M - N = \{z : z = x - y, x \in M, y \in N\}$ and $\{0\}$, i.e., the set consisting of all differences $x - y$ where $x \in M$, $y \in N$ and the set whose only element is 0 (note that the minus sign in $M - N$ does not have the usual meaning of a set difference);

2) A linear functional f separates two sets M and N if and only if it separates the sets $M - x_0 = \{z : z = x - x_0, x \in M\}$ and $N - x_0 = \{z : z = y - x, y \in N\}$ for every $x_0 \in L$.

The following theorem on the separation of convex sets in a linear space has numerous applications and is an easy consequence of the Hahn-Banach theorem:

THEOREM 6. *Let M and N be two disjoint convex sets in a real linear space L, where at least one of the sets, say M, has a nonempty interior (i.e., is a convex body). Then there exists a nontrivial linear functional f on L separating M and L.*

Proof. There is no loss of generality in assuming that the point 0 belongs to the interior of M, since otherwise we need only consider the sets $M - x_0 = \{z : z = x - x_0, x \in M\}$ and $N - x_0 = \{z : z = y - x_0, y \in N\}$, where x_0 is some point of the interior of M. Let y_0 be a point of N. Then the point $-y_0$ belongs to the interior of the set $M - N = \{z : z = x - y, x \in M, y \in N\}$, and 0 belongs to the interior of the set $M - N + y_0 = \{z : z = x - y + y_0, x \in M, y \in N\}$. Since M and N are disjoint, we have $0 \notin M - N$, $y_0 \notin M - N + y_0$. Let p be the Minkowski functional for the set $M - N + y_0$. Then $p(y_0) \geqslant 1$ since $y_0 \notin M - N + y_0$. Consider the linear functional

$$f_0(\alpha y_0) = \alpha p(y_0)$$

defined on the one-dimensional subspace of L consisting of all elements of the form αy_0. Clearly f_0 satisfies the condition

$$f_0(\alpha y_0) \leqslant p(\alpha y_0),$$

since

$$p(\alpha y_0) = \alpha p(y_0) \quad \text{if} \quad \alpha \geqslant 0,$$

while

$$f_0(\alpha y_0) = \alpha f_0(y_0) < 0 \leqslant p(\alpha y_0) \quad \text{if} \quad \alpha < 0.$$

Hence, by the Hahn-Banach theorem, the functional f_0 can be extended to a linear functional f defined on the whole space L and satisfying the condition $f(y) \leqslant p(y)$ on L. It follows that $f(y) \leqslant 1$ if $y \in M - N + y_0$, while at the same time $f(y_0) \geqslant 1$, i.e., f separates the sets $M - N + y_0$ and $\{y_0\}$. Therefore f separates the sets $M - N$ and $\{0\}$. But then f separates the sets M and N. ∎

Problem 1. Let M be the set of all points $x = (x_1, x_2, \ldots, x_n, \ldots)$ in l_2 satisfying the condition

$$\sum_{n=1}^{\infty} n^2 x_n^2 \leqslant 1.$$

Prove that M is a convex set, but not a convex body.

Problem 2. Give an example of two convex bodies whose intersection is not a convex body.

Problem 3. We say that $n + 1$ points $x_1, x_2, \ldots, x_{n+1}$ in a linear space L are "in general position" if they do not belong to any $(n - 1)$-dimensional subspace of L. The convex hull of a set of $n + 1$ points $x_1, x_2, \ldots, x_{n+1}$ in general position is called an *n-dimensional simplex*, and the points $x_1, x_2, \ldots, x_{n+1}$ themselves are called the *vertices* of the simplex. Describe the zero-dimensional, one-dimensional, two-dimensional and three-dimensional simplexes in Euclidean three-space R^3. Prove that the simplex with vertices $x_1, x_2, \ldots, x_{n+1}$ is the set of all points in L which can be represented in the form

$$x = \sum_{k=1}^{n+1} \alpha_k x_k,$$

where

$$\alpha_k \geqslant 0, \qquad \sum_{k=1}^{n+1} \alpha_k = 1.$$

Problem 4. Show that if the points $x_1, x_2, \ldots, x_{n+1}$ are in general position, then so are any $k + 1$ $(k < n)$ of them.

Comment. Hence the $k + 1$ points generate a k-dimensional simplex, called a *k-dimensional face* of the *n*-dimensional simplex with vertices $x_1, x_2, \ldots, x_{n+1}$.

Problem 5. Describe all zero-dimensional, one-dimensional and two-dimensional faces of the tetrahedron in R^3 with vertices e_1, e_2, e_3, e_4.

Problem 6. Show that in the Hahn-Banach theorem we can drop the condition that the functional p be finite.

15. Normed Linear Spaces

15.1. Definitions and examples. Chapters 2 and 3 deal with topological (in particular, metric) spaces, i.e., spaces equipped with the notion of closeness of elements, while Secs. 14 and 15 deal with linear spaces, i.e., spaces equipped with the operations of addition of elements and multiplication of elements by numbers. We now combine these two ideas, arriving at the notion of a *topological linear space*, equipped with a topology as well as with the algebraic operations characterizing a linear space. In this section and the next, we will study topological linear spaces of a particularly important type, namely *normed linear spaces* and *Euclidean spaces*. Topological linear spaces in general will be considered in Sec. 17.

DEFINITION 1. *A functional p defined on a linear space L is said to be a **norm** (in L) if it has the following properties*:
   a) *p is finite and convex*;
   b) $p(x) = 0$ *only if* $x = 0$;
   c) $p(\alpha x) = |\alpha|\, p(x)$ *for all* $x \in L$ *and all* α.

Recalling the definition of a convex functional, we see that a *norm* in L is a finite functional on L such that

1) $p(x) \geqslant 0$ for all $x \in L$, where $p(x) = 0$ if and only if $x = 0$;
2) $p(\alpha x) = |\alpha|\, p(x)$ for all $x \in L$ and all α;
3) $p(x + y) \leqslant p(x) + p(y)$ for all $x, y \in L$.

DEFINITION 2. *A linear space L, equipped with a norm $p(x) = \|x\|$, is called a **normed linear space**.*

The notation $\|x\|$ will henceforth be preferred for the norm of the element $x \in L$. In terms of this notation, properties 1)—3) take the form:

1') $\|x\| \geqslant 0$ for all $x \in L$, where $\|x\| = 0$ if and only if $x = 0$;
2') $\|\alpha x\| = |\alpha|\, \|x\|$ for all $x \in L$ and all α;
3') *Triangle inequality*: $\|x + y\| \leqslant \|x\| + \|y\|$ for all $x, y \in L$.

Every normed linear space L becomes a metric space if we set

$$\rho(x, y) = \|x - y\| \tag{1}$$

for arbitrary $x, y \in L$. The fact that (1) is a metric follows at once from properties 1')–3'). Thus everything said about metric spaces in Chap. 2 carries over to the case of normed linear spaces.

Many of the spaces considered in Chap. 2 as examples of metric spaces (or in Sec. 13 as examples of linear spaces) can be made into normed linear spaces in a natural way, as shown by the following examples (in each case, verify that the norm has all the required properties):

---

[8] One of the pioneer workers in this field was Stefan Banach (1892–1945), author of the classic *Théorie des Opérations Linéaires*, reprinted by Chelsea Publishing Co., New York (1955).

Example 1. The real line R^1 becomes a normed linear space if we set $\|x\| = |x|$ for every number $x \in R^1$.

Example 2. To make real n-space R^n into a normed linear space, we set

$$\|x\| = \sqrt{\sum_{k=1}^{n} x_k^2}$$

for every element $x = (x_1, x_2, \ldots, x_n)$ in R^n. The formula

$$\rho(x, y) = \|x - y\| = \sqrt{\sum_{k=1}^{n} (x_k - y_k)^2}$$

then defines the same metric in R^n as already considered in Example 3, p. 38.

Example 3. We can also equip real n-space with the norm

$$\|x\|_1 = \sum_{k=1}^{n} |x_k| \tag{2}$$

or the norm

$$\|x\|_0 = \max_{1 \leqslant k \leqslant n} |x_k|. \tag{3}$$

The corresponding metrics lead to the spaces R_1^n and R_0^n considered in Examples 4 and 5, p. 39.

Example 4. The formula

$$\|x\| = \sqrt{\sum_{k=1}^{n} |x_k|^2}$$

introduces a norm in complex n-space C^n. Other possible norms in C^n are given by (2) and (3).

Example 5. The space $C_{[a,b]}$ of all functions continuous on the interval $[a, b]$ can be equipped with the norm

$$\|f\| = \max_{a \leqslant t \leqslant b} |f(t)|.$$

The metric space corresponding to this norm has already been considered in Example 6, p. 39.

Example 6. Let m be the space of all bounded numerical sequences

$$x = (x_1, x_2, \ldots, x_k, \ldots),$$

and let

$$\|x\| = \sup_k |x_k|. \tag{4}$$

Then (4) obviously has all the properties of a norm. The metric "induced" by this norm is the same as that considered in Example 9, p. 41.

Example 7. A *complete* normed linear space, relative to the metric (1), is called a *Banach space.* It is easy to see that the spaces in Examples 1–6 are all Banach spaces (the details are left as an exercise).

15.2. Subspaces of a normed linear space. In Sec. 13.3 we defined a subspace of a linear space L (unequipped with any topology) as a nonempty set L_0 with the property that if $x, y \in L_0$, then $\alpha x + \beta y \in L_0$ for arbitrary α and β. The subspaces of greatest interest in a normed linear space are the *closed* subspaces, i.e., those containing all their limit points. In the case of an infinite-dimensional space, it is easy to give examples of subspaces that are not closed:[9]

Example 1. In the space of all bounded sequences, the sequences with only finitely many nonzero terms form a subspace, but not a closed subspace, since, for example, the closure of the subspace contains the sequence

$$\left(1, \frac{1}{2}, \ldots, \frac{1}{n}, \ldots\right).$$

Example 2. The set $P_{[a,b]}$ of all polynomials defined on the interval $[a, b]$ is a subspace of $C_{[a,b]}$, but obviously not a closed subspace. On the other hand, the closure of $P_{[a,b]}$ coincides with $C_{[a,b]}$, since *every* function continuous on $[a, b]$ is the limit of a uniformly convergent sequence of polynomials, by *Weierstrass' approximation theorem.*[10]

In what follows, we will be concerned as a rule with closed subspaces. Hence it is natural to modify somewhat the terminology adopted in Sec. 13.3, i.e., by a subspace of a normed normed linear space we will always mean a *closed* subspace. In particular, by the subspace generated by a set of elements $\{x_\alpha\}$ we will always mean the smallest *closed* subspace containing $\{x_\alpha\}$. This subspace will also be called the *linear closure* of $\{x_\alpha\}$. The term *linear manifold* will be reserved for a set of elements L_0 (not necessarily closed) such that $x, y \in L_0$ implies $\alpha x + \beta y \in L_0$ for arbitrary numbers α and β. A set of elements $\{x_\alpha\}$ in a normed linear space L is said to be *complete* (in L) if the linear closure of $\{x_\alpha\}$ coincides with L.

Remark. This is another meaning of the word "closed," not to be confused with its meaning in Sec. 6.4. The context will always make it clear which meaning is intended.

Example 3. By Weierstrass' approximation theorem, the set of functions $1, t, t^2, \ldots, t^n, \ldots$ is complete in $C_{[a,b]}$.

---

[9] This contingency cannot arise in a finite-dimensional subspace (see Problem 5a).

[10] See e.g., G. P. Tolstov, *Fourier Series* (translated by R. A. Silverman), Prentice-Hall, Inc., Englewood Cliffs, N.J. (1962), p. 120.

Problem 1. A subset M of a normed linear space R is said to be *bounded* if there is a constant C such that $\|x\| \leqslant C$ for all $x \in M$. Reconcile this with Problem 5, p. 65.

Problem 2. Given a Banach space R, let $\{B_n\}$ be a nested sequence of closed spheres in R (so that $B_1 \supset B_2 \supset \cdots \supset B_n \supset \cdots$). Prove that $\bigcap\limits_{n} B_n$ is nonempty (it is not assumed that the radius of B_n approaches 0 as $n \to \infty$). Give an example of a nested sequence $\{E_n\}$ of nonempty closed bounded convex sets in a Banach space R such that $\bigcap\limits_{n} E_n$ is empty (cf. Problem 6, p. 66).

Problem 3. Prove that the algebraic dimension (defined in Problem 4c, p. 128) of an infinite-dimensional Banach space is uncountable.

Problem 4. Let R be a Banach space, and let M be a closed subspace of R. Define a norm in the factor space $P = R/M$ by setting

$$\|\xi\| = \inf_{x \in \xi} \|x\|$$

for every element (residue class) $\xi \in P$. Prove that
a) $\|\xi\|$ is actually a norm in P;
b) The space P, equipped with this norm, is a Banach space.

Problem 5. Let R be a normed linear space. Prove that
a) Every finite-dimensional linear subspace of R is closed;
b) If M is a closed subspace of R and N a finite-dimensional subspace of R, then the set

$$M + N = \{z : z = x + y, \, x \in M, \, y \in N\} \tag{5}$$

is a closed subspace of R;
c) If Q is an open convex set in R and $x_0 \notin Q$, then there exists a closed hyperplane which passes through the point x_0 and does not intersect Q.

Problem 6. Let $x = (x_1, x_2, \ldots, x_k, \ldots)$ be an arbitrary element of l_2. Prove that l_2 is a normed linear space when equipped with the norm

$$\|x\| = \sqrt{\sum_{k=1}^{\infty} x_k^2}.$$

Give an example of two closed linear subspaces M and N of l_2 whose "linear sum" $M + N$ is not closed.

Problem 7. Two norms $\|\cdot\|_1$, $\|\cdot\|_2$ in a linear space R are said to be *equivalent* if there exist constants $a, b > 0$ such that

$$a \|x\|_1 \leqslant \|x\|_2 \leqslant b \|x\|_1$$

for all $x \in R$. Prove that if R is finite-dimensional, then any two norms in R are equivalent.

16. Euclidean Spaces

16.1. Scalar products. Orthogonality and bases. We begin with two key definitions:

DEFINITION 1. *By a **scalar product** in a real linear space R is meant a real function defined for every pair of elements $x, y \in R$ and denoted by (x, y), with the following properties:*
  1) $(x, x) \geqslant 0$ *where* $(x, x) = 0$ *if and only if* $x = 0$;
  2) $(x, y) = (y, x)$;
  3) $(\lambda x, y) = \lambda(x, y)$;
  4) $(x, y + z) = (x, y) + (x, z)$
(valid for all $x, y, z \in R$ and all real λ).

DEFINITION 2. *A linear space R equipped with a scalar product is called a **Euclidean space**.*

LEMMA. *Any two elements x, y of a Euclidean space R satisfy the **Schwarz inequality***
$$|(x, y)| \leqslant \|x\| \, \|y\|, \tag{1}$$
where
$$\|x\| = \sqrt{(x, x)}, \qquad \|y\| = \sqrt{(y, y)}.$$

Proof. The quadratic polynomial
$$\varphi(\lambda) = (\lambda x + y, \lambda x + y) = \lambda^2(x, x) + 2\lambda(x, y) + (y, y)$$
$$= \|x\|^2 \lambda^2 + 2(x, y)\lambda + \|y\|^2$$
is obviously nonnegative. Therefore
$$(x, y)^2 - \|x\|^2 \, \|y\|^2 \leqslant 0, \tag{2}$$
since otherwise $\varphi(\lambda)$ would become negative for some λ (why?). But (2) is equivalent to (1). ∎

We now use the scalar product in a Euclidean space R to introduce a norm in R:

THEOREM 1. *A Euclidean space R becomes a normed linear space when equipped with the norm*
$$\|x\| = \sqrt{(x, x)} \qquad (x \in R).$$

Proof. Properties 1') and 2') on p. 138 are immediate consequences of the definition of a scalar product. To prove property 3'), i.e., the triangle inequality, we note that
$$\|x + y\|^2 = (x + y, x + y) = (x, x) + 2(x, y) + (y, y)$$
$$\leqslant (x, x) + 2\,|(x, y)| + (y, y)$$
$$\leqslant \|x\|^2 + 2\,\|x\| \, \|y\| + \|y\|^2 = (\|x\| + \|y\|)^2,$$

because of the Schwarz inequality (1), and hence

$$\|x + y\| \leqslant \|x\| + \|y\|. \quad \blacksquare$$

The scalar product in R can be used to define the *angle* between two vectors as well as the length (i.e., norm) of a vector:

DEFINITION 3. *Given any two vectors x and y in a Euclidean space R, the quantity θ defined by the formula*

$$\cos \theta = \frac{(x, y)}{\|x\| \|y\|} \qquad (0 \leqslant \theta \leqslant \pi) \qquad (3)$$

*is called the **angle** between x and y.*

Remark. It follows from Schwarz's inequality (1) that the right-hand side of (3) cannot exceed 1. Therefore, given any x and y, (3) actually determines a unique angle in the interval $[0, \pi]$.

Suppose $(x, y) = 0$, so that (3) implies $\theta = \pi/2$. Then the vectors x and y are said to be *orthogonal*. A set of nonzero vectors $\{x_\alpha\}$ in R is said to be an *orthogonal system* if

$$(x_\alpha, x_\beta) = 0 \qquad \text{for} \quad \alpha \neq \beta$$

and an *orthonormal system* if

$$(x_\alpha, x_\beta) = \begin{cases} 0 & \text{for} \quad \alpha \neq \beta, \\ 1 & \text{for} \quad \alpha = \beta. \end{cases}$$

If $\{x_\alpha\}$ is an orthogonal system, then clearly

$$\left\{ \frac{x_\alpha}{\|x_\alpha\|} \right\}$$

is an orthonormal system.

THEOREM 2. *The vectors in an orthogonal system $\{x_\alpha\}$ are linearly independent.*

Proof. Suppose

$$c_1 x_{\alpha_1} + c_2 x_{\alpha_2} + \cdots + c_n x_{\alpha_n} = 0.$$

Then, taking the scalar product with x_{α_i}, we get

$$(x_{\alpha_i}, c_1 x_{\alpha_1} + c_2 x_{\alpha_2} + \cdots + c_n x_{\alpha_n}) = c_i(x_{\alpha_i}, x_{\alpha_i}) = 0,$$

by the orthogonality of $\{x_\alpha\}$. But $(x_{\alpha_i}, x_{\alpha_i}) \neq 0$, and hence

$$c_i = 0 \qquad (i = 1, 2, \ldots, n). \quad \blacksquare$$

An orthogonal system $\{x_\alpha\}$ is called an *orthogonal basis* if it is *complete*, i.e., if the smallest closed subspace containing $\{x_\alpha\}$ is the whole space R. Similarly, a complete orthonormal system is called an *orthonormal basis*.

16.2. Examples. We now give some examples of Euclidean spaces and orthogonal bases in them:

Example 1. Let R^n be real n-space, i.e., the set of all ordered n-tuples $x = (x_1, x_2, \ldots, x_n)$, $y = (y_1, y_2, \ldots, y_n)$, $\ldots$, equipped with the same algebraic operations as in Example 2, p. 119. Using the formula

$$(x, y) = \sum_{k=1}^{n} x_k y_k \tag{4}$$

to define a scalar product in R^n, we get *Euclidean n-space*.[11] The corresponding norm and distance in R^n are

$$\|x\| = \sqrt{\sum_{k=1}^{n} x_k^2}$$

and

$$\rho(x, y) = \|x - y\| = \sqrt{\sum_{k=1}^{n} (x_k - y_k)^2}. \tag{5}$$

The vectors

$$e_1 = (1, 0, 0, \ldots, 0),$$
$$e_2 = (0, 1, 0, \ldots, 0),$$
$$\cdots\cdots\cdots\cdots$$
$$e_n = (0, 0, 0, \ldots, 1)$$

form an orthonormal basis in R^n, one of infinitely many such bases.

Example 2. The space l_2 with elements $x = (x_1, x_2, \ldots, x_k, \ldots)$, $y = (y_1, y_2, \ldots, y_k, \ldots)$, $\ldots$, where

$$\sum_{k=1}^{\infty} x_k^2 < \infty, \qquad \sum_{k=1}^{\infty} y_k^2 < \infty, \ldots,$$

becomes an infinite-dimensional Euclidean space when equipped with the scalar product

$$(x, y) = \sum_{k=1}^{\infty} x_k y_k. \tag{6}$$

The convergence of the right-hand side of (6) follows from the elementary inequality

$$|x_k y_k| < (|x_k| + |y_k|)^2 < 2(x_k^2 + y_k^2),$$

and it is an easy matter to verify that (6) has all the properties of a scalar

---

[11] The term "Euclidean n-space" has already been used in Example 3, p. 38 to describe the metric space with distance (5). In so doing, we anticipated the eventual introduction of the scalar product (4).

product. The simplest orthonormal basis in l_2 consists of the vectors

$$
\begin{aligned}
e_1 &= (1, 0, 0, \ldots), \\
e_2 &= (0, 1, 0, \ldots), \\
e_3 &= (0, 0, 1, \ldots),
\end{aligned} \tag{7}
$$

$$\cdots\cdots\cdots\cdots$$

The orthonormality of the system (7) is obvious. As for the completeness of the system, given any vector $x = (x_1, x_2, \ldots, x_k, \ldots)$ in l_2, let

$$x^{(k)} = (x_1, x_2, \ldots, x_k, 0, 0, \ldots).$$

Then $x^{(k)}$ is a linear combination of the vectors $e_1, e_2, \ldots, e_k$ and $\|x^{(k)} - x\| \to 0$ as $k \to \infty$.

Example 3. The space $C^2_{[a,b]}$ consisting of all continuous functions on $[a, b]$ equipped with the scalar product

$$(f, g) = \int_a^b f(t)g(t)\, dt$$

is another example of a Euclidean space. Among the various orthogonal bases in $C^2_{[a, b]}$, one of the most important is the system of trigonometric functions

$$1, \quad \cos\frac{2\pi n t}{b - a}, \quad \sin\frac{2\pi n t}{b - a} \qquad (n = 1, 2, \ldots). \tag{8}$$

The orthogonality of this system can be verified by a simple calculation. Making the choice $a = -\pi$, $b = \pi$, we simplify (8) to

$$1, \quad \cos nt, \quad \sin nt \qquad (n = 1, 2, \ldots). \tag{8'}$$

Thus (8') is an orthogonal basis in the space $C^2_{[-\pi, \pi]}$. As for the completeness, we have

THEOREM 3. *The system* (8) *is complete in* $C^2_{[a,b]}$.

Proof. By another version of Weierstrass' approximation theorem,[12] every function φ continuous on the interval $[a, b]$ and such that $\varphi(a) = \varphi(b)$ is the limit of a uniformly convergent sequence of trigonometric polynomials, i.e., linear combinations of elements of the system (8). This sequence converges (a fortiori) to φ in the norm of the space $C^2_{[a,b]}$. But an arbitrary function $f \in C^2_{[a,b]}$ can be represented as the limit in the

---

[12] See e.g., G. P. Tolstov, *op. cit.*, Corollary 1, p. 117.

$C^2_{[a,b]}$ norm of a sequence of functions $\{\varphi_n\}$, where

$$\varphi_n(x) = \begin{cases} f(x) & \text{if } a \leqslant x \leqslant b - \dfrac{1}{n}, \\ \left[nf\left(b - \dfrac{1}{n}\right) - nf(a)\right](b-x) + f(a) & \text{if } b - \dfrac{1}{n} \leqslant x \leqslant b \end{cases}$$

FIGURE 16

coincides with f in the interval $[a, b - (1/n)]$, is linear on $[b - (1/n), b]$ and takes the same value at the point b as at the point a (see Figure 16). Hence every element of $C^2_{[a,b]}$ can be approximated arbitrarily closely (in the $C^2_{[a,b]}$ norm) by a linear combination of elements of the system (8). ∎

16.3. Existence of an orthogonal basis. Orthogonalization. From now on, we will be mainly concerned with the case of *separable* Euclidean spaces, i.e., Euclidean spaces containing a countable everywhere dense subset. For example, the spaces R^n, l_2 and $C^2_{[a,b]}$ are all separable, as shown in Sec. 6.3. An example of a nonseparable Euclidean space is given in Problem 2.

THEOREM 4. *Every orthogonal system $\{x_\alpha\}$ in a separable Euclidean space R has no more than countably many elements x_α.*

Proof. There is no loss of generality in assuming that the system $\{x_\alpha\}$ is orthonormal as well as orthogonal, since otherwise we need only replace $\{x_\alpha\}$ by

$$\left\{\frac{x_\alpha}{\|x_\alpha\|}\right\}.$$

We then have

$$\|x_\alpha - x_\beta\| = \sqrt{2} \quad \text{if } \alpha \neq \beta. \tag{9}$$

Consider the set of open spheres $S(x_\alpha, \tfrac{1}{2})$. These spheres are pairwise disjoint, because of (9). Moreover, each sphere contains at least one element from some countable subset $\{y_n\}$ everywhere dense in R. Consequently there are no more than countably many such spheres, and hence no more than countably many elements x_α. ∎

We have already exhibited an orthogonal basis in each of the spaces R^n, l_2 and $C^2_{[a,b]}$. The existence of an orthogonal basis in any separable Euclidean space is guaranteed by the following theorem and its corollary, analogous to the theorem on the existence of an orthogonal basis in any finite-dimensional Euclidean space:[13]

THEOREM 5 (*Orthogonalization theorem*). *Let*

$$f_1, f_2, \ldots, f_n, \ldots \qquad (10)$$

be any (countable) set of linearly independent elements of a Euclidean space R. Then R contains a set of elements

$$\varphi_1, \varphi_2, \ldots, \varphi_n, \ldots \qquad (11)$$

such that

1) *The system* (11) *is orthonormal*;
2) *Every element* φ_n *is a linear combination*

$$\varphi_n = a_{n1}f_1 + a_{n2}f_2 + \cdots + a_{nn}f_n \qquad (a_{nn} \neq 0)$$

of the elements $f_1, f_2, \ldots, f_n$;
3) *Every element* f_n *is a linear combination*

$$f_n = b_{n1}\varphi_1 + b_{n2}\varphi_2 + \cdots + b_{nn}\varphi_n \qquad (b_{nn} \neq 0)$$

of the elements $\varphi_1, \varphi_2, \ldots, \varphi_n$.

Moreover, every element of the system (10) *is uniquely determined by these conditions to within a factor of* ± 1.

Proof. First we construct φ_1. Setting

$$\varphi_1 = a_{11}f_1,$$

we determine a_{11} from the condition

$$(\varphi_1, \varphi_1) = a_{11}^2(f_1, f_1) = 1,$$

which implies

$$a_{11} = \frac{1}{b_{11}} = \frac{1}{\sqrt{(f_1, f_1)}}.$$

This obviously determines φ_1 uniquely (except for sign).

Next suppose elements $\varphi_1, \varphi_2, \ldots, \varphi_{n-1}$ satisfying the conditions of the theorem have already been constructed. Then f_n can be written in the form

$$f_n = b_{n1}\varphi_1 + \cdots + b_{n,n-1}\varphi_{n-1} + h_n, \qquad (12)$$

where

$$(h_n, \varphi_k) = 0 \qquad (k = 1, 2, \ldots, n - 1).$$

---

[13] See e.g., G. E. Shilov, *An Introduction to the Theory of Linear Spaces* (translated by R. A. Silverman), Prentice-Hall, Inc., Englewood Cliffs, N.J. (1961), Theorem 28, p. 142.

In fact, the coefficients b_{nk} and hence the element h_n are uniquely determined by the conditions

$$(h_n, \varphi_k) = (f_n - b_{n1}\varphi_1 - \cdots - b_{n,n-1}\varphi_{n-1}, \varphi_k)$$
$$= (f_n, \varphi_k) - b_{nk}(\varphi_k, \varphi_k) = 0,$$

i.e.,

$$b_{nk} = (f_n, \varphi_k) \qquad (k = 1, 2, \ldots, n-1).$$

Clearly $(h_n, h_n) > 0$, since $(h_n, h_n) = 0$ contradicts the assumed linear independence of the elements (10). Let

$$\varphi_n = \frac{h_n}{\sqrt{(h_n, h_n)}} \tag{13}$$

Using (12) and (13), we express h_n and hence φ_n in terms of the functions $f_1, f_2, \ldots, f_n$, i.e.,

$$\varphi_n = a_{n1}f_1 + a_{n2}f_2 + \cdots + a_{nn}f_n,$$

where

$$a_{nn} = \frac{1}{\sqrt{(h_n, h_n)}} \neq 0.$$

Moreover

$$(\varphi_n, \varphi_k) = 0 \qquad (k = 1, 2, \ldots, n-1),$$

$$(\varphi_n, \varphi_n) = 1$$

and

$$f_n = b_{n1}\varphi_1 + b_{n2}\varphi_2 + \cdots + b_{nn}\varphi_n,$$

where

$$b_{nn} = \sqrt{(h_n, h_n)} > 0.$$

Thus, starting from elements $\varphi_1, \varphi_2, \ldots, \varphi_{n-1}$ satisfying the conditions of the theorem, we have constructed elements $\varphi_1, \varphi_2, \ldots, \varphi_{n-1}, \varphi_n$ satisfying the same conditions. The proof now follows by mathematical induction. ∎

Remark. The process leading from the linearly independent elements (10) to the orthonormal system (11) is called *orthogonalization*. It is clear that the subspace generated by (10) coincides with that generated by (11). Hence the set (10) is complete if and only if the set (11) is complete.

COROLLARY. *Every separable Euclidean space R has a countable orthonormal basis.*

Proof. Let $\psi_1, \psi_2, \ldots, \psi_n, \ldots$ be a countable everywhere dense subset of R. Then a *complete* set of linearly independent elements $f_1, f_2, \ldots, f_n, \ldots$ can be selected from $\{\psi_n\}$. In fact, we need only eliminate from the sequence $\{\psi_n\}$ all elements ψ_k which can be written as linear combinations of elements ψ_i with smaller indices ($i < k$). Applying the orthogonalization process to $f_1, f_2, \ldots, f_n, \ldots$, we get an orthonormal basis. ∎

16.4. Bessel's inequality. Closed orthogonal systems. Let $e_1, e_2, \ldots, e_n$ be an orthonormal basis in R^n. Then every vector $x \in R^n$ can be written in the form

$$x = \sum_{k=1}^{n} c_k e_k,$$

where

$$c_k = (x, e_k).$$

We now show how this generalizes to the case of an infinite-dimensional Euclidean space R. Let $\varphi_1, \varphi_2, \ldots, \varphi_k, \ldots$ be an orthonormal system in R, and let f be an arbitrary element of R. Suppose that with f we associate

1) The sequence of numbers

$$c_k = (f, \varphi_k) \qquad (k = 1, 2, \ldots), \tag{14}$$

called the *components* or *Fourier coefficients* of f with respect to the system $\{\varphi_k\}$;

2) The series

$$\sum_{k=1}^{\infty} c_k \varphi_k \tag{15}$$

(for the time being, purely formal), called the *Fourier series* of f with respect to the system $\{\varphi_k\}$.

Then it is natural to ask whether the series (15) converges,[14] and if so, whether the sum of the series coincides with the original function f. To answer these questions, we first prove

THEOREM 6. *Given an orthonormal system*

$$\varphi_1, \varphi_2, \ldots, \varphi_k, \ldots \tag{16}$$

in a Euclidean space R, let f be an arbitrary element of R. Then the expression

$$\left\| f - \sum_{k=1}^{n} a_k \varphi_k \right\|$$

achieves its minimum for

$$a_k = c_k = (f, \varphi_k) \qquad (k = 1, 2, \ldots, n).$$

---

[14] More exactly, whether the sequence of partial sums corresponding to (15) converges in the metric of R.

This minimum equals

$$\|f\|^2 - \sum_{k=1}^{n} c_k^2.$$

Moreover

$$\sum_{k=1}^{\infty} c_k^2 \leqslant \|f\|^2, \tag{17}$$

*a result known as **Bessel's inequality**.*

Proof. Let

$$S_n = \sum_{k=1}^{n} a_k \varphi_k. \tag{18}$$

Then, by the orthonormality of (16),

$$\|f - S_n\|^2 = \left(f - \sum_{k=1}^{n} a_k \varphi_k, f - \sum_{k=1}^{n} a_k \varphi_k\right)$$

$$= (f,f) - 2\left(f, \sum_{k=1}^{n} a_k \varphi_k\right) + \left(\sum_{k=1}^{n} a_k \varphi_k \sum_{l=1}^{n} a_l \varphi_l\right)$$

$$= \|f\|^2 - 2\sum_{k=1}^{n} a_k c_k + \sum_{k=1}^{n} a_k^2$$

or

$$\|f - S_n\|^2 = \|f\|^2 - \sum_{k=1}^{n} c_k^2 + \sum_{k=1}^{n} (a_k - c_k)^2, \tag{19}$$

where

$$c_k = (f, \varphi_k) \qquad (k = 1, 2, \ldots, n).$$

The expression in the right-hand side of (19) obviously achieves its minimum when its last term vanishes, i.e., when

$$a_k = c_k \qquad (k = 1, 2, \ldots, n),$$

and this minimum is just

$$\|f - S_n\|^2 = \|f\|^2 - \sum_{k=1}^{n} c_k^2. \tag{20}$$

Moreover, since $\|f - S_n\|^2 \geqslant 0$, it follows from (20) that

$$\sum_{k=1}^{n} c_k^2 \leqslant \|f\|^2 \tag{21}$$

for every n. Hence the series

$$\sum_{k=1}^{\infty} c_k^2$$

is convergent. Taking the limit as $n \to \infty$ in (21), we get (17). ∎

Remark. Geometrically, Bessel's inequality (17) means that the sum of the squares of the projections of a vector f onto a set of mutually perpendicular directions cannot exceed the square of the length of the vector itself. For a geometric interpretation of the rest of Theorem 6, see Problems 5 and 6.

The case where Bessel's inequality becomes an *equality* is particularly important:

DEFINITION 4. *Suppose equality holds in* (17) *for every* $f \in R$, *i.e., suppose*

$$\sum_{k=1}^{\infty} c_k^2 = \|f\|^2 \tag{22}$$

for every $f \in R$. *Then the orthonormal system* $\varphi_1, \varphi_2, \ldots, \varphi_k, \ldots$ *is said to be* **closed**.

Remark. This is another meaning of the word "closed," not to be confused with its meaning in Sec. 6.4. The context will always make it clear which meaning is intended. Formula (22) is known as *Parseval's theorem.*

THEOREM 7. *An orthonormal system* $\varphi_1, \varphi_2, \ldots, \varphi_k, \ldots$ *in a Euclidean space R is closed if and only if every element* $f \in R$ *is the sum of its Fourier series.*

Proof. According to Definition 4, R is closed if and only if (22) holds for every $f \in R$. Taking the limit as $n \to \infty$ in (20) and using (18), we see that (22) holds for every $f \in R$ if and only if

$$\lim_{n \to \infty} \left\| f - \sum_{k=1}^{n} c_k \varphi_k \right\| = 0,$$

or equivalently

$$f = \sum_{k=1}^{\infty} c_k \varphi_k,$$

for every $f \in R$. ∎

The properties of being *complete* and being *closed* are intimately connected, as shown by

THEOREM 8. *An orthonormal system* $\varphi_1, \varphi_2, \ldots, \varphi_k, \ldots$ *in a Euclidean space R is complete if and only if it is closed.*

Proof. Suppose $\{\varphi_k\}$ is closed. Then, by Theorem 7, every element $f \in R$ is the limit of the partial sums of its Fourier series. In other words, linear combinations of elements of $\{\varphi_k\}$ are everywhere dense in R, i.e., $\{\varphi_k\}$ is complete.

Conversely, suppose $\{\varphi_k\}$ is complete. Then every element $f \in R$ can be approximated arbitrarily closely by a linear combination

$$\sum_{k=1}^{n} a_k \varphi_k$$

of elements of $\{\varphi_k\}$. But the partial sum

$$\sum_{k=1}^{n} c_k \varphi_k$$

of the Fourier series of f is at least as good an approximation. Hence f is the sum of its own Fourier series. It follows from Theorem 7 that $\{\varphi_k\}$ is closed. ∎

COROLLARY. *Every separable Euclidean space R contains a closed orthonormal system* $\varphi_1, \varphi_2, \ldots, \varphi_k, \ldots$

Proof. An immediate consequence of Theorem 8 and the corollary to Theorem 5. ∎

Example 1. The orthonormal system (7) is closed in l_2.

Remark. In introducing the concepts of Fourier coefficients and Fourier series, we assumed that the system $\{\varphi_k\}$ is *orthonormal*. More generally, suppose $\{\varphi_k\}$ is orthogonal but not orthonormal, and let

$$\psi_k = \frac{\varphi_k}{\|\varphi_k\|}.$$

Then the system $\{\psi_k\}$ is orthonormal. Given any $f \in R$, let

$$c_k = (f, \psi_k) = \frac{1}{\|\varphi_k\|}(f, \varphi_k),$$

and consider the series

$$\sum_{k=1}^{\infty} c_k \psi_k = \sum_{k=1}^{\infty} \frac{c_k}{\|\varphi_k\|} \varphi_k = \sum_{k=1}^{\infty} a_k \varphi_k,$$

where

$$a_k = \frac{c_k}{\|\varphi_k\|} = \frac{(f, \varphi_k)}{\|\varphi_k\|^2}. \tag{23}$$

Then the coefficients (23) are called the *Fourier coefficients* of the element $f \in R$ with respect to the orthogonal (but not orthonormal) system $\{\varphi_k\}$. Substituting $c_k = a_k \|\varphi_k\|$ into (17), we get the following version of Bessel's inequality for arbitrary orthogonal systems:

$$\sum_{k=1}^{\infty} a_k^2 \|\varphi_k\|^2 < \|f\|^2. \tag{17'}$$

If equality holds in (17') for every $f \in R$, the orthogonal system $\{\varphi_k\}$ is said to be *closed*, just as in Definition 4.

Example 2. The orthogonal system (8) is closed in $C^2_{[a,b]}$.

16.5. Complete Euclidean spaces. The Riesz-Fischer theorem. Given a Euclidean space R, let $\{\varphi_k\}$ be an orthonormal (but not necessarily complete) system in R. It follows from Bessel's inequality that a necessary condition for the numbers $c_1, c_2, \ldots, c_k, \ldots$ to be Fourier coefficients of an element $f \in R$ is that the series

$$\sum_{k=1}^{\infty} c_k^2$$

converge. It turns out that this condition is also sufficient if R is *complete*, as shown by

THEOREM 9 (*Riesz-Fischer*). *Given an orthonormal system $\{\varphi_k\}$ in a complete Euclidean space R, let the numbers $c_1, c_2, \ldots, c_k, \ldots$ be such that*

$$\sum_{k=1}^{\infty} c_k^2 \tag{24}$$

converges. Then there exists an element $f \in R$ with $c_1, c_2, \ldots, c_k, \ldots$ as its Fourier coefficients, i.e., such that

$$\sum_{k=1}^{\infty} c_k^2 = \|f\|^2$$

where

$$c_k = (f, \varphi_k) \qquad (k = 1, 2, \ldots).$$

Proof. Writing

$$f_n = \sum_{k=1}^{n} c_k \varphi_k,$$

we have

$$\|f_{n+p} - f_n\|^2 = \|c_{n+1}\varphi_{n+1} + \cdots + c_{n+p}\varphi_{n+p}\|^2 = \sum_{k=n+1}^{n+p} c_k^2.$$

Hence f converges to some element $f \in R$, by the convergence of (24) and the completeness of R. Moreover,

$$(f, \varphi_k) = (f_n, \varphi_k) + (f - f_n, \varphi_k), \tag{25}$$

where the first term on the right equals c_k if $n \geqslant k$ and the second term approaches zero as $n \to \infty$, since

$$|(f - f_n, \varphi_k)| \leqslant \|f - f_n\| \, \|\varphi_k\|.$$

Taking the limit as $n \to \infty$ in (25), we get

$$(f, \varphi_k) = c_k,$$

since the left-hand side is independent of n. Moreover,

$$\|f - f_n\| \to 0$$

as $n \to \infty$, and hence

$$\left(f - \sum_{k=1}^{n} c_k \varphi_k, f - \sum_{k=1}^{n} c_k \varphi_k \right) = (f, f) - \sum_{k=1}^{n} c_k^2 \to 0$$

as $n \to \infty$, i.e.,

$$\lim_{n \to \infty} \sum_{k=1}^{n} c_k^2 = \sum_{k=1}^{\infty} c_k^2 = \|f\|^2. \quad \blacksquare$$

THEOREM 10. *Let $\{\varphi_k\}$ be an orthonormal system in a complete Euclidean space. Then $\{\varphi_k\}$ is complete if and only if R contains no nonzero element orthogonal to all the elements of $\{\varphi_k\}$.*

Proof. Suppose $\{\varphi_k\}$ is complete and hence closed (by Theorem 8), and suppose f is orthogonal to all the elements of $\{\varphi_k\}$. Then all the Fourier coefficients of f vanish. Hence

$$\|f\|^2 = \sum_{k=1}^{\infty} c_k^2 = 0$$

by the Riesz-Fischer theorem, i.e., $f = 0$.

Conversely, suppose $\{\varphi_k\}$ is not complete. Then R contains an element $g \neq 0$ such that

$$\|g\|^2 > \sum_{k=1}^{\infty} c_k^2, \quad \text{where} \quad c_k = (g, \varphi_k)$$

(why?). By the Riesz-Fischer theorem, there exists an element $f \in R$ such that

$$(f, \varphi_k) = c_k, \qquad \|f\|^2 = \sum_{k=1}^{\infty} c_k^2.$$

But $f - g$ is orthogonal to all the φ_k, by construction. Moreover, it follows from

$$\|f\|^2 = \sum_{k=1}^{\infty} c_k^2 < \|g\|^2$$

that $f - g \neq 0$. $\quad \blacksquare$

16.6. Hilbert space. The isomorphism theorem. Continuing our study of complete Euclidean spaces, we concentrate our attention on infinite-dimensional spaces, since finite-dimensional spaces are considered in great detail in courses on linear algebra.

DEFINITION 5. *By a **Hilbert space**[15] is meant a Euclidean space which is complete, separable and infinite-dimensional.*

In other words, a Hilbert space is a set H of elements $f, g, \ldots$ of any kind such that

1) H is a Euclidean space, i.e., a real linear space[16] equipped with a scalar product;
2) H is complete with respect to the metric $\rho(f, g) = \|f - g\|$;
3) H is separable, i.e., H contains a countable everywhere dense subset;
4) H is infinite-dimensional, i.e., given any positive integer n, H contains n linearly independent elements.

Example. The real space l_2 is a Hilbert space (check all the properties).

DEFINITION 6. *Two Euclidean spaces R and R^* are said to be **isomorphic** (to each other) if there is a one-to-one correspondence $x \leftrightarrow x^*, y \leftrightarrow y^*$ between the elements of R and those of R^* $(x, y \in R, x^*, y^* \in R^*)$ preserving linear operations and scalar products in the sense that[17]*

$$x + y \leftrightarrow x^* + y^*, \quad \alpha x \leftrightarrow \alpha x^*, \quad (x, y) = (x^*, y^*).$$

It is well known that any two n-dimensional Euclidean spaces are isomorphic to each other, and in particular that every n-dimensional Euclidean space is isomorphic to the space R^n of Example 1, p. 144.[18] On the other hand, two infinite-dimensional Euclidean spaces need not be isomorphic. For example, the spaces l_2 and $C^2_{[a, b]}$ are not isomorphic, as can be seen from the fact that l_2 is complete while $C^2_{[a, b]}$ is not (recall Examples 4 and 5, p. 57). Nevertheless, for *Hilbert* spaces we have

THEOREM 11 (*Isomorphism theorem*). *Any two Hilbert spaces are isomorphic.*

Proof. The theorem will be proved once we manage to show that every Hilbert space H is isomorphic to l_2. Let $\{\varphi_k\}$ be any complete orthonormal system in H (such exists by the corollary to Theorem 5), and with every element $f \in H$ associate its Fourier coefficients $\{c_k\}$ with respect to $\{\varphi_k\}$. Since

$$\sum_{k=1}^{\infty} c_k^2 < \infty,$$

---

[15] Named after the celebrated German mathematician David Hilbert (1862–1943).

[16] However, see Sec. 16.9.

[17] Isomorphism of two normed linear spaces R and R^* is defined in the same way, except that preservation of scalar products is replaced by preservation of norms, i.e., by the condition $\|x\| = \|x^*\|$.

[18] See e.g., G. E. Shilov, *op. cit.*, Theorem 29, p. 144.

by Theorem 8, the sequence $(c_1, c_2, \ldots, c_k, \ldots)$ belongs to l_2. Conversely, by the Riesz-Fischer theorem, to every element $(c_1, c_2, \ldots, c_k, \ldots)$ in l_2 there corresponds an element $f \in H$ with the numbers $c_1, c_2, \ldots, c_k, \ldots$ as its Fourier coefficients. This correspondence between the elements of H and those of l_2 is obviously one-to-one. Moreover, if

$$f \leftrightarrow (c_1, c_2, \ldots, c_k, \ldots),$$
$$\tilde{f} \leftrightarrow (\tilde{c}_1, \tilde{c}_2, \ldots, \tilde{c}_k, \ldots),$$

then clearly

$$f + \tilde{f} \leftrightarrow (c_1 + \tilde{c}_1, c_2 + \tilde{c}_2, \ldots, c_k + \tilde{c}_k, \ldots),$$
$$\alpha f \leftrightarrow (\alpha c_1, \alpha c_2, \ldots, \alpha c_k, \ldots),$$

i.e., sums go into sums and scalar multiples into scalar multiples with the same factor. Finally, by Parseval's theorem,

$$(f,f) = \sum_{k=1}^{\infty} c_k^2, \qquad (\tilde{f},\tilde{f}) = \sum_{k=1}^{\infty} \tilde{c}_k^2,$$

$$(f,f) + 2(f,\tilde{f}) + (\tilde{f},\tilde{f}) = (f + \tilde{f}, f + \tilde{f}) = \sum_{k=1}^{\infty} (c_k + \tilde{c}_k)^2$$

$$= \sum_{k=1}^{\infty} c_k^2 + 2\sum_{k=1}^{\infty} c_k \tilde{c}_k + \sum_{k=1}^{\infty} \tilde{c}_k^2,$$

and hence

$$(f,\tilde{f}) = \sum_{k=1}^{\infty} c_k \tilde{c}_k,$$

so that scalar products are preserved. ∎

Remark. Theorem 11 shows that to within an isomorphism, there is only one Hilbert space (i.e., only one space with the four properties listed above, and that this space has l_2 as its "coordinate realization," just as the space of all ordered n-tuples of real numbers with the scalar product $\sum_{k=1}^{n} x_k y_k$ is the "coordinate realization" of axiomatically defined Euclidean n-space.

16.7. Subspaces. Orthogonal complements and direct sums.

In keeping with the terminology of Sec. 15.2, by a *linear manifold* in a Hilbert space H we mean a set L of elements of H such that $f, g \in H$ implies $\alpha f + \beta g \in L$ for arbitrary numbers α and β, while by a *subspace* of H we mean a *closed* linear manifold in H.

LEMMA. *If a metric space R has a countable everywhere dense subset, then so does every subset $R' \subset R$.*

Proof. Let

$$\xi_1, \xi_2, \ldots, \xi_n, \ldots$$

be a countable everywhere dense subset of R, and let

$$a_n = \inf_{\eta \in R'} \rho(\xi_n, \eta).$$

Then, given any positive integers n and p, there is a point $\eta_{np} \in R'$ such that

$$\rho(\xi_n, \eta_{np}) < a_n + \frac{1}{p}.$$

Given any $\varepsilon > 0$ and any $\eta \in R'$, let

$$\frac{1}{p} < \frac{\varepsilon}{3},$$

and choose n such that

$$\rho(\xi_n, \eta) < \frac{\varepsilon}{3}.$$

Then

$$\rho(\xi_n, \eta_{np}) < a_n + \frac{1}{p} < \frac{\varepsilon}{3} + \frac{\varepsilon}{3} = \frac{2\varepsilon}{3},$$

and hence $\rho(\eta, \eta_{np}) < \varepsilon$. In other words, R' has an everywhere dense subset $\{\eta_{np}\}$ $(n, p = 1, 2, \ldots)$ containing no more than countably many elements. ∎

THEOREM 12. *Every subspace M of a Hilbert space H is either a (complete separable) Euclidean space or itself a Hilbert space. Moreover, M has an orthonormal basis, like H itself.*

Proof. The fact that M has properties 1) and 2) of Definition 5 is obvious. The separability of M follows from the lemma. To construct an orthonormal basis in M, apply Theorem 5 to any countable everywhere dense subset of M. ∎

Subspaces of a Hilbert space H have certain special properties (not shared by subspaces of an arbitrary normed linear space), stemming from the presence of a scalar product in H and the associated concept of orthogonality:

THEOREM 13. *Let M be a subspace of a Hilbert space H, and let*

$$M' = H \ominus M$$

*denote the **orthogonal complement** of M, i.e., the set of all elements $h' \in H$ orthogonal to every $h \in M$. Then M' is also a subspace of H.*

Proof. The linearity of M' is obvious, since

$$(h_1', h) = (h_2', h) = 0$$

implies

$$(\alpha_1 h_1' + \alpha_2 h_2', h) = 0$$

for arbitrary numbers α_1 and α_2. To show that M' is closed, suppose $\{h_n'\}$ is a sequence of elements of M' converting to h'. Then, given any $h \in M$,

$$(h', h) = \lim_{n \to \infty} (h_n', h) = 0,$$

and hence $h' \in M'$. ∎

Remark. By definition, $h' \in M'$ if and only if h' is orthogonal to every $h \in M$. But then $h \in H$ if and only if h is orthogonal to every $h' \in M'$. Hence $M' = H \ominus M$ implies $M = H \ominus M'$, and we can call M and M' *(mutually) orthogonal subspaces* of H.

THEOREM 14. *Let M be a subspace of a Hilbert space H, and let $M' = H \ominus M$ be the orthogonal complement of M. Then every element $f \in H$ has a unique representation of the form*

$$f = h + h', \tag{26}$$

where $h \in M$, $h' \in M'$.

Proof. Given any $f \in H$, let $\{\varphi_k\}$ be an orthonormal basis in M, and let

$$h = \sum_{k=1}^{\infty} c_k \varphi_k, \qquad c_k = (f, \varphi_k).$$

By Bessel's inequality,

$$\sum_{k=1}^{\infty} c_k^2 < \infty,$$

and hence, by the Riesz-Fischer theorem, h exists and belongs to M. Let

$$h' = f - h.$$

Then obviously

$$(h', \varphi_k) = 0$$

for all k, and since any element $g \in M$ can be represented in the form

$$g = \sum_{k=1}^{\infty} a_k \varphi_k,$$

we have

$$(h', g) = \sum_{k=1}^{\infty} a_k (h', \varphi_k) = 0,$$

i.e., $h' \in M'$. This proves the existence of the representation (26).

To prove the uniqueness of (26), suppose there is another representation

$$f = h_1 + h_1',$$

where $h_1 \in M$, $h_1' \in M'$. Then

$$(h_1, \varphi_k) = (f, \varphi_k) = c_k$$

for all k, and hence

$$h_1 = h, \qquad h_1' = h'. \quad \blacksquare$$

COROLLARY 1. *Every orthonormal system* $\{\varphi_k\}$ *in a Hilbert space* H *can be enlarged to give a complete orthonormal system in* H.

Proof. Let M be the linear closure of $\{\varphi_k\}$, so that $\{\varphi_k\}$ is complete in M. Let $M' = H \ominus M$ be the orthogonal complement of M, and let $\{\varphi_k'\}$ be a complete orthonormal system in M' (such exists by Theorem 12, since M' is a subspace). Recalling (26), we see that the union of $\{\varphi_k\}$ and $\{\varphi_k'\}$ is a complete orthonormal system in H. $\quad \blacksquare$

COROLLARY 2. *Let* M *be a subspace of a Hilbert space* H, *and let* $M' = H \ominus M$. *Then* M' *has codimension* n *if* M *has dimension* n *and dimension* n *if* M *has codimension* n.

Proof. An immediate consequence of the representation (26) and Theorem 2, p. 122. $\quad \blacksquare$

Let M be a subspace of a Hilbert space H, with orthogonal complement $M' = H \ominus M$. If every vector $f \in H$ can be represented in the form

$$f = h + h' \qquad (h \in M, h' \in M'),$$

we say that H is the *direct sum* of the orthogonal subspaces M and M', and write

$$H = M \oplus M'.$$

The concept of a direct sum generalizes at once to the case of any finite or even countable number of subspaces: Thus H is said to be the direct sum of the subspaces $M_1, M_2, \ldots, M_n, \ldots$ and we write

$$H = M_1 \oplus M_2 \oplus \cdots \oplus M_n \oplus \cdots$$

if

1) The subspaces M_j are *pairwise orthogonal*, i.e., every element in M_j is orthogonal to every element in M_k whenever $j \neq k$;
2) Every element $f \in H$ has a representation of the form

$$f = h_1 + h_2 + \cdots + h_n + \cdots \tag{27}$$

where $h_n \in H_n$ $(n = 1, 2, \ldots)$.

It is easy to see that the representation (27) is unique if it exists and that

$$\|f\|^2 = \sum_{n=1}^{\infty} \|h_n\|^2$$

(give the details).

Besides direct sums of subspaces, we can also talk about direct sums of a finite or countable number of Hilbert spaces. Thus, given two Hilbert spaces H_1 and H_2, by the direct sum

$$H = H_1 \oplus H_2$$

is meant the set of all ordered pairs (h_1, h_2) with $h_1 \in H_1$, $h_2 \in H_2$, where linear operations and the scalar product in H are defined by

$$(h_1, h_2) + (h_1', h_2') = (h_1 + h_1', h_2 + h_2'),$$
$$\alpha(h_1, h_2) = (\alpha h_1, \alpha h_2),$$
$$((h_1, h_2), (h_1', h_2')) = (h_1, h_1') + (h_2, h_2').$$

Consider the subspace of H consisting of all pairs of the form $(h_1, 0)$ and the subspace consisting of all pairs of the form $(0, h_2)$. Then clearly these two subspaces are orthogonal and can be identified in a natural way with H_1 and H_2, respectively. More generally, given any Hilbert spaces $H_1, H_2, \ldots,$ $H_n, \ldots,$ by the direct sum

$$H = H_1 \oplus H_2 \oplus \cdots \oplus H_n \oplus \cdots$$

is meant the set of all sequences

$$h = (h_1, h_2, \ldots, h_n, \ldots) \qquad (h_n \in H_n)$$

such that

$$\sum_{n=1}^{\infty} \|h_n\|^2 < \infty,$$

with linear operations defined in the obvious way and the scalar product of two elements $h = (h_1, h_2, \ldots, h_n, \ldots), g = (g_1, g_2, \ldots, g_n, \ldots)$ defined by

$$(h, g) = \sum_{n=1}^{\infty} (h_n, g_n).$$

16.8. Characterization of Euclidean spaces. Given a normed linear space R, we now look for circumstances under which R is Euclidean. In other words, we look for extra conditions on the norm of R which guarantee that the norm be derivable from some suitably defined scalar product in R.

THEOREM 15. *A necessary and sufficient condition for a normed linear space R to be Euclidean is that*

$$\|f + g\|^2 + \|f - g\|^2 = 2(\|f\|^2 + \|g\|^2) \qquad (28)$$

for every $f, g \in R$.

Proof. Thinking of $f + g$ and $f - g$ as the "diagonals of the parallelogram in R with sides f and g," we can interpret (28) as the analogue of a familar property of parallelograms in the plane, i.e., the sum of the squares of the diagonals of a parallelogram equals the sum of the squares of its sides. The necessity of (28) is obvious, since if R is Euclidean, then

$$\begin{aligned}
\|f + g\|^2 + \|f - g\|^2 &= (f + g, f + g) + (f - g, f - g) \\
&= (f,f) + 2(f,g) + (g,g) + (f,f) \\
&\quad - 2(f,g) + (g,g) \\
&= 2(\|f\|^2 + \|g\|^2).
\end{aligned}$$

To prove the sufficiency of (28), we set

$$(f,g) = \tfrac{1}{4}(\|f + g\|^2 - \|f - g\|^2), \tag{29}$$

and show that if (28) holds, then (29) has all the properties of a scalar product listed on p. 142. Since (29) implies

$$(f,f) = \tfrac{1}{4}(\|2f\|^2 - \|f - f\|^2) = \|f\|^2, \tag{30}$$

the scalar product (29) clearly generates the given norm $\|\cdot\|$ in R. Moreover, it follows at once from (29) and (30) that
1) $(f,f) \geqslant 0$ where $(f,f) = 0$ if and only if $f = 0$;
2) $(f,g) = (g,f)$.
The proof of the linearity properties

$$(f + g, h) = (f, h) + (g, h) \tag{31}$$

and

$$(\alpha f, g) = \alpha(f, g) \tag{32}$$

requires a little work. To prove (31), consider the function of three vectors

$$\Phi(f, g, h) = 4[(f + g, h) - (f, h) - (g, h)],$$

or equivalently

$$\begin{aligned}
\Phi(f, g, h) = \|f + g + h\|^2 - \|f - g - h\|^2 - \|f + h\|^2 + \|f - h\|^2 \\
- \|g + h\|^2 + \|g - h\|^2 \tag{33}
\end{aligned}$$

after using (29). It follows from (28) that

$$\|f + g + h\|^2 = 2\|f \pm h\|^2 + 2\|g\|^2 - \|f \pm h - g\|^2. \tag{34}$$

Substituting (34) into (33), we get

$$\begin{aligned}
\Phi(f, g, h) = -\|f + h - g\|^2 + \|f - h - g\|^2 + \|f + h\|^2 \\
- \|f - h\|^2 - \|g + h\|^2 + \|g - h\|^2. \tag{35}
\end{aligned}$$

Taking half the sum (34) and (35), we find that

$$\Phi(f, g, h) = \tfrac{1}{2}(\|g + h + f\|^2 + \|g + h - f\|^2)$$
$$-\tfrac{1}{2}(\|g - h + f\|^2 + \|g - h - f\|^2)$$
$$-\|g - h\|^2 + \|g - h\|^2,$$

which becomes

$$\Phi(f, g, h) = (\|g + h\|^2 + \|f\|^2) - (\|g - h\|^2 - \|f\|^2)$$
$$-\|g + h\|^2 + \|g - h\|^2 \equiv 0$$

after applying (28) to both expressions in parentheses. But $\Phi(f, g, h) \equiv 0$ is equivalent to (31).

To prove (32), we introduce the function

$$\varphi(c) = (cf, g) - c(f, g),$$

where f and g are fixed but arbitrary elements of R. It follows at once from (29) that

$$\varphi(0) = \tfrac{1}{4}(\|g\|^2 - \|g\|^2) = 0$$

and $\varphi(-1) = 0$, since $(-f, g) = -(f, g)$. Hence, given any integer n,

$$(nf, g) = (\operatorname{sgn} n(f + \cdots + f), g) = \operatorname{sgn} n[(f, g) + \cdots + (f, g)]$$
$$= |n| \operatorname{sgn} n(f, g) = n(f, g),$$

i.e., $\varphi(n) = 0$. Moreover, given any integers p, q ($q \neq 0$),

$$\left(\frac{p}{q}f, g\right) = p\left(\frac{1}{q}f, g\right) = \frac{p}{q}q\left(\frac{1}{q}f, g\right) = \frac{p}{q}(f, g),$$

i.e., $\varphi(c) = 0$ for all rational c. But $\varphi(c)$ is a continuous function of c (why?), and hence $\varphi(c) \equiv 0$, which is equivalent to (32). ∎

Example 1. The n-dimensional space R_p^n, equipped with the norm

$$\|x\|_p = \left(\sum_{k=1}^n |x_k|^p\right)^{1/p},$$

is a normed linear space if $p > 1$ (see Example 10, p. 41) and a Euclidean space if $p = 2$ (see Example 1, p. 144). However, R_p^n fails to be Euclidean if $p \neq 2$. In fact, for the two vectors

$$f = (1, 1, 0, \ldots, 0),$$
$$g = (1, -1, 0, \ldots, 0),$$

we have

$$f + g = (2, 0, 0, \ldots, 0),$$
$$f - g = (0, 2, 0, \ldots, 0),$$

and hence
$$\|f\|_p = \|g\|_p = 2^{1/p}, \qquad \|f + g\| = \|f - g\| = 2.$$
Therefore the "parallelogram condition" (28) fails if $p \neq 2$.

Example 2. Consider the space $C_{[0,\pi/2]}$ of all functions continuous on the interval $[0, \pi/2]$, and let
$$f(t) = \cos t, \qquad g(t) = \sin t.$$
Then
$$\|f\| = \|g\| = 1,$$
and
$$\|f + g\| = \max_{0 \leqslant t \leqslant \pi/2} |\cos t + \sin t| = \sqrt{2},$$
$$\|f - g\| = \max_{0 \leqslant t \leqslant \pi/2} |\cos t - \sin t| = 1.$$
Therefore
$$\|f + g\|^2 + \|f - g\|^2 \neq 2(\|f\|^2 + \|g\|^2).$$

It follows that the norm in $C_{[0,\pi/2]}$ cannot be generated by any scalar product whatsoever, i.e., the space $C_{[0,\pi/2]}$ fails to be Euclidean. It is easy to see that the same is true of the space $C_{[a,b]}$ for any a and b ($a < b$).

16.9. Complex Euclidean spaces. Besides real Euclidean spaces, we can also consider *complex* Euclidean spaces, i.e., complex linear spaces equipped with a scalar product. However, we must now modify the properties of the scalar product listed on p. 142, since in the complex case these properties are contradictory as they stand. In fact, it follows from properties 2) and 3), p. 142 that
$$(\lambda x, \lambda x) = \lambda^2 (x, x),$$
and hence, after choosing $\lambda = i$, that
$$(ix, ix) = -(x, x),$$
i.e., the norms of the vectors x and ix cannot both be positive, contrary to property 1). To remedy this difficulty, we define the scalar product in a complex Euclidean space R as a *complex-valued* function (x, y), defined for every pair of elements $x, y \in R$, with the following properties:

1') $(x, x) \geqslant 0$ where $(x, x) = 0$ if and only if $x = 0$;
2') $(x, y) = \overline{(y, x)}$;
3') $(\lambda x, y) = \lambda(x, y)$;
4') $(x, y + z) = (x, z) + (y, z)$

(valid for all $x, y, z \in R$ and all complex λ). It follows from 2') and 3') that
$$(x, \lambda y) = \overline{(\lambda y, x)} = \overline{\lambda} \overline{(y, x)} = \bar{\lambda}(x, y)$$
(as usual, the overbar denotes the complex conjugate).

Example 1. The space C^n introduced in Example 2, p. 119 becomes a complex Euclidean space if we define the scalar product of two elements $x = (x_1, \ldots, x_n), y = (y_1, \ldots, y_n)$ in C^n as

$$(x, y) = \sum_{k=1}^{n} x_k \bar{y}_k.$$

Example 2. The *complex* space l_2 with elements $x = (x_1, x_2, \ldots, x_k, \ldots)$, $y = (y_1, y_2, \ldots, y_k, \ldots), \ldots$, where

$$\sum_{k=1}^{\infty} |x_k|^2 < \infty, \qquad \sum_{k=1}^{\infty} |y_k|^2 < \infty, \ldots,$$

becomes an infinite-dimensional complex Euclidean space when equipped with the scalar product

$$(x, y) = \sum_{k=1}^{\infty} x_k y_k.$$

Example 3. The space $C_{[a,b]}^2$ of all *complex-valued* functions continuous on the interval $[a, b]$, equipped with the scalar product

$$(f, g) = \int_a^b f(t)\overline{g(t)} \, dt,$$

is another example of an infinite-dimensional complex Euclidean space.

The *norm* (length) of a vector in a complex Euclidean space is defined by the same formula

$$\|x\| = \sqrt{(x, x)}$$

as in the real case. However, the notion of the angle between two vectors x and y plays no role in the complex case, since the quantity

$$\frac{(x, y)}{\|x\| \, \|y\|}$$

is in general complex and hence cannot be the cosine of a real angle. On the other hand, the notion of orthogonality is defined in the same way as before, i.e., two elements x and y of a complex Euclidean space are said to be *orthogonal* if $(x, y) = 0$.

Let $\{\varphi_k\}$ be any orthogonal system in a complex Euclidean space R, and let f be any element of R. Then, just as in the real case, the numbers

$$a_k = \frac{1}{\|\varphi_k\|^2} (f, \varphi_k)$$

and the series

$$\sum_{k=1}^{\infty} a_k \varphi_k$$

are called the *Fourier coefficients* and the *Fourier series* of the function f, with respect to the system $\{\varphi_k\}$. In the complex case, Bessel's inequality (17') becomes

$$\sum_{k=1}^{\infty} |a_k|^2 \|\varphi_k\|^2 \leqslant \|f\|^2.$$

If the system $\{\varphi_k\}$ is orthonormal, the Fourier coefficients become

$$a_k = c_k = (f, \varphi_k),$$

and Bessel's inequality simplifies to

$$\sum_{k=1}^{\infty} |c_k|^2 \leqslant \|f\|^2.$$

By a *complex Hilbert space* is meant a complex Euclidean space which is complete, separable and infinite-dimensional. Theorem 11 carries over at once to the complex case, with isomorphism being defined exactly as in Definition 6:

THEOREM 11' (*Isomorphism theorem*). *Any two complex Hilbert spaces are isomorphic.*

Proof. This time show that every complex Hilbert space is isomorphic to the *complex* space l_2, the "coordinate realization" of a complex Hilbert space. ∎

Remark. As an exercise, the reader should state and prove the complex analogues of all the other theorems of Sec. 16.

Problem 1. Prove that in a Euclidean space, the operations of addition, multiplication by numbers and the formation of scalar products are all continuous. More exactly, prove that if $x_n \to x$, $y_n \to y$ (in the sense of norm convergence) and $\lambda_n \to \lambda$ (in the sense of ordinary convergence), then

$$x_n + y_n \to x + y, \quad \lambda_n x_n \to \lambda x, \quad (x_n, y_n) \to (x, y).$$

Hint. Use Schwarz's inequality.

Problem 2. Let R be the set of all functions f defined on the interval $[0, 1]$ such that

1) $f(t)$ is nonzero at no more than countably many points $t_1, t_2, \ldots$;

2) $\sum_{i=1}^{\infty} f^2(t_i) < \infty.$

Define addition of elements and multiplication of elements by scalars in the ordinary way, i.e., $(f + g)(t) = f(t) + g(t)$, $(\alpha f)(t) = \alpha f(t)$. If f and g are two elements of R, nonzero only at the points $t_1, t_2, \ldots$ and $t_1', t_2', \ldots,$

respectively, define the scalar product of f and g as

$$(f, g) = \sum_{i,j=1}^{\infty} f(t_i)g(t_j').$$

Prove that this scalar product makes R into a Euclidean space. Prove that R is nonseparable, i.e., that R contains no countable everywhere dense subset.

Problem 3. Give an example of a (nonseparable) Euclidean space which has no orthonormal basis. Prove that a complete Euclidean space (not necessarily separable) always has an orthonormal basis.

Problem 4. Prove that every nested sequence of nonempty closed bounded convex sets in a complete Euclidean space (not necessarily separable) has a nonempty intersection.

Comment. Cf. Problem 6, p. 66 and Problem 2, p. 141.

Problem 5. Given a Euclidean space R, let $\varphi_1, \varphi_2, \ldots, \varphi_k, \ldots$ be an orthonormal basis in R and f an arbitrary element of R. Prove that the element

$$f - \sum_{k=1}^{n} a_k \varphi_k$$

is orthogonal to all linear combinations of the form

$$\sum_{k=1}^{n} b_k \varphi_k$$

if and only if

$$a_k = (f, \varphi_k) \qquad (k = 1, 2, \ldots, n).$$

Problem 6. According to elementary geometry, the length of the perpendicular dropped from a point P to a line L or plane Π is smaller than the length of any other line segment joining P to L or Π. What is the natural generalization of this fact to the case of an arbitrary Euclidean space?

Hint. Use Theorem 6 and the result of the preceding problem.

Problem 7. Let R be a complete Euclidean space (not necessarily separable), so that R has an orthonormal basis $\{\varphi_\alpha\}$, by Problem 3. Prove that every vector $f \in R$ satisfies the formulas

$$f = \sum_{\alpha} (f_\alpha, \varphi)\varphi_\alpha, \qquad \|f\|^2 = \sum_{\alpha} |(f, \varphi_\alpha)|^2,$$

where neither sum contains more than countably many nonzero terms.

Problem 8. Give an example of a Euclidean space R and an orthonormal system $\{\varphi_n\}$ in R such that R contains no nonzero element orthogonal to every φ_n, even though $\{\varphi_n\}$ fails to be complete.

Comment. By Theorem 10, R cannot be complete.

Problem 9. Given a Euclidean space R, not necessarily complete, let R^* be the completion of R as defined in Sec. 7.4. Define linear operations and the scalar product in R^* by "continuous extension" of those in $R \subset R^*$. More exactly, if $x_n \to x$, $y_n \to y$ where $x_n, y_n \in R$, let

$$x + y = \lim_{n \to \infty} (x_n + y_n), \quad \alpha x = \lim_{n \to \infty} \alpha x_n, \quad (x, y) = \lim_{n \to \infty} (x_n, y_n).$$

Prove that

a) These limits exist and are independent of the choice of the sequences $\{x_n\}$, $\{y_n\}$ in R converging to x and y;

b) R^* is itself a Euclidean space.

Complete $C^2_{[a, b]}$ in this way, and show that the resulting space is a Hilbert space.

Comment. The elements belonging to the completion of $C^2_{[a, b]}$ but not to $C^2_{[a, b]}$ are themselves functions, in fact discontinuous functions whose squares are Lebesgue-integrable on $[a, b]$, as defined in Sec. 29.

Problem 10. Prove that each of the following sets is a subspace of the Hilbert space l_2:

a) The set of all $(x_1, x_2, \ldots, x_k, \ldots) \in l_2$ such that $x_1 = x_2$;

b) The set of all $(x_1, x_2, \ldots, x_k, \ldots) \in l_2$ such that $x_k = 0$ for all even k.

Problem 11. Show that every complex Euclidean space of finite dimension n is isomorphic to the space C^n of Example 1, p. 164. Generalize Problem 9 to the case where $C^2_{[a, b]}$ is the complex space of Example 3, p. 164.

17. Topological Linear Spaces

17.1. Definitions and examples. Specification of a norm is only one way of introducing a topology into a linear space. There are many situations in analysis, notably in the theory of generalized functions (to be discussed in Sec. 21), where it is desirable to use other methods of equipping a linear space with a topology:

DEFINITION 1. *By a **topological linear space** is meant a set E with the following properties:*

1) *E is a linear space;*

2) *E is a topological space;*

3) *The operations of addition of elements of E and multiplication of elements of E by numbers (real or complex) are continuous with respect to the topology in E, in the sense that*

a) *If $z_0 = x_0 + y_0$, then, given any neighborhood U of the point z_0, there are neighborhoods V and W of the points x_0 and y_0, respectively, such that $x + y \in U$ whenever $x \in V$, $y \in W$;*

b) *If $\alpha_0 x_0 = y_0$, then, given any neighborhood U of the point y_0, there is a neighborhood V of the point x_0 and a number $\varepsilon > 0$ such that $\alpha x \in U$ whenever $x \in V$, $|\alpha - \alpha_0| < \varepsilon$.*

THEOREM 1. *Let E be a topological linear space, and let U be any neighborhood of zero. Then the set*

$$U + x_0 = \{y : y = x + x_0, x \in U\}$$

is a neighborhood of x_0. Moreover, every neighborhood of x_0 is a set of this form, i.e., some neighborhood of zero "shifted by the vector x_0."

Proof. It follows from property 3a) that the mapping $f(x) = x - x_0$ carrying E into itself is continuous. Hence, by Theorem 10, p. 87, the preimage $f^{-1}(U)$ of any neighborhood U of the point zero is itself a neighborhood. But $f^{-1}(U) = U + x_0$. Therefore $U + x_0$ is a neighborhood, obviously of the point x_0. Similarly, given any neighborhood V of the point x_0, let $U = V - x_0 = V + (-x_0)$. Then U is a neighborhood of zero, by the continuity of the mapping $g(x) = x + x_0$. But clearly $U + x_0 = V$. ∎

Remark. Thus the topology in E is determined by giving a *neighborhood base at zero*, i.e., a system $\mathcal{N}_0$ of neighborhoods of zero with the property that, given any open set $G \subset E$ containing the point zero, there is a neighborhood $N \in \mathcal{N}_0$ contained in G. In fact, the mapping $f(x) = x + x_0$ carries a neighborhood base at zero into a neighborhood base at x_0. Hence $\mathcal{N}_0$ and its "translates," i.e., the system of all sets of the form $\{V : V = U + x, U \in \mathcal{N}_0, x \in E\}$, constitute a base for the topology in E. In this sense, $\mathcal{N}_0$ "generates" the topology in E.

Example 1. Every normed linear space is clearly a topological linear space. In fact, it is an immediate consequence of the properties of a norm that the operations of addition of vectors and multiplication of vectors by scalars are continuous with respect to the topology "induced" by the norm.

Example 2. Let R^∞ be the linear space of all numerical sequences $x = (x_1, \ldots, x_k, \ldots)$, real or complex, and let $\mathcal{N}_0$ consist of all sets of the form

$$U_{k_1, \ldots, k_r; \varepsilon} = \{x : x \in R^\infty, |x_{k_1}| < \varepsilon, \ldots, |x_{k_r}| < \varepsilon\}$$

for some number $\varepsilon > 0$ and positive integers $k_1, \ldots, k_r$. Then R^∞ becomes a topological linear space when equipped with the topology generated by $\mathcal{N}_0$.[19]

---

[19] As an exercise, verify that $\mathcal{N}_0$ and its translates satisfy Theorem 2 (or Theorem 3) of Sec. 9.3 and that the linear operations in R^∞ are continuous with respect to the topology generated by $\mathcal{N}_0$.

Example 3. Let $K_{[a,b]}$ be the linear space of all infinitely differentiable functions on the interval $[a, b]$,[20] and let $\mathcal{N}_0$ consist of all sets of the form

$$U_{r,\varepsilon} = \{\varphi : \varphi \in K_{[a,b]}, |\varphi^{(0)}(x)| < \varepsilon, \dots, |\varphi^{(r)}(x)| < \varepsilon \text{ for all } x \in [a, b]\}$$

for some number $\varepsilon > 0$ and positive integer r. Then $K_{[a,b]}$ becomes a topological linear space when equipped with the topology generated by this neighborhood base (again supply some missing details).

DEFINITION 2. *A subset M of a topological linear space E is said to be* **bounded** *if, given any neighborhood U of zero, there is a number $\alpha > 0$ such that $M \subset \alpha U = \{z : z = \alpha x, x \in U\}$.[21]*

DEFINITION 3. *A topological linear space E is said to be* **locally bounded** *if it contains at least one nonempty bounded open set.*

THEOREM 2. *Every normed linear space E is locally bounded.*

Proof. Given any $\varepsilon > 0$, the set of all $x \in E$ such that $\|x\| < \varepsilon$ is obviously nonempty, bounded and open. ∎

DEFINITION 4. *A topological linear space E is said to be* **locally convex** *if every nonempty open set in E contains a nonempty convex open subset.*

THEOREM 3. *Every normed linear space E is locally convex.*

Proof. Merely note that every nonempty open set in E contains an open sphere. ∎

Remark. It follows from Theorems 2 and 3 that every normed linear space is both locally bounded and locally convex. Conversely, it can be shown that every locally bounded and locally convex topological linear space satisfying the first axiom of separation is *normable*, in the sense that E can be equipped with a norm $\|\cdot\|$ generating the given topology in E, via the metric $\rho(x, y) = \|x - y\|$.

17.2. Historical remarks. For some time it was thought that the concept of a normed linear space (introduced in the thirties, notably in the work of Banach) was general enough to serve all the concrete needs of analysis. However, it subsequently became apparent that this was not so and that there are a number of problems involving such spaces as the space of infinitely differentiable functions, the space R^∞ of all numerical sequences, etc., in which the natural topology cannot be specified in terms of any norm whatsoever. Thus topological linear spaces, as opposed to normed linear

---

[20] A function φ is said to be *infinitely differentiable* if it has derivatives $\varphi^{(k)}$ of all orders $k = 0, 1, 2, \dots$ (the zeroth derivative $\varphi^{(0)}$ is just the function φ itself).

[21] A sequence $\{x_n\}$ of points in E is said to be bounded if the *set* $\{x_1, x_2, \dots, x_n, \dots\}$, consisting of all terms of the sequence, is bounded.

spaces, are by no means "exotic" or "pathological." On the contrary, some of these spaces are no less natural and important a generalization of finite-dimensional Euclidean space than, say, Hilbert space.

Problem 1. Reconcile Definition 2 with Problem 1, p. 141 in the case where E is a normed linear space.

Problem 2. Let E be a topological linear space. Prove that

a) If U and V are open sets, then so is $U + V = \{z : z = x + y, x \in U, y \in V\}$;
b) If U is open, then so is $\alpha U = \{z : z = \alpha x, x \in U\}$ provided that $\alpha \neq 0$;
c) If $F \subset E$ is closed, then so is αF for arbitrary α.

Problem 3. Prove that a topological linear space is a T_1-space if and only if the intersection of all neighborhoods of zero contains no nonzero elements.

Problem 4. Prove that a topological linear space E automatically has the following separation property: Given any point $x \in E$ and any neighborhood U of x, there is another neighborhood V of x such that $[V] \subset U$.

Hint. If U is a neighborhood of zero, then, by the continuity of subtraction, there is a neighborhood V of zero such that [22]

$$V - V = \{z : z = x - y, x \in V, y \in V\} \subset U.$$

Suppose $y \in [V]$. Then every neighborhood of y, in particular $V + y$, contains a point of V. Hence there is a point $z \in V$ such that $z + y \in V$. It follows that $y \in V - V \subset U$.

Problem 5. Prove that a topological space T has the separation property figuring in Problem 4 if and only if for each point $x \in T$ and each closed set $F \subset T$ not containing x, there is an open set O_1 containing x and an open set O_2 containing F such that $O_1 \cap O_2 = \varnothing$.

Comment. Thus, for T_1-spaces, this separation property is "halfway between" that of a Hausdorff space and that of a normal space.

Problem 6. Given a topological linear space E, prove that

a) If $\{x_n\}$ is a convergent sequence of points in E, then the set $M = \{x_1, x_2, \ldots, x_n, \ldots\}$ is bounded;
b) A subset $M \subset E$ is bounded if and only if, given any sequence $\{x_n\}$ of points in M and any sequence $\{\varepsilon_n\}$ of positive numbers converging to zero, the sequence $\{\varepsilon_n x_n\}$ also converges to zero.

---

[22] Here the minus sign in $V - V$ does not have the usual meaning of a set difference (the same kind of notation was used in Sec. 14.5).

Problem 7. Prove that

a) The space R^∞ of Example 2, p. 168 is not locally bounded;
b) Every locally bounded topological linear space satisfies the first axiom of countability.

Problem 8. Let x be any point of a locally convex topological linear space E, and let U be any neighborhood of x. Prove that x has a convex neighborhood contained in U.

Hint. It is enough to consider the case $x = 0$. Suppose U is a neighborhood of zero. Then there is a neighborhood V of zero such that $V - V \subset U$, where $V - V$ is the same as in the hint to Problem 4. Since E is locally convex, there is a nonempty convex open set $V' \subset V$. If $x_0 \in V'$, then $V' - x_0$ is a convex neighborhood of zero contained in U.

Problem 9. Prove that an open set U in a topological linear space is convex if and only if $U + U = 2U$.

Problem 10. Given a linear space E, a set $U \subset E$ is said to be *symmetric* if $x \in U$ implies $-x \in U$. Let $\mathscr{B}$ be the set of all convex symmetric subsets of E such that each coincides with its own interior. Prove that

a) $\mathscr{B}$ is a system of neighborhoods of zero determining a locally convex topology τ in E which satisfies the first axiom of separation;
b) The topology τ is the strongest locally convex topology compatible with the linear operations in E;
c) Every linear functional on E is continuous with respect to τ.

Problem 11. Two norms $\|\cdot\|_1$ and $\|\cdot\|_2$ in a linear space E are said to be *compatible* if, whenever a sequence $\{x_k\}$ in E is fundamental with respect to both norms and converges to a limit $x \in E$ with respect to one of them, it also converges to the same limit x with respect to the other norm. A linear space E equipped with a countable system of compatible norms $\|\cdot\|_n$ is said to be *countably normed*. Prove that every countably normed linear space becomes a topological linear space when equipped with the topology generated by the neighborhood base consisting of all sets of the form

$$U_{r, \varepsilon} = \{x : x \in E, \|x\|_1 < \varepsilon, \ldots, \|x\|_r < \varepsilon\} \tag{1}$$

for some number $\varepsilon > 0$ and positive integer r.

Problem 12. Prove that each of the following spaces is countably normed, i.e., in each case verify the compatibility of the given system of norms $\|\cdot\|_n$:

a) The space $K_{[a,b]}$ of infinitely differentiable functions on $[a, b]$, equipped with the norms

$$\|f\|_n = \sup_{\substack{a \leqslant t \leqslant b \\ 0 \leqslant k \leqslant n}} |f^{(k)}(t)| \qquad (n = 0, 1, 2, \ldots); \tag{2}$$

b) The space S_∞ of all infinitely differentiable functions $f(t)$ on $(-\infty, \infty)$ such that $f(t)$ and all its derivatives approach zero as $|t| \to \infty$ faster than any power of $1/|t|$ (i.e., such that $t^p f^{(q)}(t) \to 0$ as $|t| \to \infty$ for arbitrary p and q), equipped with the norms

$$\|f\|_n = \sup_{p,q \leqslant n} |t^p f^{(q)}(t)| \qquad (n = 0, 1, 2, \ldots);$$

c) The space Φ of all numerical sequences $x = (x_1, \ldots, x_k, \ldots)$ such that

$$\sum_{k=1}^{\infty} k^n x_k^2$$

converges for all $n = 0, 1, 2, \ldots$, equipped with the norms

$$\|x\|_n = \sqrt{\sum_{k=1}^{\infty} k^n x_k^2} \qquad (n = 0, 1, 2, \ldots).$$

Show that (1) and (2) define the same topology in $K_{[a,b]}$ as in Example 3, p. 169.

Comment. Φ might be called the space of "rapidly decreasing sequences."

Problem 13. A norm $\|\cdot\|_1$ is said to be *stronger* than a norm $\|\cdot\|_2$ if there is a constant $c > 0$ such that $\|x\|_1 \geqslant c \|x\|_2$ for all $x \in E$ (then $\|\cdot\|_2$ is said to be *weaker* than $\|\cdot\|_1$). Discuss the norms (2) in this language.

Comment. Two norms are said to be *comparable* if one is stronger than the other, and *equivalent* if one is both stronger and weaker than the other (cf. Problem 7, p. 141).

Problem 14. Prove that every countably normed space satisfies the first axiom of countability.

Hint. Replace the system of neighborhoods $U_{r,\varepsilon}$ by the subsystem such that ε takes only the values

$$1, \frac{1}{2}, \ldots, \frac{1}{n}, \ldots$$

(this can be done without changing the topology).

Comment. Thus the topology in E can be described in terms of convergent sequences (recall Sec. 9.4).

Problem 15. Prove that the topology in a countably normed space can be specified in terms of the metric

$$\rho(x, y) = \sum_{n=1}^{\infty} \frac{1}{2^n} \frac{\|x - y\|_n}{1 + \|x - y\|_n} \qquad (x, y \in E). \tag{3}$$

First verify that $\rho(x, y)$ has all the properties of a metric, and is invariant under shifts in the sense that $\rho(x + z, y + z) = \rho(x, y)$ for all $x, y, z \in E$.

Comment. A countably normed space is said to be *complete* if it is complete with respect to the metric (3).

Problem 16. Prove that a sequence $\{x_k\}$ in a countably normed space is fundamental with respect to the metric (4) if and only if it is fundamental with respect to each of the norms $\|\cdot\|_n$. Prove that $\{x_k\}$ converges to an element $x \in E$ with respect to the metric (3) if and only if it converges to x with respect to each of the norms $\|\cdot\|_n$.

Comment. Thus, in particular, a countably normed space E is said to be complete if a sequence $\{x_k\}$ in E converges whenever it is fundamental with respect to each of the norms $\|\cdot\|_n$.

Problem 17. An infinite-dimensional separable linear space H equipped with a countable system of scalar products $(\cdot, \cdot)_n$ is said to be *countably Hilbert* if the norms

$$\|x\|_n = \sqrt{(x, x)_n} \qquad (x \in H)$$

generated by these scalar products are compatible and if the space H is complete. Prove that the space Φ of Problem 12c is countably Hilbert when equipped with the scalar products

$$(x, y)_n = \sum_{k=1}^{\infty} k^n x_k y_k \qquad (n = 0, 1, 2, \ldots),$$

where $x = (x_1, \ldots, x_k, \ldots)$, $y = (y_1, \ldots, y_k, \ldots)$ are any two elements of Φ.

Problem 18. The norms $\|\cdot\|_n$ in a countably normed space E can be assumed to satisfy the condition

$$\|x\|_k \leqslant \|x\|_l \qquad \text{if} \quad k < l, \tag{4}$$

since otherwise we can replace $\|\cdot\|_n$ by

$$\| \cdot \|_n' = \sup \{\| \cdot \|_1, \| \cdot \|_2, \ldots, \| \cdot \|_n\}.$$

(Prove that this does not change the topology in E.) Let E_n denote the completion of E with respect to the norm $\|\cdot\|_n$. Using (4), prove that

$$E_1 \supset E_2 \supset \cdots \supset E_n \supset \cdots.$$

Clearly,

$$E \subset \bigcap_{n=1}^{\infty} E_n.$$

Prove that E is complete if and only if

$$E = \bigcap_{n=1}^{\infty} E_n.$$

Problem 19. Let $C_{[a,b]}^{(n)}$ be the space of all functions defined on the interval $[a, b]$ with continuous derivatives up to order n inclusive, equipped with the norm

$$\|f\|_n = \sup_{\substack{a \leqslant t \leqslant b \\ 0 \leqslant k \leqslant n}} |f^{(k)}(t)|$$

(note that $C_{[a,b]}^{(0)} = C_{[a,b]}$). Prove that $C_{[a,b]}^{(n)}$ is complete. Prove that $K_{[a,b]}$ equals the intersection

$$\bigcap_{n=0}^{\infty} C_{[a,b]}^{(n)},$$

and hence is complete (by Problem 18).

5

LINEAR FUNCTIONALS

---

18. Continuous Linear Functionals

18.1. Continuous linear functionals on a topological linear space. A (real) functional f defined on a topological linear space E is said to be *linear* on E if

$$f(\alpha x + \beta y) = \alpha f(x) + \beta f(y)$$

for all $x, y \in E$ and arbitrary numbers α, β (recall Sec. 13.5), and *continuous at the point* $x_0 \in E$ if, given any $\varepsilon > 0$, there is a neighborhood U of x_0 such that

$$|f(x) - f(x_0)| < \varepsilon \tag{1}$$

for all $x \in U$ (recall Sec. 9.6). We say that the functional f is *continuous (on E)* if it is continuous at every point $x_0 \in E$.

THEOREM 1. *Let f be a linear functional on a topological linear space E, and suppose f is continuous at some point $x_0 \in E$. Then f is continuous on E, i.e., at every point of E.*

Proof. Given any point $y \in E$ and any number $\varepsilon > 0$, let U be a neighborhood of x_0 such that $x \in U$ implies (1). Then

$$V = U + (y - x_0) = \{z : z = x + y - x_0, x \in U\}$$

is a neighborhood of y, by Theorem 1, p. 168. Moreover, $x \in V$ implies $x + x_0 - y \in U$ and hence

$$|f(x) - f(y)| = |f(x + x_0 - y) - f(x_0)| < \varepsilon,$$

i.e., f is continuous at y. ∎

COROLLARY. *The continuity of a linear functional on a topological linear space need only be checked at a single point, for example, at the point zero.*

THEOREM 2. *Let f be a linear functional on a topological linear space E. Then f is continuous on E if and only if f is bounded in some neighborhood of zero.*[1]

Proof. Suppose f is continuous on E, in particular at the point zero. Then, given any $\varepsilon > 0$, there is a neighborhood of zero in which $|f(x)| < \varepsilon$. Obviously, f is bounded in this neighborhood.

Conversely, suppose f is bounded in some neighborhood U of zero, so that $|f(x)| < C$ for all $x \in U$, where C is a suitable constant. Then, given any $\varepsilon > 0$, we have $|f(x)| < \varepsilon$ for all x in the neighborhood

$$\frac{\varepsilon}{C} U = \left\{ z : z = \frac{\varepsilon}{C} x, x \in U \right\},$$

i.e., f is continuous at zero and hence on all of E. ∎

THEOREM 3. *A necessary condition for a linear functional f to be continuous on a topological linear space E is that f be bounded on every bounded set. The condition is also sufficient if E satisfies the first axiom of countability.*

Proof. To prove the necessity, suppose f is continuous on E. Then f is bounded in some neighborhood U of zero:

$$|f(x)| < C \qquad (x \in U).$$

Let $M \subset E$ be any bounded set, as defined in Definition 2, p. 169. Then $M \subset \alpha U$ for some $\alpha > 0$, and hence

$$|f(x)| < C\alpha \qquad (x \in M),$$

i.e., f is bounded on M.

As for the sufficiency, let $\{U_n\}$ be a countable neighborhood base at the point zero such that

$$U_1 \supset U_2 \supset \cdots \supset U_n \supset \cdots$$

(cf. the proof of Theorem 7, p. 84). If f fails to be continuous on E, it cannot be bounded on any of these neighborhoods of zero. Therefore in each U_n there is a point x_n such that $|f(x_n)| > n$. The sequence $\{x_n\}$ is bounded (recall footnote 21, p. 169), and even converges to zero, while the sequence $\{f(x_n)\}$ is unbounded. But then f fails to be bounded on the bounded set $\{x_1, x_2, \ldots, x_n, \ldots\}$, contrary to hypothesis. ∎

Guided by Theorem 3, we introduce

---

[1] Recall footnote 14, p. 110.

DEFINITION 1. *Given a linear functional f on a topological linear space E, suppose f is bounded on every bounded subset of E. Then f is said to be a **bounded linear functional**.*

Remark. In general, a bounded linear functional need not be continuous.

18.2. Continuous linear functionals on a normed linear space. Suppose E is a normed linear space, so that in particular E satisfies the first axiom of countability (recall the remark on p. 83). Then, by Theorem 3, a linear functional on E is continuous if and only if it is bounded. But by a bounded set in a normed linear space we mean a set contained in some closed sphere $\|x\| \leqslant C$ (recall Problem 1, p. 141). Therefore a linear functional f on a normed linear space is bounded (and hence continuous) if and only if it is bounded on every closed sphere $\|x\| \leqslant C$, or equivalently on the closed unit sphere $\|x\| \leqslant 1$, because of the linearity of f. In other words, f is bounded if and only if the number

$$\|f\| = \sup_{\|x\| \leqslant 1} |f(x)| \tag{2}$$

is finite.

DEFINITION 2. *Given a bounded linear functional f on a normed linear space E, the number* (2), *equal to the least upper bound of* $|f(x)|$ *on the closed unit sphere* $\|x\| \leqslant 1$, *is called the **norm** of f.*

THEOREM 4. *The norm* $\| f \|$ *has the following two properties:*

$$\|f\| = \sup_{x \neq 0} \frac{|f(x)|}{\|x\|}, \tag{3}$$

$$|f(x)| \leqslant \|f\| \, \|x\| \quad for\ all \quad x \in E. \tag{4}$$

Proof. Clearly,

$$\|f\| = \sup_{\|x\| \leqslant 1} |f(x)| = \sup_{\|x\| = 1} |f(x)|$$

(why?). But the set of all vectors in E of norm 1 coincides with the set of all vectors

$$\frac{x}{\|x\|} \quad (x \in E, x \neq 0), \tag{5}$$

and hence

$$\|f\| = \sup_{\|x\| = 1} |f(x)| = \sup_{x \neq 0} \left| f\left(\frac{x}{\|x\|} \right) \right| = \sup_{x \neq 0} \frac{|f(x)|}{\|x\|},$$

which proves (3). Moreover, since the vectors (5) all have norm 1, it follows from (2) that

$$\left| f\left(\frac{x}{\|x\|} \right) \right| = \frac{|f(x)|}{\|x\|} \leqslant \|f\| \quad (x \in E, x \neq 0),$$

which implies (4) for $x \neq 0$. The validity of (4) for $x = 0$ is obvious. ∎

Example 1. Let R^n be Euclidean n-space, and let a be any fixed nonzero vector in R^n. Then the scalar product

$$f(x) = (x, a) \qquad (x \in R^n)$$

defines a functional on R^n which is obviously linear. By Schwarz's inequality,

$$|f(x)| = |(x, a)| < \|x\| \, \|a\|. \tag{6}$$

Therefore f is bounded and hence continuous on R^n. It follows from (6) that

$$\frac{|f(x)|}{\|x\|} < \|a\| \qquad (x \neq 0). \tag{7}$$

The right-hand side of (7) is independent of x, and hence

$$\sup_{x \neq 0} \frac{|f(x)|}{\|x\|} < \|a\|,$$

i.e.,

$$\|f\| < \|a\|.$$

But choosing $x = a$, we get

$$|f(a)| = |(a, a)| = \|a\|^2,$$

or equivalently

$$\frac{|f(a)|}{\|a\|} = \|a\|.$$

It follows from (3) that

$$\|f\| = \|a\|.$$

Example 2. More generally, let R be an arbitrary Euclidean space, and let a be a fixed element of R. Then the same argument as in the preceding example shows that the scalar product

$$f(x) = (x, a) \qquad (x \in R)$$

defines a bounded linear functional on R, with norm

$$\|f\| = \|a\|.$$

Example 3. The integral

$$I(x) = \int_a^b x(t) \, dt$$

is a linear functional on the space $C_{[a,b]}$. Since

$$|I(x)| = \left| \int_a^b x(t) \, dt \right| < \max_{a \leqslant t \leqslant b} |x(t)| \, (b - a) = \|x\| \, (b - a),$$

where the equality holds if $x(t) \equiv$ const, we see that the functional I is bounded, with norm

$$\|I\| = b - a. \tag{8}$$

Example 4. More generally, let $y_0(t)$ be a fixed function in $C_{[a,b]}$, and let

$$I(x) = \int_a^b x(t)y_0(t)\,dt.$$

Then I is a linear functional on $C_{[a,b]}$. Since

$$|I(x)| = \left| \int_a^b x(t)y_0(t)\,dt \right| \leqslant \|x\| \int_a^b |y_0(t)|\,dt,$$

where the equality holds if $x(t) \equiv \text{const}$, the functional I is bounded, with norm

$$\|I\| = \int_a^b |y_0(t)|\,dt. \tag{9}$$

Note that (9) reduces to (8) in the case $y_0(t) \equiv 1$.

Example 5. As in Example 3, p. 124, let

$$\delta_{t_0}(x) = x(t_0)$$

be the linear functional on $C_{[a,b]}$ which assigns to each function $x(t) \in C_{[a,b]}$ its value at some fixed point $t_0 \in [a, b]$. Clearly

$$|x(t_0)| \leqslant \max_{a \leqslant t \leqslant b} |x(t)| = \|x\|,$$

where equality holds if $x(t) \equiv \text{const}$. Hence δ_{t_0} is bounded, with norm

$$\|\delta_{t_0}\| = 1.$$

The concept of the norm of a bounded linear functional on a normed linear space can be given a simple geometric interpretation. As shown in Theorem 4, p. 127, every nontrivial linear functional f can be associated with a hyperplane

$$M_f = \{x : f(x) = 1\}.$$

Let d be the distance from the hyperplane M_f to the point $x = 0$, defined as

$$d = \inf_{f(x)=1} \|x\|$$

(cf. Problem 9, p. 54). Since, as always

$$|f(x)| \leqslant \|f\|\,\|x\|,$$

$f(x) = 1$ implies

$$\|x\| \geqslant \frac{1}{\|f\|} \qquad (x \in M_f),$$

i.e.,

$$d \geqslant \frac{1}{\|f\|}. \tag{10}$$

On the other hand, it follows from (3) that, given any $\varepsilon > 0$, there is an element x_ε such that $f_\varepsilon(x) = 1$ and

$$(\|f\| - \varepsilon)\,\|x_\varepsilon\| < 1.$$

Therefore

$$d = \inf_{f(x)=1} \|x\| < \frac{1}{\|f\| - \varepsilon}\,,$$

and hence

$$d \leqslant \frac{1}{\|f\|}\,, \tag{11}$$

since $\varepsilon > 0$ is arbitrary. Comparing (10) and (11), we get

$$d = \frac{1}{\|f\|}\,,$$

i.e., the norm of the linear functional f equals the reciprocal of the distance between the hyperplane $f(x) = 1$ and the point $x = 0$.

18.3. The Hahn-Banach theorem for a normed linear space. Let $f_0(x)$ be a linear functional defined on a subset L of a linear space E, satisfying the condition

$$|f_0(x)| \leqslant p(x), \tag{12}$$

where p is a finite convex functional on E. Then, according to the Hahn-Banach theorem (Theorem 5, p. 132), f_0 can be extended onto the whole space E without violating the condition (12) As applied to bounded linear functionals on a normed linear space E, this result can be formulated as follows:

THEOREM 5 (*Hahn-Banach*). *Given a real normed linear space E, let L be a subspace of E and f_0 a bounded linear functional on L Then f_0 can be extended to a bounded linear functional f on the whole space E without increasing its norm, i.e.,*

$$\|f\|_{\text{on } E} = \|f_0\|_{\text{on } L}.$$

Proof. We need only choose the functional p in Theorem 5, p. 132 to be the convex functional $k\,\|x\|$, where

$$k = \|f_0\|_{\text{on } L}. \qquad \blacksquare$$

This form of the Hahn-Banach theorem has a simple geometric interpretation. The equation

$$f_0(x) = 1 \tag{13}$$

specifies a hyperplane in the subspace L, at distance

$$\frac{1}{\|f_0\|}$$

from the origin (the point $x = 0$) The fact that the functional f_0 can be extended onto the whole space E without increasing its norm means that the hyperplane (13) can be extended to a larger hyperplane in the whole space E in such a way that the distance between the larger hyperplane and the origin is the same as the distance between the hyperplane (13) and the origin.

In the same way, starting from the complex version of the Hahn-Banach theorem (Theorem 5′, p. 134), we get

THEOREM 5′. *Given a complex normed linear space E, let L be a subspace of E and f_0 a bounded linear functional on L. Then f_0 can be extended to a bounded linear functional f on the whole space E without increasing its norm, i.e.,* $\|f\|_{\text{on } E} = \|f_0\|_{\text{on } L}$.

In the case of an arbitrary topological linear space E, a nontrivial continuous linear functional on E may not even exist. However, by imposing suitable restrictions on E, we can guarantee the existence of "sufficiently many" continuous linear functionals on E.[2]

DEFINITION 3. *A topological linear space E is said to have **sufficiently many** continuous linear functionals if for each pair of distinct points $x_1, x_2 \in E$ there exists a continuous linear functional f on E such that $f(x_1) \neq f(x_2)$, or equivalently, if for each nonzero element $x_0 \in E$ there exists a continuous linear functional on E such that $f(x_0) \neq 0$.*

THEOREM 6. *Every normed linear space E has sufficiently many continuous linear functionals.*

Proof. Given any nonzero element $x_0 \in E$, we define a linear functional
$$f_0(\lambda x_0) = \lambda$$
on the set L of all elements of the form λx_0. We then use the Hahn-Banach theorem to extend f_0 onto the whole space E. This gives a continuous linear functional on E such that $f(x_0) = 1 \neq 0$. ∎

Problem 1. Prove that a functional f on a T_1-space E is continuous at a point $x \in E$ if and only if $x_n \to x$ implies $f(x_n) \to f(x)$.

Problem 2. Prove that every linear functional on a finite-dimensional topological linear space is automatically continuous.

Problem 3. Let E be a topological linear space. Prove that a linear functional f on E is continuous if and only if

a) Its null space $\{x : f(x) = 0\}$ is closed in E;
b) There exists an open set $U \subset E$ and a number t such that $t \notin f(U)$.

---

[2] See Theorem 6 and Problems 7–8.

Problem 4. Given a topological linear space E, prove that

a) If every linear functional on E is continuous, then the topology in E is the topology τ of Problem 10, p. 171;

b) If E is infinite-dimensional and normable, then there exists a non-continuous linear functional on E;

c) If E has a neighborhood base at zero whose power does not exceed the algebraic dimension of E, then there exists a noncontinuous linear functional on E.

Hint. In b) use the existence of a Hamel basis in E (recall Problem 4, p. 128, where algebraic dimension is also defined).

Problem 5. Prove that

$$f(x) = ax(0) + bx(1),$$

$$g(x) = \int_0^{1/2} x(t)\, dt - \int_{1/2}^1 x(t)\, dt$$

are both bounded linear functionals on the space $C_{[0,1]}$. What are their norms?

Problem 6. As in Problem 11, p. 171, let E be a countably normed space with norms $\|\cdot\|_n$, where

$$\|x\|_1 \leqslant \|x\|_2 \leqslant \cdots \leqslant \|x\|_n \leqslant \cdots \tag{14}$$

(as in Problem 18, p. 173, this condition entails no loss of generality). Let E^* be the set of all continuous linear functionals on E, and let E_n^* be the set of all linear functionals on E which are continuous with respect to the norm $\|\cdot\|_n$. Prove that

$$E_1^* \subset E_2^* \subset \cdots \subset E_n^* \subset \cdots$$

and

$$E^* = \bigcup_{n=1}^{\infty} E_n^*. \tag{15}$$

Hint. If f is a continuous linear functional on E, then, by Theorem 2, there is a neighborhood U of zero in which f is bounded. It follows from (14) and the definition of the topology in E that there is a number $\varepsilon > 0$ and a positive integer k such that the open sphere $\|x\|_k < \varepsilon$ is contained in U. Being bounded on this sphere, f is bounded and continuous with respect to the norm $\|\cdot\|_k$.

Comment. Let f be a continuous linear functional on E, i.e., let $f \in E^*$. Then by the *order* of f is meant the smallest integer n for which $f \in E_n^*$. It follows from (15) that every continuous linear functional on E is of finite order.

Problem 7. Prove that every countably normed space E has sufficiently many continuous linear functionals.

Hint. Given any nonzero element $x_0 \in E$, use Theorem 6 to construct a linear functional f continuous with respect to the norm $\|\cdot\|_1$ such that $f(x_0) \neq 0$.

Problem 8. Show that every real locally convex topological linear space E satisfying the first axiom of separation has sufficiently many continuous linear functionals.

Hint. Given any nonzero element $x_0 \in E$, show that there is a convex symmetric[3] neighborhood U of zero such that $x_0 \notin U$. Let p_U be the Minkowski functional of U. Then, as in the proof of Theorem 6, p. 136, p_U is a finite convex functional on E such that $p_U(-x) = p_U(x)$ and

$$p_U(x) < 1 \quad \text{if} \quad x \in U, \qquad p_U(x_0) \geqslant 1.$$

Define a linear functional $f_0(\lambda x_0) = \lambda$ on the set L of all elements of the form λx_0. Clearly $|f_0(x)| \leqslant p_0(x)$ on L and $f_0(x_0) = 1$. Now use the Hahn-Banach theorem to extend f_0 onto the whole space E.

Comment. The importance of locally convex spaces is mainly due to this property (which continues to hold in the complex case).

19. The Conjugate Space

19.1. Definition of the conjugate space. The operations of addition of functionals and multiplication of functionals by numbers are defined in the obvious way:

DEFINITION 1. *Let f and g be two functionals defined on a topological linear space E, and let α be any number. Then by the **sum** of f and g, denoted by $f + g$, is meant the functional whose value at every point $x \in E$ is the sum of the values of f and g at x, while by the **product** of α and f, denoted by αf, is meant the functional whose value at every point $x \in E$ is the product of α and the value of f at x. More concisely,*

$$(f + g)(x) = f(x) + g(x),$$
$$\alpha f(x) = \alpha f(x)$$

for every $x \in E$.

Clearly, if f and g are linear functionals, then so are $f + g$ and αf. Moreover, if f and g are bounded (and hence continuous), so are $f + g$ and αf.

---

[3] Recall Problem 10, p. 171.

Let E^* be the set of all continuous linear functionals on E. Then the space E^*, called the *conjugate space* of E, is itself a linear space, when equipped with the operations of addition of functionals and multiplication of functionals by numbers. This can be seen at once by verifying the three axioms in Definition 1, p. 118. Note that the zero element in E^* is the functional $f = 0$, equal to zero for all $x \in E$.

The next step is to introduce a topology in E^*, besides the linear operations just described. This can be done in various ways. First we consider the particularly simple case where the original space E is a normed linear space.

19.2. The conjugate space of a normed linear space. Let f be a continuous linear functional on a normed linear space E. In Sec. 18.2 we introduced the concept of the norm of f, equal to

$$\|f\| = \sup_{x \neq 0} \frac{|f(x)|}{\|x\|}$$

(recall Theorem 4, p. 177). This quantity clearly has all the properties of a norm, as listed on p. 138. In fact,

1) $\|f\| \geqslant 0$ where $\|f\| = 0$ if and only if $f = 0$;
2) $\|\alpha f\| = |\alpha| \, \|f\|$;
3) $\|f + g\| \leqslant \|f\| + \|g\|$, since obviously

$$\sup_{x \neq 0} \frac{|f(x) + g(x)|}{\|x\|} \leqslant \sup_{x \neq 0} \frac{|f(x)|}{\|x\|} + \sup_{x \neq 0} \frac{|g(x)|}{\|x\|}.$$

Hence the space E^* conjugate to E can be made into a normed linear space by simply equipping each functional $f \in E^*$ with its norm $\|f\|$. The corresponding topology in E^* is called the *strong topology* in E^*. In cases where we want to emphasize that E^* is equipped with the norm $\|\cdot\|$, we will write $(E^*, \|\cdot\|)$ instead of E^*.

Example 1. Let E be Euclidean n-space (real or complex), and let $e_1, \ldots, e_n$ be any basis in E, so that every vector $x \in E$ has a unique representation of the form

$$x = \sum_{k=1}^{n} x_k e_k.$$

If f is a linear functional on E, then clearly

$$f(x) = \sum_{k=1}^{n} f(e_k) x_k. \tag{1}$$

Thus a linear functional on E is uniquely determined by its values on the basis vectors $e_1, \ldots, e_n$, where these values can be assigned arbitrarily.

Consider the linear functionals $f_1, \ldots, f_n$ defined by

$$f_j(e_k) = \begin{cases} 1 & \text{if } j = k, \\ 0 & \text{if } j \neq k. \end{cases}$$

It is clear that these functionals are linearly independent, and moreover that

$$f_j(x) = x_j.$$

Hence we can write (1) in the form

$$f(x) = \sum_{k=1}^{n} f(e_k) f_k(x).$$

Thus the functionals $f_1, \ldots, f_n$ form a basis in the space E^*, called the *dual* of the basis $e_1, \ldots, e_n$ in the original space E. Therefore E^* is itself an n-dimensional linear space. Of course, different norms in E "induce" different norms in E^* (see Problem 1).

Example 2. Let c_0 be the space of all sequences $x = (x_1, \ldots, x_k, \ldots)$ converging to zero, with norm

$$\|x\| = \sup_k \|x_k\|.$$

Then the space $(c_0^*, \|\cdot\|)$ conjugate to c_0 is isomorphic (see footnote 17, p. 155) to the space l_1 of all absolutely summable sequences $f = (f_1, \ldots, f_k, \ldots)$,[4] with norm

$$\|f\| = \sum_{k=1}^{\infty} |f_k|,$$

To prove this, we first note that, given any element $f = (f_1, \ldots, f_k, \ldots) \in l_1$, the formula

$$\tilde{f}(x) = \sum_{k=1}^{\infty} x_k f_k \tag{2}$$

defines a functional $\tilde{f}$ on the space c_0, where $\tilde{f}$ is clearly linear. Moreover, it follows from (2) that

$$|\tilde{f}(x)| \leqslant \|x\| \sum_{k=1}^{\infty} |f_k|,$$

and hence

$$\|\tilde{f}\| \leqslant \|f\|. \tag{3}$$

---

[4] A sequence $\{f_k\}$, or $f = (f_1, \ldots, f_k, \ldots)$ in "point notation," is said to be *absolutely summable* if

$$\sum_{k=1}^{\infty} |f_k| < \infty.$$

Consider the vectors

$$e_1 = (1, 0, 0, \ldots),$$
$$e_2 = (0, 1, 0, \ldots),$$
$$e_3 = (0, 0, 1, \ldots),$$
$$\cdots \cdots \cdots \cdots$$

in c_0, and let

$$x^{(n)} = \sum_{k=1}^{n} \frac{f_k}{|f_k|} e_k$$

(if $f_k = 0$, we set $f_k/|f_k| = 0$). Then $x^{(n)} \in c_0$, and

$$\|x^{(n)}\| \leqslant 1. \tag{4}$$

Moreover

$$\tilde{f}(x^{(n)}) = \sum_{k=1}^{n} \frac{f_k}{|f_k|} \tilde{f}(e_k) = \sum_{k=1}^{n} |f_k|,$$

so that

$$\lim_{n \to \infty} \tilde{f}(x^{(n)}) = \sum_{k=1}^{\infty} |f_k| = \|f\|. \tag{5}$$

It follows from (4) and (5) that

$$\|\tilde{f}\| \geqslant \|f\| \tag{6}$$

(why?). Comparing (3) and (6), we get

$$\|\tilde{f}\| = \|f\|.$$

Thus the mapping carrying f into $\tilde{f}$ is a "norm-preserving" mapping of l_1 into c_0^*. We must still verify that this mapping is one-to-one and "onto" (see p. 5), i.e., that every functional $\tilde{f} \in c_0^*$ has a unique representation of the form (2), where $f = (f_1, \ldots, f_k, \ldots) \in l_1$. Let $x = (x_1, \ldots, x_k, \ldots) \in c_0$. Then

$$x = \sum_{k=1}^{\infty} x_k e_k,$$

where the series on the right converges in c_0 to the element x, since

$$\left\| x - \sum_{k=1}^{n} x_k e_k \right\| = \sup_{k > n} |x_k| \to 0$$

as $n \to \infty$. Since the functional $\tilde{f} \in c_0^*$ is continuous,

$$\tilde{f}(x) = \sum_{k=1}^{\infty} x_k \tilde{f}(e_k)$$

(where is the continuity used?). Hence $\tilde{f}$ has a unique representation of the form (2), and we need only verify that

$$\sum_{k=1}^{\infty} |\tilde{f}(e_k)| < \infty. \tag{7}$$

This time let

$$x^{(n)} = \sum_{k=1}^{n} \frac{\tilde{f}(e_k)}{|\tilde{f}(e_k)|} e_k.$$

Noting that $x^{(n)} \in c_0$ and $\|x^{(n)}\| \leqslant 1$, we find that

$$\sum_{k=1}^{n} |\tilde{f}(e_k)| = \sum_{k=1}^{n} \frac{\tilde{f}(e_k)}{|\tilde{f}(e_k)|} \tilde{f}(e_k) = \tilde{f}(x^{(n)}) < \|\tilde{f}\|.$$

But this implies (7), since n can be made arbitrarily large.

Whether or not the original space E is complete, we have

THEOREM 1. *The conjugate space* $(E^*, \|\cdot\|)$ *is complete.*

Proof. Let $\{f_n\}$ be a fundamental sequence of functionals in E^*. Then, given any $\varepsilon > 0$, there is an integer N such that $n, n' > N$ implies

$$\|f_n - f_{n'}\| < \varepsilon,$$

so that

$$|f_n(x) - f_{n'}(x)| \leqslant \|f_n - f_{n'}\| \|x\| < \varepsilon \|x\|$$

for every $x \in E$. Therefore the sequence $\{f_n(x)\}$ is fundamental and hence convergent for every $x \in E$. Let

$$f(x) = \lim_{n \to \infty} f_n(x).$$

Then f is linear, since

$$f(\alpha x + \beta y) = \lim_{n \to \infty} f_n(\alpha x + \beta y)$$

$$= \lim_{n \to \infty} [\alpha f_n(x) + \beta f_n(y)] = \alpha f(x) + \beta f(y).$$

Moreover, choosing n so large that $\|f_n - f_{n+p}\| < 1$ for all $p \geqslant 0$, we have $\|f_{n+p}\| < \|f_n\| + 1$ for all $p \geqslant 0$, and hence

$$|f_{n+p}(x)| \leqslant (\|f_n\| + 1) \|x\|.$$

It follows that

$$\lim_{p \to \infty} |f_{n+p}(x)| = |f(x)| \leqslant (\|f_n\| + 1) \|x\|,$$

so that f is bounded and hence continuous.

To complete the proof, we now show that the functional f is the limit of the sequence $\{f_n\}$, i.e., that

$$\lim_{n \to \infty} \|f_n - f\| = 0. \tag{8}$$

Given any $\varepsilon > 0$, let n be so large that

$$\|f_n - f_{n+p}\| < \frac{\varepsilon}{3} \tag{9}$$

for all $p \geqslant 0$. By the definition of the norm in E^*, there is a nonzero element $x_{n,\varepsilon} \in E$ such that

$$\|f_n - f\| \leqslant \frac{|f_n(x_{n,\varepsilon}) - f(x_{n,\varepsilon})|}{\|x_{n,\varepsilon}\|} + \frac{\varepsilon}{3} = |f_n(u_{n,\varepsilon}) - f(u_{n,\varepsilon})| + \frac{\varepsilon}{3},$$

where

$$u_{n,\varepsilon} = \frac{x_{n,\varepsilon}}{\|x_{n,\varepsilon}\|}.$$

Therefore

$$\|f_n - f\| \leqslant |f_n(u_{n,\varepsilon}) - f_{n+p}(u_{n,\varepsilon})| + |f_{n+p}(u_{n,\varepsilon}) - f(u_{n,\varepsilon})| + \frac{\varepsilon}{3}$$

$$\leqslant \|f_n - f_{n+p}\| \, \|u_{n,\varepsilon}\| + |f_{n+p}(u_{n,\varepsilon}) - f(u_{n,\varepsilon})| + \frac{\varepsilon}{3},$$

or

$$\|f_n - f\| \leqslant |f_{n+p}(u_{n,\varepsilon}) - f(u_{n,\varepsilon})| + \frac{2\varepsilon}{3} \tag{10}$$

after using (9) and the fact that $\|u_{n,\varepsilon}\| = 1$. But

$$\lim_{p \to \infty} f_{n+p}(u_{n,\varepsilon}) = f(u_{n,\varepsilon}),$$

by the very definition of f. Hence, taking the limit as $p \to \infty$ in (10), we get

$$\|f_n - f\| < \varepsilon,$$

which implies (8), since $\varepsilon > 0$ is arbitrary. ∎

Next we examine the structure of the space conjugate to a Hilbert space:

THEOREM 2. *Let H be a real Hilbert space. Then, given any $x_0 \in H$, the formula*

$$f(x) = (x, x_0) \qquad (x \in H) \tag{11}$$

defines a continous linear functional on H, with $\|f\| = \|x_0\|$. Conversely, given any continuous linear functional f on H, there is a unique element $x_0 \in H$ such that (11) holds, with $\|x_0\| = \|f\|$.

Proof. Given any $x_0 \in H$, formula (11) obviously defines a linear functional on H. By Schwarz's inequality,

$$|f(x)| = |(x, x_0)| \leqslant \|x\| \, \|x_0\|, \tag{12}$$

so that f is bounded and hence continuous. Moreover $\|f\| = \|x_0\|$, because of (12) and the fact that $f(x_0) = \|x_0\|^2$.

Conversely, let f be any continuous linear functional on H. If $f = 0$, then f obviously has the representation (11) with $x_0 = 0$ (in this case $\|x_0\| = \|f\| = 0$). Otherwise, let

$$H_0 = \{x : f(x) = 0\}$$

be the null space of f. Since f is continuous, H_0 is a *closed* subspace of H. According to Theorem 3, Corollary 2, p. 126, the codimension of the null space of any nontrivial linear functional f equals 1. Therefore, by Theorem 14, Corollary 2, p. 159, the orthogonal complement H_0' of the space H_0 is one-dimensional, i.e., there exists a nonzero vector y_0 orthogonal to H_0 such that every vector $x \in H$ has a unique representation of the form

$$x = y + \lambda y_0, \tag{13}$$

where $y \in H_0$. Clearly, there is no loss of generality in assuming that $\|y_0\| = 1$. Now let

$$x_0 = f(y_0)y_0. \tag{14}$$

Then, given any $x \in H$, we have

$$f(x) = f(y + \lambda y_0) = \lambda f(y_0)$$

because of (13), and

$$(x, x_0) = \lambda(y_0, x_0) = \lambda f(y_0)(y_0, y_0) = \lambda f(y_0)$$

because of (14). Therefore (11) holds for all $x \in H$. To prove the uniqueness of x_0, suppose

$$f(x) = (x, x_0') \qquad (x \in H). \tag{11'}$$

Then, subtracting (11') from (11), we get

$$(x, x_0 - x_0') = 0 \qquad (x \in H),$$

which immediately implies $x_0' = x_0$ after choosing $x = x_0 - x_0'$. ∎

COROLLARY. *The correspondence $x_0 \leftrightarrow f$ is an isomorphism between H and H^*, regarded as normed linear spaces.*

Proof. If

$$f(x) = (x, x_0), \qquad g(x) = (x, y_0),$$

then

$$\alpha f(x) + \beta g(x) = (x, \alpha x_0 + \beta y_0).$$

Moreover $\|x_0\| = \|f\|$. ∎

19.3. The strong topology in the conjugate space. Let E be a normed linear space. Then as we have seen, the conjugate space E^* is itself a normed linear space, and a neighborhood of zero in E^* means the set of all continuous linear functionals on E satisfying the condition $\|f\| < \varepsilon$ for some $\varepsilon > 0$. In other words, for a neighborhood base at zero in the space E^* we can take the set of all functionals in E^* such that $|f(x)| < \varepsilon$ when x ranges over the closed unit sphere $\|x\| \leqslant 1$ in the space E. Suppose E is a topological linear space, but not a normed linear space. Then in defining the topology in E^* it seems natural to start from an arbitrary bounded set $A \subset E$, since there is no longer a "unit sphere." This suggests

DEFINITION 2. *Let* E *be a topological linear space, with conjugate space* E^*. *Then by the* **strong topology**[5] *in* E^* *is meant the topology generated by the neighborhood base at zero consisting of all sets of the form*

$$U_{A,\varepsilon} = \{f \colon |f(x)| < \varepsilon \text{ for all } x \in A\} \tag{15}$$

for some number $\varepsilon > 0$ *and bounded set* $A \subset E$.[6]

Regardless of the topology in the original set E, we have

THEOREM 3. *The conjugate space* E^*, *equipped with the strong topology, is a locally convex* T_1-*space*.

Proof. If $f_0 \in E^*$ and $f_0 \neq 0$, then there is an element $x_0 \in E$ such that $f_0(x_0) \neq 0$. Let

$$\varepsilon = \tfrac{1}{2}|f(x_0)|, \qquad A = \{x_0\}.$$

Then clearly $f_0 \notin U_{A,\varepsilon}$, and hence E^* is a T_1-space. To verify that the strong topology in E^* is locally convex, we need only note that $U_{A,\varepsilon}$ is a convex set in E^* for any $\varepsilon > 0$ and any bounded set $A \subset E$. ∎

Remark. The strong topology in E^* will be denoted by the symbol b. In cases where we want to emphasize that E^* is equipped with the strong topology, we will write (E^*, b) instead of E^*.

19.4. The second conjugate space. Since the set of all continuous linear functionals on a topological linear space E is itself a topological linear space, namely the conjugate space (E^*, b), we can also talk about the *second conjugate space* $E^{**} = (E^*)^*$, i.e., set of all continuous linear functionals on E^*, the *third conjugate space* $E^{***} = (E^{**})^*$, and so on.

THEOREM 4. *Given a topological linear space* E *with conjugate space* E^*, *let* x_0 *be any fixed element of* E. *Then*

$$\psi_{x_0}(f) = f(x_0)$$

---

[5] As opposed to the weak topology in E^*, to be discussed in Sec. 20.3.
[6] See Problem 8.

is a continuous linear functional on E^*.

Proof. The linearity is obvious, since

$$\psi_{x_0}(\alpha f + \beta g) = \alpha f(x_0) + \beta g(x_0) = \alpha\psi_{x_0}(f) + \beta\psi_{x_0}(g) \qquad (f, g \in E^*).$$

As for the continuity, given any $\varepsilon > 0$, let A be a bounded subset of E containing x_0, and let $U_{A,\varepsilon}$ be the neighborhood (15). Then

$$|\psi_{x_0}(f)| = |f(x_0)| < \varepsilon \qquad \text{if } f \in U_{A,\varepsilon},$$

i.e., the functional ψ_{x_0} is continuous at 0 and hence continuous on the whole space E^*. ∎

Thus the mapping

$$\pi(x) = \psi_x(f),$$

called the *natural mapping of E into E^**, is a mapping of the whole space E onto some subset $\pi(E)$ of the second conjugate space E^{**}. Clearly π is linear, in the sense that

$$\pi(\alpha x + \beta y) = f(\alpha x + \beta y) = \alpha f(x) + \beta f(y) = \alpha\pi(x) + \beta\pi(y).$$

Suppose E has sufficiently many continuous linear functionals, e.g., suppose E is a normed linear space or a locally convex topological linear space satisfying the first axiom of separation.[7] Then π is one-to-one, since, given any two distinct elements $x_1, x_2 \in E$, there is a functional $f \in E^*$ such that $f(x_1) \neq f(x_2)$ and hence $\pi(x_1) \neq \pi(x_2)$. Being the conjugate space of (E^*, b), E^{**} can also be equipped with a strong topology (introduced by the obvious analogue of Definition 2), which we denote by b^*.

If $\pi(E) = E^{**}$, the space E is said to be *semireflexive*. It can be shown (see Problem 9) that the inverse mapping π^{-1} carrying $\pi(E)$ into E is always continuous. If E is semireflexive and if π (as well as π^{-1}) is continuous, the space E is said to be *reflexive* and π then establishes a homeomorphism between the space E and (E^{**}, b^*). In this case, each element $x \in E$ can be identified with the corresponding element $\pi(x) \in E^{**}$, and hence it is convenient to denote the value of a functional $f \in E^*$ at the point $x \in E$ by the more symmetric notation

$$f(x) = (f, x).$$

Thus (f, x) can be regarded as a functional on E for each fixed $f \in E^*$, and as a functional on E^* for each fixed $x \in E$ (in the latter case, x also acts like an element of E^{**}).

THEOREM 5. *If E is a normed linear space (so that in particular E^* and E^{**} are also normed linear spaces), then the natural mapping of E into E^{**} is an isometry.*

---

[7] Recall Problem 8, p. 183.

Proof. Given an element $x \in E$, let $\|x\|$ denote the norm of x in E and $\|x\|_2$ the norm of its image in E^{**}. We want to show that $\|x\| = \|x\|_2$. To this end, let f be any element of E^*. Then

$$|(f, x)| \leqslant \|f\| \, \|x\|,$$

i.e.,

$$\|x\| \geqslant \frac{|(f, x)|}{\|f\|} \qquad (f \neq 0),$$

and since the left-hand side is independent of f,

$$\|x\| \geqslant \sup_{f \in E^*} \frac{|(f, x)|}{\|f\|} = \|x\|_2. \tag{16}$$

On the other hand, by the Hahn-Banach theorem, for every $x_0 \in E$ there is a linear functional f_0 such that

$$|(f_0, x_0)| = \|f_0\| \, \|x_0\|. \tag{17}$$

In fact, to construct such a functional, we need only set $f_0(x) = \lambda$ for any element of the form λx_0, and then extend f_0 to a functional on the whole space E (without changing its norm). It follows from (17) that

$$\|x\|_2 = \sup_{f \in E^*} \frac{|(f, x)|}{\|f\|} \geqslant \|x\|. \tag{18}$$

Comparing (16) and (18), we get

$$\|x\| = \|x\|_2. \quad \blacksquare$$

COROLLARY. *The concepts of semireflexivity and reflexivity coincide for a normed linear space.*

Proof. If the natural mapping π is an isometry, then obviously both π and π^{-1} are continuous. $\quad \blacksquare$

Remark. According to Theorem 5, every normed linear space E is isometric to the linear manifold $\pi(E) \subset E^{**}$[8]. Identifying E with $\pi(E)$, we can assert that $E \subset E^{**}$ in general, and $E = E^{**}$ if E is reflexive (or semireflexive).

THEOREM 6. *Every reflexive normed linear space is complete.*

Proof. If E is reflexive, then $E = E^{**}$. But $E^{**} = (E^*)^*$ is complete, by Theorem 1, p. 187. $\quad \blacksquare$

---

[8] The set $\pi(E)$ need not be closed.

Example 1. Finite-dimensional Euclidean spaces and Hilbert space are the simplest examples of reflexive spaces (in fact, for such spaces $E = E^*$). This follows from Theorem 2 (cf. Problem 5).

Example 2. The space c_0 of all sequences $x = (x_1, \ldots, x_k, \ldots)$ converging to zero is an example of a complete nonreflexive space. In fact, as we saw in Example 2, p. 185, the conjugate space of c_0 is the space l_1 of all absolutely summable sequences, which in turn has the space m of all bounded sequences (not necessarily converging to zero) as its conjugate space (see Problem 2c).

Example 3. It can be shown that the space $C_{[a,b]}$ of all continuous functions on $[a, b]$ is nonreflexive, and even that there is *no* normed linear space with $C_{[a,b]}$ as its conjugate space.

Example 4. The space l_p, where $1 < p \neq 2$, is an example of a reflexive space which does not coincide with its conjugate space. In fact, $l_p^* = l_q$, where

$$\frac{1}{p} + \frac{1}{q} = 1,$$

and hence $l_p^{**} = l_q^* = l_p$.

Problem 1. Let E be Euclidean n-space (real or complex), and let $e_1, \ldots, e_n$ be a basis in E. Let $x_1, \ldots, x_n$ be the coordinates of a vector $x \in E$ with respect to the basis $e_1, \ldots, e_n$, and let $f^1, \ldots, f^n$ be the coordinates of a functional $f \in E^*$ with respect to the dual basis $f_1, \ldots, f_n$. Prove that in each of the following pairs, the norm in E^* is the norm "induced" by the corresponding norm in E:

a) $\|x\| = \left(\sum_{k=1}^{n} |x_k|^2\right)^{1/2}$, $\|f\| = \left(\sum_{k=1}^{n} |f^k|^2\right)^{1/2}$;

b) $\|x\| = \left(\sum_{k=1}^{n} |x_k|^p\right)^{1/p}$, $\|f\| = \left(\sum_{k=1}^{n} |f^k|^q\right)^{1/q}$

   where $\dfrac{1}{p} + \dfrac{1}{q} = 1$ $(p, q > 1)$;

c) $\|x\| = \sup_{1 \leqslant k \leqslant n} |x_k|$, $\|f\| = \sum_{k=1}^{n} |f^k|$;

d) $\|x\| = \sum_{k=1}^{n} x_k$, $\|f\| = \sup_{0 \leqslant k \leqslant n} |f^k|$.

Problem 2. Let l_p be the normed linear space of all sequences $x = (x_1, \ldots, x_k, \ldots)$ with norm

$$\|x\| = \left(\sum_{k=1}^{\infty} |x_k|^p\right)^{1/p} < \infty \qquad (p \geqslant 1).$$

Prove that

  a) If $p > 1$, the space l_p^* conjugate to l_p is isomorphic to the space l_q, where

$$\frac{1}{p} + \frac{1}{q} = 1;$$

  b) If $p > 1$, the general form of a continuous linear functional on l_p is

$$\tilde{f}(x) = \sum_{k=1}^{\infty} x_k f_k,$$

  where $x = (x_1, \ldots, x_k, \ldots) \in l_p, f = (f_1, \ldots, f_k, \ldots) \in l_q;$

  c) If $p = 1$, l_1^* is isomorphic to the space m of all bounded sequences $x = (x_1, \ldots, x_k, \ldots)$ with norm $\|x\| = \sup_k |x_k|$.

Problem 3. Let E be an incomplete normed linear space, with completion $\bar{E}$. Prove that the conjugate spaces E^* and $(\bar{E})^*$ are isomorphic.

Hint. Given any $f \in E^*$, extend f by continuity to a functional $\tilde{f} \in (\bar{E})^*$. Conversely, given any $\tilde{f} \in (\bar{E})^*$, let f be the restriction of $\tilde{f}$ to E, namely the functional $f(x) = \tilde{f}(x)$ for all $x \in E$. Show that $f \leftrightarrow \tilde{f}$ is the desired isomorphism (with $\|f\| = \|\tilde{f}\|$).

Problem 4. Let E be an incomplete Euclidean space with the Hilbert space H as its completion. Prove that E^* and H are isomorphic.

Problem 5. Particularize Theorem 2 to the case of a finite-dimensional Euclidean space.

Problem 6. Generalize Theorem 2 to the case of a complex Hilbert space.

Hint. Write $x_0 = \overline{f(y_0)}y_0$ instead of (14). The isomorphism of H and H^* associating the functional $f(x) = (x, x_0)$ with x_0 is then "conjugate-linear" in the sense that $\bar{\alpha}f$ is associated with αx_0.

Problem 7. Let Φ be the same countably normed space of "rapidly decreasing sequences" as in Problem 12c, p. 172. Find the conjugate space Φ^*.

Hint. Use Problem 6, p. 182.

Ans. Φ^* is the space of all functionals $\tilde{f}$ of the form

$$\tilde{f}(x) = \sum_{k=1}^{n} x_k f_k,$$

where $f = (f_1, \ldots, f_k, \ldots)$ is any sequence satisfying the condition

$$\sum_{k=1}^{\infty} k^{-n} f_k^2 < \infty$$

for some nonnegative integer n.

Problem 8. Let E, E^*, and $U_{A,\varepsilon}$ be the same as in Definition 2. Verify that the system $U_{A,\varepsilon}$ actually generates a topology b in E^* such that the linear operations in E^* are continuous with respect to b. Prove that if E is a normed linear space, then b coincides with the "norm topology" of Sec. 19.2.

Problem 9. Let E be a topological linear space, and let b^* be the strong topology in E^{**} and π the natural mapping of E into E^{**}. Prove that π^{-1} is continuous.

Hint. The topology b^* induces a topology $\pi^{-1}(b^*)$ in the space E, in which a set $G \subset E$ is said to be open if its image $\pi(G)$ is the intersection of $\pi(E)$ with an open subset of (E^{**}, b^*). Show that $\pi^{-1}(b^*)$ is stronger than the original topology in E.

Problem 10. Prove that every closed subspace of a reflexive space is itself reflexive.

20. The Weak Topology and Weak Convergence

20.1. The weak topology in a topological linear space. Let E be a topological linear space, with conjugate space E^*. Given any $\varepsilon > 0$ and any finite set of continuous linear functionals $f_1, \ldots, f_r \in E^*$, the set

$$U = U_{f_1, \ldots, f_n; \varepsilon} = \{x : |f_1(x)| < \varepsilon, \ldots, |f_r(x)| < \varepsilon\} \tag{1}$$

is open in E and contains the point zero, i.e., U is a neighborhood of zero. Let $\mathcal{N}_0$ be the system of all sets of the form (1). Then $\mathcal{N}_0$ is a neighborhood base at zero, generating a topology in E which is again the topology of a topological linear space (the details are left as an exercise). This topology is called the *weak topology* in E. Every subset of E which is open in the weak topology is also open in the original topology of E, but the converse may not be true, i.e., $\mathcal{N}_0$ may not be a neighborhood base at zero for the original topology in E. In other words, the weak topology is weaker (as defined on p. 80) than the original topology, as anticipated by the terminology. Clearly, the weak topology in E is the weakest topology τ with the property that every linear functional continuous with respect to the original topology is also continuous with respect to τ.

20.2. Weak convergence. The weak topology in E may not satisfy the first axiom of countability, even in the case where E is a normed linear space. Hence the weak topology cannot in general be described in the language of convergent sequences. Nevertheless, the weak topology determines an important kind of convergence in E, called *weak convergence*. By contrast, the convergence in E determined by the original topology (by the norm, if E is a normed linear space) is called *strong convergence*.

THEOREM 1. *A sequence $\{x_n\}$ of elements in a topological linear space E is weakly convergent to an element $x_0 \in E$ if and only if the numerical sequence $\{f(x_n)\}$ converges to $f(x_0)$ for every $f \in E^*$, i.e., for every continuous linear functional f on E.*

Proof. Clearly, there is no loss of generality in assuming that $x_0 = 0$. Suppose $f(x_n) \to 0$ for every $f \in E^*$. Then, given any "weak neighborhood" (1), let N_i be such that $|f_i(x_n)| < \varepsilon$ for all $n > N_i$ ($i = 1, \ldots, r$), and let $N = \max \{N_1, \ldots, N_r\}$. Then $x_n \in U$ for all $n > N$, i.e., $\{x_n\}$ converges to x_0 in the weak topology.

Conversely, suppose that for each neighborhood (1), there is an integer $N = N(U)$ such that $x_n \in U$ for all $n > N$. Then obviously $f(x_n) \to 0$ for any given $f \in E^*$, as we see by choosing f to be one of the functionals $f_1, \ldots, f_r$ figuring in the definition of U. ∎

Specializing to the case where E is a normed linear space, we have

THEOREM 2. *Let $\{x_n\}$ be a weakly convergent sequence of elements in a normed linear space E. Then $\{x_n\}$ is bounded, i.e., there is a constant C such that*

$$\|x_n\| \leqslant C \qquad (n = 1, 2, \ldots).$$

Proof. Suppose $\{x_n\}$ is unbounded. Then $\{x_n\}$ is unbounded on every closed sphere

$$S[f_0, \varepsilon] = \{f : \|f - f_0\| \leqslant \varepsilon\}$$

in E^*, in the sense that the set of numbers

$$\{(f, x_n) : f \in S[f_0, \varepsilon], n = 1, 2, \ldots\}$$

is unbounded for every $S[f_0, \varepsilon] \subset E^*$. In fact, if the sequence $\{x_n\}$ is bounded on $S[f_0, \varepsilon]$, then it is also bounded on the sphere

$$S[0, \varepsilon] = \{g : \|g\| \leqslant \varepsilon\},$$

since if $g \in S[0, \varepsilon]$, then

$$f_0 + g \in S[f_0, \varepsilon],$$

$$(g, x_n) = (f_0 + g, x_n) - (f_0, x_n),$$

where the numbers (f_0, x_n) are bounded, by the weak convergence of $\{x_n\}$. But if $|(g, x_n)| \leqslant C$ for all $g \in S[0, \varepsilon]$, then, by the isometry of the natural mapping of E into E^{**},

$$\|x_n\| = \sup_{\|g\| \leqslant 1} |(g, x_n)| = \frac{1}{\varepsilon} \sup_{\|g\| \leqslant \varepsilon} |(g, x_n)| < \frac{C}{\varepsilon} \qquad (n = 1, 2, \ldots),$$

so that $\{x_n\}$ is unbounded, contrary to assumption. It follows that if $\{x_n\}$ is unbounded, then $\{x_n\}$ is unbounded on every closed sphere in E^*.

Next, choosing any closed sphere $S_0 \subset E^*$, we find an integer n_1 and an element $f \in S_0$ such that

$$|(f, x_{n_1})| > 1. \qquad (2)$$

Since (f, x) depends continuously on x, the inequality (2) holds for all f belonging to some closed sphere $S_1 \subset S_0$. Repeating this argument, we find an integer n_2 and a closed sphere $S_2 \subset S_1$ such that

$$|(f, x_{n_2})| > 2$$

for all $f \in S_2$, and so on, where in general there is an integer n_k and a closed sphere $S_k \subset S_{k-1}$ such that

$$|(f, x_{n_k})| > k$$

for all $f \in S_k$. At the same time, we can obviously see to it that the radius of the sphere S_k approaches zero as $k \to \infty$. Since E^* is complete, by Theorem 1, p. 187, it follows from the nested sphere theorem (Theorem 2, p. 60) that there is an element $\tilde{f}$ contained in all the spheres S_k. But then

$$|(\tilde{f}, x_{n_k})| > k \qquad (k = 1, 2, \ldots),$$

contrary to the assumed weak convergence of the sequence $\{x_n\}$. ∎

COROLLARY 1. *Let $\{x_n\}$ be a sequence of elements in a normed linear space E such that the numerical sequence $\{(f, x_n)\}$ is bounded for every $f \in E^*$. Then $\{x_n\}$ is bounded.*

Proof. In proving Theorem 2, the weak convergence of $\{x_n\}$ was invoked only to infer the boundedness of the sequence $\{(f_0, x_n)\}$. ∎

Generalizing Corollary 1, we get

COROLLARY 2. *Let M be a weakly bounded subset of a normed linear space E, i.e., a subset bounded in the weak topology. Then M is strongly bounded, i.e., M is contained in some closed sphere.*

Proof. Suppose M contains a sequence $\{x_n\}$ such that $\|x_n\| \to \infty$, and let M' be the set of all points x_n $(n = 1, 2, \ldots)$. Since M is weakly bounded, so is M'. This means that M' is "absorbed" by any weak neighborhood of zero, in particular by any neighborhood

$$U = \{x : |(f, x)| < 1, f \in E^*\},$$

in the sense that there is a number $\alpha > 0$ such that $M' \subset \alpha U$. But then $|(f, x_n)| < \alpha$ for all n, which, by Corollary 1, contradicts the assumption that $\|x_n\| \to 0$. ∎

COROLLARY 3. *A necessary and sufficient condition for a subset M of a normed linear space E to be (strongly) bounded is that every continuous linear functional f ∈ E* be bounded on M.*

Proof. The necessity follows at once from the inequality

$$|(f, x)| \leqslant \|f\| \, \|x\|,$$

while the sufficiency is an immediate consequence of Corollary 2 and the meaning of weak boundedness. ∎

A useful test for weak convergence of a sequence is given by

THEOREM 3. *A bounded sequence $\{x_n\}$ of elements in a normed linear space E is weakly convergent to an element $x \in E$ if $f(x_n) \to f(x)$ for every $f \in \Delta$, where Δ is any set whose linear hull is everywhere dense in E*.*

Proof. Let φ be an arbitrary element of E^*, and let $\{\varphi_k\}$ be a sequence of linear combinations of elements of Δ converging to φ (such a sequence exists, since Δ is everywhere dense in E^*). Let C be such that

$$\|x\| \leqslant C, \qquad \|x_n\| \leqslant C \qquad (n = 1, 2, \ldots).$$

Moreover, given any $\varepsilon > 0$, choose k so large that $\|\varphi - \varphi_k\| < \varepsilon$ (this is possible, since $\varphi_k \to \varphi$). Then

$$
\begin{aligned}
|\varphi(x_n) - \varphi(x)| &\leqslant |\varphi(x_n) - \varphi_k(x_n)| + |\varphi_k(x_n) - \varphi_k(x)| \\
&\quad + |\varphi_k(x) - \varphi(x)| \\
&\leqslant C\varepsilon + C\varepsilon + |\varphi_k(x_n) - \varphi_k(x)|.
\end{aligned}
\tag{3}
$$

But $\varphi_k(x_n) \to \varphi_k(x)$ as $n \to \infty$, since φ_k is a linear combination of elements of Δ, and $f(x_n) \to f(x)$ for every $f \in \Delta$, by hypothesis. Therefore we can make the right-hand side of (3) as small as we please, by choosing ε sufficiently small and n sufficiently large. It follows that $\varphi(x_n) \to \varphi(x)$ for every $\varphi \in E^*$, i.e., $\{x_n\}$ converges weakly to x. ∎

The meaning of weak convergence in various spaces is illustrated by the following examples:

Example 1. Given a finite-dimensional Euclidean space R^n, let $e_1, \ldots, e_n$ be any orthonormal basis in R^n, and let $\{x^{(k)}\}$ be a sequence in R^n converging weakly to a vector $x = (x_1, \ldots, x_n) \in R^n$. Then

$$(x^{(k)}, e_j) = x_j^{(k)} \to (x, e_j) = x_j \qquad (j = 1, \ldots, n),$$

i.e., for every j the sequence $\{x^{(k)}\}$ of components of the vectors $x^{(k)}$ converges to the corresponding component of the limit vector x. But then

$$\rho(x^{(k)}, x) = \sqrt{\sum_{k=1}^{n} (x_j^{(k)} - x_j)^2} \to 0$$

as $k \to \infty$, so that $\{x^{(k)}\}$ converges strongly to x. On the other hand, strong convergence obviously implies weak convergence in any space. Thus we see that *weak convergence and strong convergence are equivalent concepts in* R^n.

Example 2. Let $\{x^{(k)}\}$ be a (strongly) bounded sequence of elements of l_2. Then $\{x^{(k)}\}$ converges weakly to an element $x \in l_2$ if

$$(x^{(k)}, e_j) = x_j^{(k)} \to (x, e_j) = x_j \qquad (j = 1, 2, \ldots),$$

where

$$e_1 = (1, 0, 0, \ldots), \qquad e_2 = (0, 1, 0, \ldots), \ldots$$

is an orthonormal basis in l_2. This follows from Theorem 3, since linear combinations of the elements $e_1, e_2, \ldots$ are everywhere dense in l_2, which coincides with its own conjugate space (recall Problem 2a, p. 194). Thus weak convergence in l_2 has the same interpretation in terms of components as in R^n, i.e., for every j the sequence $\{x_j^{(k)}\}$ of components of the vectors $x^{(k)}$ converges to the corresponding component of the limit vector x. However, the concepts of weak convergence and strong convergence no longer coincide in l_2. In fact, although obviously not strongly convergent, the sequence of basis vectors $\{e_k\}$ converges weakly to zero. To see this, we note that by Theorem 2, p. 188, every continuous linear functional f on l_2 can be written as a scalar product

$$f(x) = (x, a)$$

of a variable vector $x \in l_2$ with a fixed vector $a = (a_1, \ldots, a_n, \ldots) \in l_2$, so that in particular

$$f(e_k) = a_k.$$

But $a_k \to 0$ as $k \to \infty$ for every $a \in l_2$, and hence $f(e_k) \to 0 = f(0)$.

Example 3. Consider the space $C_{[a,b]}$ of all functions continuous on $[a, b]$, and let $\{x_n(t)\}$ be a sequence of functions in $C_{[a,b]}$ converging weakly to a function $x(t) \in C_{[a,b]}$. Among the continuous linear functionals on $C_{[a,b]}$, we have the functionals $\delta_{t_0}, a \leqslant t_0 \leqslant b$ (see Example 5, p. 179), where δ_{t_0} assigns to each function $x(t) \in C_{[a,b]}$ its value at the fixed point t_0. Clearly,

$$\delta_{t_0}(x_n) \to \delta_{t_0}(x)$$

means that

$$x_n(t_0) \to x(t_0).$$

Hence, if the sequence $\{x_n(t)\}$ is weakly convergent, then

1) $\{x_n(t)\}$ is *uniformly bounded* on $[a, b]$, i.e., there is a constant C such that $|x_n(t)| \leqslant C$ for all $n = 1, 2, \ldots$ and all $t \in [a, b]$;[9]
2) $\{x_n(t)\}$ is *pointwise convergent* on $[a, b]$, i.e., $\{x_n(t)\}$ is a convergent numerical sequence for every fixed $t \in [a, b]$.

---

[9] This follows from Theorem 2.

20.3. The weak topology and weak convergence in a conjugate space. Let E be a topological linear space, with conjugate space E^*. Suppose that in Definition 2, p. 190, we require A to be *finite* instead of *bounded*. Then the resulting topology, generated by the neighborhood base at zero consisting of all sets of the form

$$U_{A,\varepsilon} = \{f : |f(x)| < \varepsilon \text{ for all } A\} \tag{4}$$

for some number $\varepsilon > 0$ and finite set $A \subset E$, is called the *weak topology* in E^* instead of the *strong topology*. Clearly, the set (4) can also be written as

$$U_{x_1, \ldots, x_n; \varepsilon} = U_{A,\varepsilon} = \{f : |f(x_1)| < \varepsilon, \ldots, |f(x_n)| < \varepsilon\} \tag{4'}$$

for some $\varepsilon > 0$ and points $x_1, \ldots, x_n \in E$. Since every finite set $A \subset E$ is bounded, while in general there are bounded infinite sets in E, the weak topology in E^* is in fact weaker than the strong topology in E^* (and in general does not coincide with the strong topology).

The weak topology in E^* determines a kind of convergence in E^*, called *weak convergence (of functionals)*. Weak convergence of functionals plays an important role in many problems of functional analysis, in particular in the theory of generalized functions (to be discussed in the next section). Obviously, a sequence $\{f_n\}$ of functionals $f_n \in E^*$ is weakly convergent to a functional $f \in E^*$ if and only if $\{f_n(x)\}$ converges to $f(x)$ for every $x \in E$.

For weakly convergent sequences of functionals, we have the following analogues of Theorems 2 and 3:

THEOREM 2'. *Let $\{f_n\}$ be a weakly convergent sequence of continuous linear functionals on a Banach space E. Then $\{f_n\}$ is bounded, i.e., there is a constant C such that*

$$\|f_n\| < C \qquad (n = 1, 2, \ldots).$$

Proof. The proof is the exact analogue of that of Theorem 2. Note that this time we must specify that E is a *complete* normed linear space (i.e., a Banach space). ∎

THEOREM 3'. *A bounded sequence $\{f_n\}$ of continuous linear functionals on a Banach space E is weakly convergent to a functional $f \in E^*$ if $f_n(x) \to f(x)$ for every $x \in \Delta$, where Δ is any set whose linear hull is everywhere dense in E.*

Proof. The exact analogue of the proof of Theorem 3. ∎

Example. Let E be the space $C_{[a,b]}$ of all functions continuous on $[a, b]$, and consider the functional

$$\delta_{t_0}(x) = x(t_0), \tag{5}$$

as in Example 3 above. For simplicity (and without loss of generality), we assume that $t_0 = 0 \in (a, b)$, so that (5) becomes

$$\delta_0(x) = x(0). \tag{6}$$

Let $\{f_n(t)\}$ be a sequence of functions continuous on $[a, b]$ such that[10]

1) $f_n(t)$ is positive if $|t| < \dfrac{1}{n}$ and zero if $|t| \geqslant \dfrac{1}{n}$,

2) $\displaystyle\int_a^b f_n(t)\, dt = 1$ for all $n = 1, 2, \ldots,$

and let

$$\delta_0^{(n)}(x) = \int_a^b f_n(t)x(t)\, dt.$$

Then $\delta_0^{(n)}$ is a continuous linear functional on $C_{[a,b]}$ (recall Example 4, p. 179). Moreover, given any function $x(t) \in C_{[a,b]}$, we have

$$\delta_0^{(n)}(x) = \int_a^b f_n(t)x(t)\, dt = \int_{-1/n}^{1/n} f_n(t)x(t)\, dt = x(\tau)\int_{-1/n}^{1/n} f_n(t)\, dt = x(\tau)$$

for some $\tau \in [-1/n, 1/n]$, by the mean value theorem for integrals, and hence

$$\delta_0^{(n)}(x) \to x(0) = \delta_0(x) \tag{7}$$

as $n \to \infty$. Thus the sequence of functionals $\{\delta_0^{(n)}\}$ converges weakly to the functional δ_0. Suppose we write (6) in the form

$$\delta_0(x) = \int_a^b \delta(t)x(t)\, dt,$$

in terms of the "delta function" $\delta(t)$, as in Example 3, p. 124. Then, loosely speaking, (7) says that "the generalized function $\delta(t)$ is the weak limit of the sequence of ordinary functions $f_n(t)$."

20.4. The weak* topology. There are two ways of regarding the space E^* of continuous linear functionals on a given space E, either as the space conjugate to the original space E, or else as an "original space" in its own right, with conjugate space E^{**}. Correspondingly, there are two ways of introducing a weak topology into E^*, either by using neighborhoods of the form (4'), or else by using the values of functionals in E^{**} on the space E^*, as in Sec. 20.1. Clearly, the two topologies will be the same if and only if E is reflexive (why?). Suppose E is nonreflexive. Then, to avoid confusion, the weak topology determined in E^* with the aid of E^{**} will be called simply the *weak topology*, while the topology determined in E^* with the aid of E

---

[10] As an exercise, give an explicit example of such a sequence $\{f_n(t)\}$.

will be called the *weak* topology*.[11] Clearly, the weak* topology in E^* is weaker than the weak topology in E^*, i.e., the weak* topology has fewer open sets than the weak topology. Note that weak convergence as defined in Sec. 20.3 now means weak* convergence.

The following theorem is important in various applications of the concept of weak convergence of functionals:

THEOREM 4. *Every bounded sequence* $\{f_n\}$ *of functionals in the space* E^* *conjugate to a separable normed linear space* E *contains a weakly* convergent subsequence.*

Proof. Since E is separable, there is a countable set of points $x_1, x_2, \ldots, x_n, \ldots$ everywhere dense in E. Suppose the sequence $\{f_n\}$ of functionals in E^*, i.e., continuous linear functionals on E, is bounded (in norm). Then the numerical sequence

$$f_1(x_1), f_2(x_1), \ldots, f_n(x_1), \ldots$$

is bounded, and hence, by the Bolzano-Weierstrass theorem (see p. 101), $\{f_n\}$ contains a subsequence

$$f_1^{(1)}, f_2^{(1)}, \ldots, f_n^{(1)}, \ldots$$

such that the numerical sequence

$$f_1^{(1)}(x_1), f_2^{(1)}(x_1), \ldots, f_n^{(1)}(x_1), \ldots$$

converges. By the same token, the subsequence $\{f_n^{(1)}\}$ in turn contains a subsequence

$$f_1^{(2)}, f_2^{(2)}, \ldots, f_n^{(2)}, \ldots$$

such that the sequence

$$f_1^{(2)}(x_2), f_2^{(2)}(x_2), \ldots, f_n^{(2)}(x_2), \ldots$$

converges. Continuing this construction, we get a system of subsequences $\{f_n^{(k)}\}$, $k = 1, 2, \ldots$ such that

1) $\{f_n^{(k+1)}\}$ is a subsequence of $\{f_n^{(k)}\}$ for all $k = 1, 2, \ldots$;

2) $\{f_n^{(k)}\}$ converges at the points $x_1, x_2, \ldots, x_k$.

Hence, taking the "diagonal sequence"

$$f_1^{(1)}, f_2^{(2)}, \ldots, f_n^{(n)}, \ldots,$$

we get a sequence of continuous linear functionals on E such that

$$f_1^{(1)}(x_n), f_2^{(2)}(x_n), \ldots$$

---

[11] Read "weak*" as "weak star."

converges for all n. But then, by Theorem 3′, the sequence

$$f_1^{(1)}(x), f_2^{(2)}(x), \ldots$$

converges for all $x \in E$. ∎

COROLLARY 1. *Every bounded set in the space E^* conjugate to a separable normed linear space E is relatively countably compact in the weak* topology.*

Proof. An immediate consequence of Theorem 4 and the meaning of relative countable compactness (see Sec. 10.4). ∎

COROLLARY 2. *A subset of the space E^* conjugate to a separable Banach space E is bounded if and only if it is relatively countably compact in the weak* topology.*

Proof. An immediate consequence of Theorem 2′ and Corollary 1. ∎

As we will see in a moment, the word "countably" is superfluous in Corollaries 1 and 2. First we need

THEOREM 5. *Given a separable normed linear space E, let S be the closed unit sphere in E and S^* the closed unit sphere in the conjugate space E^*. Then the topology induced in S^* by the weak* topology in E^* is the same as that induced by the metric*

$$\rho(f, g) = \sum_{n=1}^{\infty} 2^{-n} |(f - g, x_n)|,$$

where $\{x_1, \ldots, x_n, \ldots\}$ is any countable set everywhere dense in S.

Proof. Clearly, $\rho(f, g)$ has all the properties of a metric, and moreover is invariant under shifts, in the sense that

$$\rho(f + h, g + h) = \rho(f, g).$$

Hence we need only verify that

1) Every "open sphere"

$$Q_\varepsilon = \{f : \rho(f, 0) < \varepsilon\}$$

contains the intersection of S^* with some weak neighborhood of zero in E^*;

2) Every weak neighborhood of zero in E^* contains the intersection of S with some Q_ε.

Let N be such that $2^{-N} < \varepsilon/2$, and consider the weak neighborhood of zero

$$U = U_{x_1, \ldots, x_N; \varepsilon/2} = \left\{ f : |(f, x_1)| < \frac{\varepsilon}{2}, \ldots, |(f, x_N)| < \frac{\varepsilon}{2} \right\}.$$

Then $f \in S^* \cap U$ implies

$$\rho(f, 0) = \sum_{n=1}^{N} 2^{-n} |(f, x_n)| + \sum_{n=N+1}^{\infty} 2^{-n} |(f, x_n)|$$

$$\leqslant \frac{\varepsilon}{2} \sum_{n=1}^{N} 2^{-n} + \sum_{n=N+1}^{\infty} 2^{-n} < \varepsilon,$$

and hence $S^* \cap U \subset Q_\varepsilon$. This proves 1).

To prove 2), this time let

$$U = U_{y_1, \ldots, y_m; \delta} = \{ f : |(f, y_1)| < \delta, \ldots, |(f, y_m)| < \delta \}$$

be any weak neighborhood of zero in E^*, where it can clearly be assumed that $\|y_1\| < 1, \ldots, \|y_m\| < 1$. Since $\{x_1, \ldots, x_n, \ldots\}$ is everywhere dense in S, there are indices $n_1, \ldots, n_m$ such that

$$\|y_k - x_{n_k}\| < \frac{\delta}{2} \qquad (k = 1, \ldots, m).$$

Let

$$N = \max \{ n_1, \ldots, n_m \}, \qquad \varepsilon = \frac{\delta}{2^{N+1}}.$$

Then $f \in S^* \cap Q_\varepsilon$ implies

$$\sum_{n=1}^{\infty} 2^{-n} |(f, x_n)| < \varepsilon$$

and hence

$$|(f, x_n)| < 2^n \varepsilon,$$

in particular

$$|(f, x_{n_k})| < 2^{n_k} \varepsilon \leqslant 2^N \varepsilon = \frac{\delta}{2}.$$

Therefore $f \in S^* \cap Q_\varepsilon$ implies

$$|(f, y_k)| \leqslant |(f, x_{n_k})| + |(f, y_k - x_{n_k})| < \frac{\delta}{2} + \|f\| \, \|y_k - x_{n_k}\| < \delta,$$

so that $S^* \cap Q_\varepsilon \subset U$. ∎

We can now drop the word "countably" in Corollaries 1 and 2:

COROLLARY 1'. *Every bounded set in the space E^* conjugate to a separable normed linear space E is relatively compact in the weak* topology.*

Proof. Use Theorem 5 and the fact that compactness and countable compactness are equivalent concepts in a metric space (see Sec. 11.2.). ∎

COROLLARY 2'. *A subset of the space E^* conjugate to a separable Banach space E is bounded if and only if it is relatively compact in the weak* topology.*

Proof. Identical with that of Corollary 1'. ∎

Finally we prove

THEOREM 6. *Every closed sphere in the space (E^*, b) conjugate to a separable normed linear space E is compact in the weak* topology.*

Proof. Every closed sphere in the space (E^*, b) is closed in the weak* topology. In fact, since a shift in E^* carries every closed set (in the weak* topology) into another closed set, we need only prove the assertion for every sphere of the form

$$S_c = \{f : \|f\| \leqslant c\}.$$

Suppose $f_0 \notin S_c$. Then, by the definition of the norm of the functional f_0, there is an element $x \in E$ such that $\|x\| = 1$ and

$$f_0(x) = \alpha > c.$$

But then the set

$$U = \{f : f(x) > \tfrac{1}{2}(\alpha + c)\}$$

is a weak* neighborhood of f_0 containing no elements of S_c. Therefore S_c is closed in the weak* topology, and hence compact in the weak* topology, by Corollary 1'. ∎

Remark. Theorem 6 is a special case of the following more general theorem, which will not be proved here: *Every bounded subset of the space (E^*, b) conjugate to a locally convex topological linear space E is relatively compact in the weak* topology.*

Problem 1. Given a topological linear space E, suppose E has sufficiently many continuous linear functionals. Prove that E is a Hausdorff space, when equipped with the weak topology.

Problem 2. Let $\{x_n\}$ be a sequence of elements in a Hilbert space H such that

1) $\{x_n\}$ converges weakly to an element $x \in H$;
2) $\|x_n\| \to \|x\|$ as $n \to \infty$.

Prove that $\{x_n\}$ converges strongly to x, i.e., $\|x_n - x\| \to 0$ as $n \to \infty$.

Problem 3. Prove that the conclusion of the preceding problem remains valid if the condition 2) is replaced by either of the following conditions:

2') $\|x_n\| \leqslant \|x\|$ for all n;

2") $\varlimsup_{n \to \infty} \|x_n\| \leqslant \|x\|$.

Problem 4. Let H be a (separable) Hilbert space and M a bounded subset of H. Prove that the topology in M induced by the weak topology in H can be specified by a metric.

Problem 5. Prove that every closed convex subset of a Hilbert space H is closed in the weak topology (so that, in particular, every closed linear subspace of H is weakly closed). Give an example of a closed set in H which is not weakly closed.

Problem 6. Show that the two conditions in Example 3, p. 199 are sufficient as well as necessary for weak convergence of a sequence $\{x_n(t)\}$ in $C_{[a,b]}$. Give an example of a weakly convergent sequence in $C_{[a,b]}$ which is not strongly convergent.

21. Generalized Functions

21.1. Preliminary remarks. The degree of generality attaching to the notion of "function" varies from problem to problem. Some problems involve continuous functions, others involve functions differentiable one or more times, and so on. However, there are a number of situations in which the classical notion of a function turns out to be inadequate, even when understood in the most general sense (i.e., as an arbitrary rule f assigning a number $f(x)$ to each element x in the domain of definition of f). Here are two such cases:

1) A linear mass distribution can be conveniently characterized by giving the density of the distribution. However, no "ordinary" function can specify the density corresponding to one or more points with positive mass.

2) In many problems, situations arise in which various mathematical operations cannot be carried out. For example, a function with no derivative (at certain, possibly all, points) cannot be differentiated if the derivative is interpreted in the usual way, as an "ordinary" function. Of course, such difficulties can be avoided without relinquishing classical definitions, by suitably restricting the class of "admissible functions," for example, by considering only analytic functions. However, restricting the class of admissible functions in

this way is often quite undesirable. Fortunately, it turns out that difficulties of this kind can be overcome, and just as successfully at that, by *enlarging* (rather than restricting) the class of admissible functions, i.e., by introducing the notion of a "generalized function," not encountered in classical analysis. In doing so, a key role will be played by the concept of a conjugate space, considered earlier in this chapter.

Remark. It cannot be emphasized too strongly that the introduction of generalized functions is motivated by the need to solve perfectly concrete problems of analysis, and not merely by a desire to see how far the notion of function can be pushed.

Before going into details, we indicate the basic idea behind the theory of generalized functions. Let f be a fixed function on the real line, integrable on every finite interval, and let φ be any continuous function vanishing outside some finite interval (such a function φ is said to be *finite*[12]). Suppose each φ is assigned the number

$$(f, \varphi) = \int_{-\infty}^{\infty} f(x)\varphi(x)\, dx, \tag{1}$$

involving the given function f, where the integration is in effect only over a finite interval, because of the finiteness of φ. In other words, the function f can be regarded as a functional (a linear functional, because of the basic properties of the integral) defined on some space K of finite functions. However, there are many other linear functionals on K besides functionals of the form (1). For example, by assigning each function φ its value at the point $x = 0$, we get a linear functional which cannot be represented in the form (1). In this sense, the functions f can be regarded as part of a much larger set, namely the set of all possible linear functionals on K. The space K of "test functions" φ can be chosen in various ways. For example, K might consist of all continuous finite functions, as above. However, as will soon be apparent, it makes sense to require the test functions to satisfy rather stringent smoothness conditions (besides being continuous and finite).

21.2. The test space and test functions. Generalized functions. Turning now to details, let K be the set of all finite functions φ on $(-\infty, \infty)$ with continuous derivatives of all orders (equivalently, the set of all infinitely differentiable functions), where every function $\varphi \in K$, being finite, vanishes outside some interval depending on the choice of φ. Clearly K is a linear

---

[12] Do not confuse the notion of a *finite* function (which vanishes outside some finite interval) with the notion of a *bounded* function (whose range is contained in some finite interval). Finite functions are often called "functions of finite (or compact) support."

space, when equipped with the usual operations of addition of functions and multiplication of functions by numbers. Although the space K is not normable, there is a natural way of introducing the notion of convergence in K:

DEFINITION 1. *A sequence $\{\varphi_n\}$ of functions in K is said to **converge to** a function $\varphi \in K$ if*
1) *There exists an interval outside which all the functions φ_n vanish;*
2) *The sequence $\{\varphi_n^{(k)}\}$ of derivatives of order k converges uniformly on this interval to $\varphi^{(k)}$ for every $k = 0, 1, 2, \ldots$.*[13]

The linear space K equipped with this notion of convergence is called the **test space** *(or* **fundamental space**)*, and the functions in K are called* **test functions** *(or* **fundamental functions**)*.*

DEFINITION 2. *Every continuous linear functional $T(\varphi)$ on the test space K is called a **generalized function** on $(-\infty, \infty)$, where continuity of $T(\varphi)$ means that $\varphi_n \to \varphi$ in K implies $T(\varphi_n) \to T(\varphi)$.*

Let $f(x)$ be a *locally integrable* function, i.e., a function integrable on every finite interval. Then $f(x)$ generates a generalized function via the expression

$$T_f(\varphi) = (f, \varphi) = \int_{-\infty}^{\infty} f(x)\varphi(x)\,dx, \tag{2}$$

which is clearly a continuous linear functional on K. Generalized functions of this type will be called *regular*, and all other generalized functions, i.e., those not representable in the form (2), will be called *singular*. The following are all examples of singular generalized functions:

Example 1. The "delta function"

$$T(\varphi) = \varphi(0) \tag{3}$$

is a continuous linear functional on K, i.e., a generalized function in the sense of Definition 2. This functional can be written in the form

$$T(\varphi) = \int_{-\infty}^{\infty} \delta(x)\varphi(x)\,dx, \tag{4}$$

where $\delta(x)$ is a "fictitious" function,[14] equal to zero everywhere except at $x = 0$ and such that

$$\int_{-\infty}^{\infty} \delta(x)\,dx = 1$$

---

[13] As always, $\varphi_n^{(0)} = \varphi_n$, $\varphi^{(0)} = \varphi$.

[14] The term "delta function" will be applied to both the generalized function $T(\varphi)$ and the fictitious function $\delta(x)$ generating $T(\varphi)$ via the representation (4).

(these properties are of course paradoxical), since then we have, purely formally,

$$T(\varphi) = \int_{-\infty}^{\infty} \delta(x)\varphi(x)\,dx = \varphi(0) \int_{-\infty}^{\infty} \delta(x)\,dx = \varphi(0).$$

The advantage of regarding the delta function as a functional on the test space K rather than on the space $C_{[a,b]}$ as in Example 3, p. 124 will soon be apparent.

Example 2. Generalizing (3) and (4), we can write the functional

$$T(\varphi) = \varphi(a) \tag{3'}$$

in the form

$$T(\varphi) = \int_{-\infty}^{\infty} \delta(x - a)\varphi(x)\,dx, \tag{4'}$$

in terms of the "shifted delta function" $\delta(x - a)$.

21.3. Operations on generalized functions. Addition of generalized functions and multiplication of generalized functions by numbers are defined in the same way as for linear functionals in general, i.e., by the obvious analogue of Definition 1, p. 183 (with φ and K playing the roles of x and E). In the case of regular generalized functions, these are just the operations associated with the corresponding operations for "ordinary" functions. More exactly, if

$$T_f(\varphi) = \int_{-\infty}^{\infty} f(x)\varphi(x)\,dx, \qquad T_g(\varphi) = \int_{-\infty}^{\infty} g(x)\varphi(x)\,dx,$$

where f and g are locally integrable and $\varphi \in K$, then clearly

$$(T_f + T_g)(\varphi) = T_f(\varphi) + T_g(\varphi) = T_{f+g}(\varphi)$$

and

$$(\alpha T_f)(\varphi) = \alpha T_f(\varphi) = T_{\alpha f}(\varphi)$$

for any number α.

DEFINITION 3. *A sequence of generalized functions* $\{T_n\}$ *is said to* **converge** *to a generalized function T if $T_n(\varphi) \to T(\varphi)$ for every $\varphi \in K$. The space of generalized functions equipped with this notion of convergence is denoted by K^*.*

Remark. In other words, convergence of generalized functions is just weak* convergence of continuous linear functionals on K.

We will often denote a generalized function by the symbol f, as if a representation of the form

$$(f, \varphi) = \int_{-\infty}^{\infty} f(x)\varphi(x)\,dx \tag{5}$$

existed, even in the case where the generalized function is singular. Let f be a regular generalized function, and let $\alpha = \alpha(x)$ be an infinitely differentiable "ordinary" function. Then (5) implies

$$(\alpha f, \varphi) = \int_{-\infty}^{\infty} \alpha(x) f(x) \varphi(x)\, dx$$
$$= \int_{-\infty}^{\infty} f(x) \alpha(x) \varphi(x)\, dx = (f, \alpha\varphi),$$

where $\alpha\varphi$ obviously belongs to K. Carrying this over to the singular case, we get

DEFINITION 4. *The product αf of an infinitely differentiable function α and a generalized function f is the functional defined by the formula*

$$(\alpha f, \varphi) = (f, \alpha\varphi). \tag{6}$$

Remark. It follows from (6) that the functional αf is linear and continuous, and hence itself a generalized function.

Again let T be a regular generalized function of the form

$$T(\varphi) = \int_{-\infty}^{\infty} f(x) \varphi(x)\, dx, \tag{5'}$$

and suppose the derivative f' exists and is locally integrable. Then it is natural to define the derivative of T as the functional

$$\frac{dT}{dx}(\varphi) = \int_{-\infty}^{\infty} f'(x) \varphi(x)\, dx. \tag{7}$$

Integrating (7) by parts and using the fact that every test function φ vanishes outside some finite interval, we find at once that

$$\frac{dT}{dx}(\varphi) = -\int_{-\infty}^{\infty} f(x) \varphi'(x)\, dx, \tag{8}$$

thereby obtaining an expression for dT/dx which does not involve the derivative of f. Carrying this over to the singular case, we get

DEFINITION 5. *The derivative dT/dx of a generalized function T is the functional defined by the formula*

$$\frac{dT}{dx}(\varphi) = -T(\varphi'). \tag{9}$$

Remark 1. The functional (9) is obviously linear and continuous, and hence itself a generalized function. Second, third and higher-order derivatives are defined in the same way.

Remark 2. If a generalized function is denoted by the symbol f, as in (6), then its derivative is denoted by f', and (9) takes the form

$$(f', \varphi) = -(f, \varphi'). \tag{9'}$$

It is an immediate consequence of Definition 5 that
1) Every generalized function has derivatives of all orders;
2) If a sequence of generalized functions $\{f_n\}$ converges to a generalized function f (in the sense of Definition 3), then the sequence of derivatives $\{f_n'\}$ converges to the derivative f' of the limit function.[15]

Example 1. If f is a regular generalized function whose derivative exists and is locally integrable (in particular, continuous or piecewise continuous), then the derivative of f as a generalized function coincides with its derivative in the ordinary sense. In fact, integrating (8) by parts, we get back (7).

Example 2. As in Example 1, p. 208, consider the delta function

$$T(\varphi) = \int_{-\infty}^{\infty} \delta(x)\varphi(x)\, dx.$$

It follows from Definition 5 that

$$\frac{dT}{dx}(\varphi) = -\int_{-\infty}^{\infty} \delta(x)\varphi'(x)\, dx = -\varphi'(0).$$

Example 3. Consider the "step function"

$$f(x) = \begin{cases} 0 & \text{if } x < 0, \\ 1 & \text{if } x \geqslant 0, \end{cases} \tag{10}$$

defining the linear functional

$$T(\varphi) = \int_{-\infty}^{\infty} f(x)\varphi(x)\, dx = \int_{0}^{\infty} \varphi(x)\, dx.$$

It follows from Definition 5 that

$$\frac{dT}{dx}(\varphi) = -\int_{0}^{\infty} \varphi'(x)\, dx = \varphi(0),$$

since φ vanishes at infinity. Hence the derivative of (10) is just the delta function $\delta(x)$.

21.4. Differential equations and generalized functions. The development of the theory of generalized functions was to a large extent motivated by

---

[15] Equivalently, every convergent series of generalized functions can be differentiated term by term any number of times.

problems involving differential equations, particularly partial differential equations. We now discuss a few simple ideas concerning generalized functions and ordinary differential equations. The application of generalized functions to partial differential equations is a subject lying beyond the scope of this book.[16]

LEMMA 1. *A test function φ_0 can be represented as the derivative of another test function φ_1 if and only if*

$$\int_{-\infty}^{\infty} \varphi_0(x)\,dx = 0. \tag{11}$$

Proof. If $\varphi_0(x) = \varphi_1'(x)$, where φ_1 is a test function, then

$$\int_{-\infty}^{\infty} \varphi_0(x)\,dx = \varphi_1(x)\Big|_{-\infty}^{\infty} = 0.$$

Conversely,

$$\varphi_1(x) = \int_{-\infty}^{x} \varphi_0(t)\,dt$$

is an infinitely differentiable function, with derivative $\varphi_0(x)$, and in fact a finite function if (11) holds, since then φ_0 and φ_1 vanish outside the same interval. ∎

LEMMA 2. *Let φ_1 be a fixed test function such that*

$$\int_{-\infty}^{\infty} \varphi_1(x)\,dx = 1. \tag{12}$$

Then an arbitrary test function φ can be represented in the form

$$\varphi = \varphi_0 + c\varphi_1,$$

where c is a constant and φ_0 is a test function which is the derivative of another test function.

Proof. Let

$$c = \int_{-\infty}^{\infty} \varphi(x)\,dx, \qquad \varphi_0(x) = \varphi(x) - \varphi_1(x)\int_{-\infty}^{\infty} \varphi(x)\,dx.$$

Then

$$\int_{-\infty}^{\infty} \varphi_0(x)\,dx = 0,$$

and the proof follows from Lemma 1. ∎

---

[16] See e.g., A. Friedman, *Generalized Functions and Partial Differential Equations,* Prentice-Hall, Inc., Englewood Cliffs, N.J. (1963). A key role in the development of the theory of generalized functions was played by the pioneer work of L. Schwartz, *Théorie des Distributions,* Hermann et Cie., Paris, *Volume 1* (1957), *Volume 2* (1959).

THEOREM 1. *Every solution of the differential equation*

$$y' = 0 \tag{13}$$

(in the space K^* of generalized functions) is a constant.

Proof. Equation (13) means that

$$(y', \varphi) = (y, -\varphi') = 0 \tag{14}$$

for every $\varphi \in K$. This determines the value of the functional

$$(y, \varphi) = \int_{-\infty}^{\infty} y \varphi(x)\, dx$$

for every function in the space $K' \subset K$ of all test functions which are derivatives of other test functions. In fact,

$$(y, \varphi_0) = 0$$

for every $\varphi_0 \in K'$. Let φ be an arbitrary test function. By Lemma 2, $\varphi = \varphi_0 + c\varphi_1$, where $\varphi_0 \in K'$ and φ_1 is a fixed test function satisfying the condition (12). We are free to give (y, φ_1) any value at all, without violating (14). Let

$$(y, \varphi_1) = \alpha = \text{const.}$$

Then

$$(y, \varphi) = (y, \varphi_0 + c\varphi_1) = (y, \varphi_0) + c(y, \varphi_1) = \alpha c = \text{const,}$$

and moreover y satisfies the differential equation (13). In fact, $\varphi \in K$ implies $-\varphi' \in K'$ and hence

$$(y', \varphi) = (y, -\varphi') = 0. \quad \blacksquare$$

COROLLARY. *If two generalized functions f and g have the same derivative, then $f = g + \text{const}$.*

Proof. Obvious, since $(f - g)' = 0$. $\quad \blacksquare$

THEOREM 2. *Given any generalized function f, there is another generalized function y satisfying the differential equation*

$$y' = f(x). \tag{15}$$

Proof. Any generalized function satisfying (15) is called an *antiderivative* of f. Equation (15) means that

$$(y', \varphi) = (y, -\varphi') = (f, \varphi) = \left(f, \int_{-\infty}^{x} \varphi'(t)\, dt \right) \tag{16}$$

for every $\varphi \in K$. This determines the value of the functional (y, φ) for every function in the space $K' \subset K$ of all test functions which are derivatives of other test functions. In fact,

$$(y, \varphi_0) = \left(f, -\int_{-\infty}^{x} \varphi_0(t)\, dt\right)$$

for every $\varphi_0 \in K'$. Let φ be an arbitrary test function. By Lemma 2, $\varphi = \varphi_0 + c\varphi_1$, where $\varphi_0 \in K'$ and φ_1 is a fixed test function satisfying (12). We are free to give (y, φ_1) any value at all, without violating (16). Let

$$(y, \varphi_1) = \alpha = \text{const.}$$

Then y satisfies the differential equation (15). In fact, $\varphi \in K$ implies $-\varphi' \in K'$ and hence

$$(y', \varphi) = (y, -\varphi') = \left(f, \int_{-\infty}^{x} \varphi'(t)\, dt\right) = (f, \varphi). \quad \blacksquare$$

COROLLARY. *Any two antiderivatives of a generalized function f differ only by a constant.*

Proof. Obvious by construction or from the corollary to Theorem 1. $\blacksquare$

21.5. Further developments. We now sketch some of the many extensions and modifications of the notion of generalized functions.

a) *Generalized functions of several variables.* Let K^n be the set of all functions $\varphi(x_1, \ldots, x_n)$ of n variables with partial derivatives of all orders with respect to all arguments, such that every $\varphi \in K^n$ vanishes outside some parallelepiped

$$a_i \leqslant x_i \leqslant b_i \qquad (i = 1, \ldots, n) \tag{17}$$

in n-space. Then K^n is a linear space, with addition of functions and multiplication of functions by numbers defined in the usual way. We introduce convergence in K^n by the natural generalization of Definition 1, i.e., a sequence $\{\varphi_k\}$ of functions in K^n is said to converge to a function $\varphi \in K^n$ if

1) There exists a parallelepiped (17) outside which all the functions φ_k vanish;
2) The sequence of partial derivatives

$$\left\{\frac{\partial^r \varphi_k}{\partial x_1^{\alpha_1} \cdots \partial x_n^{\alpha_n}}\right\} \qquad \left(\sum_{i=1}^{n} \alpha_i = r\right)$$

converges uniformly on this parallelepiped to the partial derivative

$$\frac{\partial^r \varphi}{\partial x_1^{\alpha_1} \cdots \partial x_n^{\alpha_n}}$$

for all $r, \alpha_1, \ldots, \alpha_n$.

Every continuous linear functional on K^n is then called a *generalized function of n variables*, and moreover every "ordinary" function $f(x_1, \ldots, x_n)$ of n variables integrable on every parallelepiped can be regarded as a generalized function, in fact the one giving rise to the functional

$$(f, \varphi) = \int f(x)\varphi(x) \, dx,$$

where

$$x = (x_1, \ldots, x_n), \qquad dx = dx_1 \cdots dx_n$$

and the integral is over all of n-space. Convergence of generalized functions is defined by the obvious analogue of Definition 3, while partial derivatives of generalized functions are defined by the formula

$$\left(\frac{\partial^r f(x)}{\partial x_1^{\alpha_1} \cdots \partial x_n^{\alpha_n}}, \varphi(x)\right) = (-1)^r \left(f(x), \frac{\partial^r \varphi(x)}{\partial x_1^{\alpha_1} \cdots \partial x_n^{\alpha_n}}\right).$$

It is clear that every generalized function of n variables has partial derivatives of all orders.

b) *Complex generalized functions.* So far we have only considered real generalized functions. Suppose the test functions are now allowed to be complex-valued, but still finite and infinitely differentiable. Then every continuous linear functional on the corresponding test space K is called a *complex generalized function.* If (f, φ) is such a functional, then

$$(f, \alpha\varphi) = \alpha(f, \varphi).$$

We can also consider conjugate-linear functionals on K, satisfying the condition (cf. p. 123)

$$(f, \alpha\varphi) = \bar{\alpha}(f, \varphi),$$

where the overbar denotes the complex conjugate. If f is an "ordinary" complex-valued function on the line, there are two natural ways of associating linear functionals with f, i.e.,

$$(f, \varphi)_1 = \int_{-\infty}^{\infty} f(x)\varphi(x) \, dx,$$

$$(f, \varphi)_2 = \int_{-\infty}^{\infty} \overline{f(x)}\varphi(x) \, dx,$$

and two natural ways of associating conjugate-linear functionals with f:

$$(f, \varphi)_3 = \int_{-\infty}^{\infty} f(x)\overline{\varphi(x)} \, dx,$$

$$(f, \varphi)_4 = \int_{-\infty}^{\infty} \overline{f(x)}\overline{\varphi(x)} \, dx.$$

Each of these four choices corresponds to a possible way of embedding the space of "ordinary" functions in the space of generalized functions. Operations on complex generalized functions are defined by analogy with the real case.

c) *Generalized functions on the circle.* Sometimes it is convenient to consider generalized functions defined on a bounded set. As a simple example, consider generalized functions on a circle C, choosing the test space K_C to be the set of all infinitely differentiable functions on C, equipped with the usual operations of addition of functions and multiplication of functions by numbers. (Note that the test functions are now automatically finite, since C is bounded.) Then every continuous linear functional on K_C is called a *generalized function on the circle.* Every "ordinary" function on C can be regarded as a periodic function on the line. In the same way, we regard every generalized function on the circle as a periodic generalized function, where a generalized function f is said to be *periodic*, with *period a*, if

$$(f(x),\ \varphi(x - a)) = (f(x),\ \varphi(x))$$

for every test function $\varphi \in K$.

d) *Other test spaces.* There are many possible choices of the test space other than the space of infinitely differentiable finite functions. For example, we can choose the test space to be the somewhat larger space S_∞ of all infinitely differentiable functions which, together with all their derivatives, approach zero faster than any power of $1/|x|$. More exactly, a function φ belongs to S_∞ if and only if, given any $p, q = 0, 1, 2, \ldots$, there is a constant C_{pq} (depending on p, q and φ) such that[17]

$$|x^p \varphi^{(q)}(x)| < C_{pq} \qquad (-\infty < x < \infty).$$

A sequence $\{\varphi_n\}$ of functions in S_∞ is said to converge to a function $\varphi \in S_\infty$ if

1) The sequence $\{\varphi_n^{(q)}\}$ converges uniformly to $\varphi^{(q)}$ on every finite interval;
2) The constants C_{pq} in the inequalities

$$|x^p \varphi_n^{(q)}(x)| < C_{pq}$$

can be chosen independently of n.

There are somewhat fewer continuous linear functionals on S_∞ than on K. For example, the function $f(x) = e^{x^2}$ corresponds to a continuous linear functional (f, φ) on K but not on S_∞.

Remark. As the theory of generalized functions has evolved, it has become apparent that there is no need to commit oneself once and for all to any definite choice of test space. Rather it is best to choose a test space

---

[17] As an exercise, verify that this is the same space S_∞ as in Problem 12b, p. 172 .

which is most suitable for solving the class of problems at hand. In general, the smaller the test space, the greater the freedom in carrying out various analytical operations (differentiation, passage to the limit, etc.) and the larger the number of continuous linear functionals on the space (why?). However, we must make sure not to make the test space too small, i.e., we must require not only that the test functions be "sufficiently smooth" but also that there be "sufficiently many" of them (in the sense of Problem 9) to allow us to "tell ordinary functions[18] apart."

Problem 1. In the test space K of all infinitely differentiable finite functions, let $\mathcal{N}_0$ be the neighborhood base at zero consisting of all sets of the form

$$U_{\gamma_0,\ldots,\gamma_n} = \{\varphi : \varphi \in K, |\varphi(x)| < \gamma_0(x), \ldots, |\varphi^{(n)}(x)| < \gamma_n(x) \text{ for all } x\}$$

for some positive functions $\gamma_0, \ldots, \gamma_n$ continuous on $(-\infty, \infty)$. Prove that the topology generated in K by $\mathcal{N}_0$ leads to the same kind of convergence in K as in Definition 1.

Comment. There are other topologies in K leading to the same convergence.

Problem 2. Let K be the test space of all infinitely differentiable finite functions, and let K_m be the subspace of K consisting of all functions $\varphi \in K$ vanishing outside the interval $[-m, m]$. We can make K_m into a countably normed space by setting

$$\|\varphi\|_n = \sup_{\substack{0 \leqslant k \leqslant n \\ |x| \leqslant m}} |\varphi^{(k)}(x)| \qquad (n = 0, 1, 2, \ldots)$$

(cf. Problem 12a, p. 171). Verify that the topology induced in K_m by the system of norms $\|\cdot\|_n$ coincides with the topology induced in K_m by the topology of Problem 1. Verify that the convergence in K_m induced in K_m by the norms $\|\cdot\|_n$ coincides with the convergence induced in K_m by the convergence in Definition 1. Clearly $K_1 \subset K_2 \subset \cdots \subset K_m \subset \cdots$, and

$$K = \bigcup_{m=1}^{\infty} K_m.$$

Show that a set $Q \subset K$ is bounded with respect to the topology in K if and only if there is an integer m such that Q is a bounded subset of the countably normed space K_m.

Problem 3. Let K and K_m be the same as in Problem 2, and let T be a linear functional on K. Prove that the following four conditions are

---

[18] More exactly, regular generalized functions.

equivalent:

   a) T is continuous with respect to the topology of the space K;
   b) T is bounded on every bounded subset $Q \subseteq K$;
   c) If $\varphi_n \in K$ and $\varphi_n \to 0$, then $T(\varphi_n) \to 0$ (provided convergence of sequences is defined as in Definition 1);
   d) The restriction T_m of the functional T to the space $K_m \subseteq K$ is a continuous functional on K_m for every $m = 1, 2, \ldots$

Problem 4. Let

$$T(\varphi) = \int_{-\infty}^{\infty} \frac{1}{x}\, \varphi(x)\, dx \qquad (18)$$

for every φ in the test space K. Prove that $T(\varphi)$ is a generalized function if the integral is understood in the sense of the Cauchy principal value.

Hint. If φ vanishes outside the interval $[a, b]$, write

$$\int_{-\infty}^{\infty} \frac{1}{x}\, \varphi(x)\, dx = \int_a^b \frac{\varphi(x) - \varphi(0)}{x}\, dx + \int_a^b \frac{\varphi(0)}{x}\, dx.$$

Problem 5. Prove that the delta function and its derivative are singular generalized functions. Prove that the same is true of (18).

Problem 6. Prove that addition of two generalized functions and multiplication of a generalized function by an infinitely differentiable function α (in particular, a constant) are continuous operations in the sense that $f_n \to f$, $\tilde{f}_n \to \tilde{f}$ implies $f_n + \tilde{f}_n \to f + \tilde{f}$, $\alpha f_n \to \alpha f$. Prove that there is no way of similarly defining a continuous product of two generalized functions, unless the functions are regular, in which case the appropriate definition is $T_{fg} = T_f T_g$ where

$$T_f(\varphi) = \int_{-\infty}^{\infty} f(x)\varphi(x)\, dx, \qquad T_g(\varphi) = \int_{-\infty}^{\infty} g(x)\varphi(x)\, dx,$$

$$T_{fg}(\varphi) = \int_{-\infty}^{\infty} f(x)g(x)\varphi(x)\, dx.$$

Problem 7. Let f be a piecewise continuous function on $(-\infty, \infty)$, differentiable everywhere except at the points $x_1, x_2, \ldots, x_n, \ldots$, where it has jumps

$$f(x_n + 0) - f(x_n - 0) = h_n \qquad (n = 1, 2, \ldots).$$

Prove that the generalized derivative of f (i.e., the derivative of f regarded as a generalized function) is the sum of its ordinary derivative (at the points where it exists) and the generalized function

$$g(x) = \sum_{n=1}^{\infty} h_n \delta(x - x_n).$$

Comment. Note that (g, φ) reduces to a finite sum for every test function φ.

Problem 8. Find the generalized derivative of the function of period 2π equal to

$$f(x) = \begin{cases} \dfrac{\pi - x}{2} & \text{if } 0 < x \leqslant \pi, \\ 0 & \text{if } x = 0, \\ -\dfrac{\pi + x}{2} & \text{if } -\pi \leqslant x < 0 \end{cases} \tag{19}$$

in the interval $[-\pi, \pi]$.

Ans. $f'(x) = -\frac{1}{2} + \pi \sum\limits_{n=-\infty}^{\infty} \delta(x - 2n\pi)$.

Comment. The function (19) is the sum of the trigonometric series

$$\sum_{n=1}^{\infty} \frac{\sin nx}{n}. \tag{20}$$

Differentiating (20) term by term, we get the divergent series

$$\sum_{n=1}^{\infty} \cos nx.$$

Hence the concept of a generalized function allows us to ascribe a definite meaning to a series that diverges in the ordinary sense. The same can be done for many divergent integrals (like those encountered in quantum field theory and other branches of theoretical physics).

Problem 9. Prove that the test space K of all infinitely differentiable finite functions has "sufficiently many" functions in the sense that, given any two distinct continuous functions f_1 and f_2, there exists a function $\varphi \in K$ such that

$$\int_{-\infty}^{\infty} f_1(x)\varphi(x)\,dx \neq \int_{-\infty}^{\infty} f_2(x)\varphi(x)\,dx.$$

Hint. Since $f(x) = f_1(x) - f_2(x) \not\equiv 0$, there is a point x_0 such that $f(x_0) \neq 0$, and hence an interval $[\alpha, \beta]$ in which $f(x)$ does not change sign. Let

$$\varphi(x) = \begin{cases} e^{-1/(x-\alpha)^2} e^{-1/(x-\beta)^2} & \text{if } \alpha < x < \beta, \\ 0 & \text{otherwise.} \end{cases}$$

Then $\varphi \in K$ and

$$\int_{-\infty}^{\infty} f(x)\varphi(x)\,dx = \int_{\alpha}^{\beta} f(x)\varphi(x)\,dx \neq 0.$$

Comment. This result can be extended to functions more general than continuous functions, with the help of the concept of the Lebesgue integral (introduced in Sec. 29).

Problem 10. Consider the homogeneous system of n linear differential equations

$$y_i' = \sum_{k=1}^{n} a_{ik}(x)y_k \qquad (i = 1, \ldots, n) \tag{21}$$

in n unknowns $y_1, \ldots, y_n$, where the a_{ik} are infinitely differentiable functions. Prove that every solution of (21) in the class K^* of generalized functions is a set of "ordinary" (in fact, infinitely differentiable) functions.

Comment. This can be expressed by saying that every "generalized solution" of (21) is also a "classical solution."

Problem 11. Consider the nonhomogeneous system of n linear differential equations

$$y_i' = \sum_{k=1}^{n} a_{ik}(x)y_k + f_i(x) \qquad (i = 1, \ldots, n), \tag{22}$$

where the a_{ik} are infinitely differentiable functions and the f_i are generalized functions. Prove that (22) has a generalized solution, which is unique to within a solution of the homogeneous system (21). What happens if the f_i are "ordinary" functions?

Problem 12. Interpret

$$f(x) = \sum_{n=1}^{\infty} \cos nx$$

as a periodic generalized function.

Hint. Recall Problem 8.

Problem 13. Show that S_∞ becomes a countably normed space when equipped with the system of norms

$$\|\varphi\|_n = \sum_{p+q=n} \sup_{\substack{-\infty < x < \infty \\ 0 \leqslant i \leqslant p \\ 0 \leqslant j \leqslant q}} |(1 + |x|^i)\varphi^{(j)}(x)|.$$

Prove that convergence of sequences in this countably normed space is equivalent to convergence of sequences in S_∞ as defined on p. 216.

6

LINEAR OPERATORS

---

22. Basic Concepts

22.1. Definitions and examples. Given two topological linear spaces E and E_1, any mapping

$$y = Ax \qquad (x \in E, y \in E_1)$$

of a subset of E (possibly E itself) into E_1 is called an *operator* (from E to E_1). The operator A is said to be *linear* if

$$A(\alpha x_1 + \beta x_2) = \alpha A x_1 + \beta A x_2.$$

Let D_A be the set of all $x \in E$ for which A is defined. Then D_A is called the *domain (of definition)* of the operator A. Although in general D_A need not equal E, we will always assume that D_A is a linear subspace of E, i.e., that $x, y \in D_A$ implies $\alpha x + \beta y \in D_A$ for all α and β.

The operator A is said to be *continuous at the point* $x_0 \in D$ if, given any neighborhood V of the point $y_0 = A x_0$, there is a neighborhood U of the point x_0 such that $Ax \in V$ for all $x \in U \cap D_A$. We say that the operator A is *continuous* if it is continuous at every point $x_0 \in D_A$.

Remark 1. Suppose E and E_1 are normed linear spaces. Then it is easy to see that A is continuous if and only if, given any $\varepsilon > 0$, there is a $\delta > 0$ such that

$$\|x' - x''\| < \delta \qquad (x', x'' \in D_A)$$

implies

$$\|Ax' - Ax''\| < \varepsilon.$$

Remark 2. In the case where E_1 is the real line, the concept of a linear operator reduces to that of a linear functional, and the definition of continuity reduces to that given on p. 175. As we will see below, much of the theory of linear functionals carries over in a straightforward way to the case of linear operators.

Example 1. Given a topological linear space E, let $Ix = x$ for all $x \in E$. Then I is a continuous linear operator, called the *identity* (or *unit*) *operator*, carrying each element of E into itself.

Example 2. Let E and E_1 be arbitrary topological linear spaces, and let $Ox = 0$ for all $x \in E$, where 0 is the zero element of the space E_1. Then O is a continuous linear operator, called the *zero operator*.

Example 3. Suppose A is a linear operator mapping the m-dimensional space R^m with basis $e_1, \ldots, e_m$ into the n-dimensional space R^n with basis $e_1', \ldots, e_n'$. If x is an arbitrary vector in R^m, then

$$x = \sum_{j=1}^{m} x_j e_j,$$

and hence, by the linearity of A,

$$y = Ax = \sum_{j=1}^{m} x_j A e_j.$$

Thus the operator A is completely determined once we know the vectors in R^n into which A carries the basis vectors $e_1, \ldots, e_m$. Suppose we expand each vector Ae_j with respect to the basis $e_1', \ldots, e_n'$, obtaining

$$Ae_j = \sum_{i=1}^{n} a_{ij} e_i'.$$

Then

$$y = \sum_{i=1}^{n} y_i e_i' = \sum_{j=1}^{m} x_j A e_j = \sum_{j=1}^{m} x_j \sum_{i=1}^{n} a_{ij} e_i'$$

and hence

$$y_i = \sum_{j=1}^{m} a_{ij} x_j,$$

i.e., the operator A is completely determined by the matrix $\|a_{ij}\|$ made up of the coefficients a_{ij}.

Example 4. Let H_1 be any subspace of a Hilbert space H, and let $H_2 = H \ominus H_1$ be the orthogonal complement of H_1, so that an arbitrary element $h \in H$ has a unique representation of the form

$$h = h_1 + h_2 \qquad (h_1 \in H_1, h_2 \in H_2)$$

(see Theorem 14, p. 158). Let

$$Ph = h_1.$$

Then P is a continuous linear operator, called a *projection operator*. Interpreted geometrically, P "projects the whole space H onto the subspace H_1."

22.2. Continuity and boundedness. A linear operator mapping E into E_1 is said to be *bounded* if it maps every bounded subset of E into a bounded subset of E_1. The operator analogue of Theorem 3, p. 176 for functionals is given by

THEOREM 1. *A necessary condition for a linear operator A to be continuous on a topological linear space E is that A be bounded. The condition is also sufficient if E satisfies the first axiom of countability.*

Proof. To prove the necessity, suppose A is continuous and suppose there is a bounded set M in E_1 whose image $AM = \{y : y = Ax, x \in M\}$ is unbounded in E_1. Then there is a neighborhood V of zero in E_1 such that none of the sets

$$\frac{1}{n} AM \qquad (n = 1, 2, \ldots)$$

is contained in V. Hence there is a sequence $\{x_n\}$ of elements of M such that none of the elements

$$\frac{1}{n} Ax_n \qquad (n = 1, 2, \ldots)$$

belongs to V. But then the sequence

$$\left\{ \frac{1}{n} x_n \right\}$$

converges to zero in E (recall Problem 6b, p. 170), while the sequence

$$\left\{ \frac{1}{n} Ax_n \right\}$$

fails to converge to zero in E_1, contrary to the assumption that A is continuous.

As for the sufficiency, let $\{U_n\}$ be a countable neighborhood base at zero in E such that

$$U_1 \supset U_2 \supset \cdots \supset U_n \supset \cdots.$$

If A fails to be continuous on E, then, by the operator analogue of Theorem 1, p. 175,[1] there is a neighborhood V of zero in E_1 and a sequence $\{x_n\}$ in E such that

$$x_n \in \frac{1}{n} U_n, \quad Ax_n \notin V \qquad (n = 1, 2, \ldots).$$

---

[1] As an exercise, state and prove this analogue.

The sequence $\{nx_n\}$ is bounded in E (and even converges to zero), while the sequence $\{nAx_n\}$ is unbounded in E_1, since it is contained in none of the sets nV. But then A fails to be bounded on the bounded set $\{x_1, x_2, \ldots, x_n, \ldots\}$, contrary to hypothesis. ∎

Next we consider the operator analogues of Definition 2 and Theorem 4, p. 177. Suppose E and E_1 are both normed linear spaces, so that in particular, E satisfies the first axiom of countability. Then, by Theorem 1, a linear operator A mapping E into E_1 is continuous if and only if it is bounded. But by a bounded set in a normed linear space we mean a set contained in some closed sphere $\|x\| \leqslant C$. Therefore a linear operator A on a normed linear space is bounded (and hence continuous) if and only if it is bounded on every closed sphere $\|x\| \leqslant C$, or equivalently on the closed unit sphere $\|x\| \leqslant 1$, because of the linearity of A. In other words, A is bounded if and only if the number

$$\|A\| = \sup_{\|x\| \leqslant 1} \|Ax\| \tag{1}$$

is finite.

DEFINITION. *Given a bounded linear operator mapping a normed linear space E into another normed linear space E_1, the number* (1), *equal to the least upper bound of $\|Ax\|$ on the closed unit sphere $\|x\| \leqslant 1$, is called the* **norm** *of A.*

THEOREM 2. *The norm $\|A\|$ has the following two properties*:

$$\|A\| = \sup_{x \neq 0} \frac{\|Ax\|}{\|x\|}, \tag{2}$$

$$\|Ax\| \leqslant \|A\| \, \|x\| \text{ for all } x \in E. \tag{3}$$

Proof. Clearly,

$$\|A\| = \sup_{\|x\| \leqslant 1} \|Ax\| = \sup_{\|x\| = 1} \|Ax\|$$

(why?). But the set of all vectors in E of norm 1 coincides with the set of all vectors

$$\frac{x}{\|x\|} \qquad (x \in E, \, x \neq 0), \tag{4}$$

and hence

$$\|A\| = \sup_{\|x\| = 1} \|Ax\| = \sup_{x \neq 0} \left\| A\left(\frac{x}{\|x\|}\right) \right\| = \sup_{x \neq 0} \frac{\|Ax\|}{\|x\|},$$

which proves (2). Moreover, since the vectors (4) all have norm 1, it follows from (1) that

$$\left\| A\left(\frac{x}{\|x\|}\right) \right\| = \frac{\|Ax\|}{\|x\|} \leqslant \|A\| \qquad (x \in E, \, x \neq 0),$$

which implies (3) for $x \neq 0$. The validity of (3) for $x = 0$ is obvious. ∎

22.3. Sums and products of operators. Let A and B be two operators from one topological linear space E to another topological linear space E_1. Then by the *sum* of A and B, denoted by $A + B$, we mean the operator assigning the element

$$y = Ax + Bx \in E_1$$

to each $x \in E$. The domain D_C of the sum $C = A + B$ is just the intersection $D_A \cap D_B$ of the domains of A and B. It is clear that C is linear if A and B are linear, and continuous if A and B are continuous. Let E and E_1 be normed linear spaces, and suppose A and B are bounded operators. Then $C = A + B$ is also bounded, with norm

$$\|C\| \leqslant \|A\| + \|B\|,$$

since, by Theorem 2 and Problem 10,

$$\|Cx\| = \|Ax + Bx\| \leqslant \|Ax\| + \|Bx\| \leqslant (\|A\| + \|B\|) \, \|x\|$$

for every $x \in E$.

Next, given three topological linear spaces E, E_1 and E_2, let A be an operator from E to E_1 and B an operator from E_1 to E_2. Then by the *product* of A and B, denoted by BA (in that order), we mean the operator assigning the element

$$z = B(Ax) \in E_2$$

to each $x \in E$. The domain D_C of the product $C = BA$ consists of those $x \in D_A$ for which $Ax \in D_B$. Again it is clear that C is linear if A and B are linear, and continuous if A and B are continuous. Let E, E_1 and E_2 be normed linear spaces, and suppose A and B are bounded operators. Then $C = BA$ is also bounded, with norm

$$\|C\| \leqslant \|A\| \, \|B\|,$$

since

$$\|Cx\| = \|B(Ax)\| \leqslant \|B\| \, \|Ax\| \leqslant \|B\| \, \|A\| \, \|x\|.$$

Remark 1. Sums and products of three or more operators are defined in the natural way, e.g.,

$$CBA = C(BA) = (CB)A,$$

$$A + B + C = A + (B + C) = (A + B) + C.$$

Note that addition of operators is associative and commutative, while multiplication of operators is associative but in general not commutative (give an example where $AB \neq BA$).

Remark 2. By the product αA of the operator A and the number α is meant the operator assigning the element αAx to each $x \in E$. Let $\mathscr{L}(E, E_1)$ be the set of all continuous linear operators mapping E into E_1. Then $\mathscr{L}(E, E_1)$ is clearly a linear space when equipped with the operations of addition of operators and multiplication of operators by numbers.

Problem 1. Prove that every linear operator on a finite-dimensional space is automatically continuous (cf. Problem 2, p. 181).

Problem 2. Let A be a linear operator mapping m-space R^m into n-space R^n. Prove that the image of R^m, i.e., the set $\{y : y = Ax, x \in R^m\}$, has dimension no greater than m.

Problem 3. Let $C_{[a,b]}$ be the linear space of functions continuous on the interval $a < x < b$, equipped with the norm

$$\|f\| = \max_{a \leqslant x \leqslant b} |f(x)|.$$

Let $K(x, y)$ be a fixed function of two variables, continuous on the square $a < x < b, a < y < b$, and let A be the operator defined by

$$g(x) = Af(x) = \int_a^b K(x, y) f(y) \, dy.$$

Prove that A is a continuous linear operator mapping $C_{[a,b]}$ into itself.

Problem 4. Let $C_{[a,b]}^2$ be the space of functions continuous on $[a, b]$, equipped with the norm

$$\|f\| = \sqrt{\int_a^b f^2(x) \, dx},$$

and let A be the same as in the preceding problem. Prove that A is a continuous linear operator mapping $C_{[a,b]}^2$ into itself.

Problem 5. Given a fixed function $\varphi(x)$ continuous on $[a, b]$, let A be the mapping defined by

$$g(x) = Af(x) = \varphi(x) f(x).$$

Prove that A is a continuous linear operator on both spaces $C_{[a,b]}$ and $C_{[a,b]}^2$, mapping each space into itself.

Problem 6. Let $C_{[a,b]}^{(1)}$ be the set of all continuously differentiable functions on $[a, b]$, and let D be the differentiation operator, defined by

$$Df(x) = f'(x)$$

for all $f \in C_{[a,b]}^{(1)}$. Prove that

a) $C_{[a,b]}^{(1)}$ is a linear space;
b) D is a linear operator mapping $C_{[a,b]}^{(1)}$ onto $C_{[a,b]}$;
c) D is not continuous on $C_{[a,b]}$;
d) D is continuous with respect to the norm

$$\|f\|_1 = \max_{a \leqslant x \leqslant b} |f(x)| + \max_{a \leqslant x \leqslant b} |f'(x)|.$$

Problem 7. Let $K_{[a,b]}$ be the space of infinitely differentiable functions on $[a, b]$, equipped with the topology generated by the countable system of norms

$$\|f\|_n = \sup_{\substack{a \leqslant x \leqslant b \\ 0 \leqslant k \leqslant n}} |f^{(k)}(x)|$$

(cf. Problem 12a, p. 171). Prove that the differentiation operator D is a continuous linear operator on $K_{[a,b]}$, mapping $K_{[a,b]}$ onto itself.

Problem 8. Interpret the differentiation operator as a continuous linear operator on the space of all generalized functions.

Hint. Take continuity to mean that if a sequence of generalized functions $\{f_n(x)\}$ converges to a generalized function $f(x)$, then $\{f_n'(x)\}$ converges to $f'(x)$.

Problem 9. Prove that

a) The operators in Problems 3–7 and Examples 1–4, p. 222 are all bounded;

b) A linear operator on a countably normed space is continuous if and only if it is bounded.

Problem 10. Let A be a bounded linear operator mapping a normed linear space E into another normed linear space E_1. Suppose $\|A\|$ is defined as the smallest number C such that $\|Af\| \leqslant C \|f\|$ for all $x \in E$. Prove that $\|A\|$ is the same number as in the definition on p. 224. Particularize this to the case of a bounded linear functional on E.

Problem 11. Let E and E_1 be normed linear spaces, and let $\mathscr{L}(E, E_1)$ be the same as in Remark 2 above. Prove that

a) $\mathscr{L}(E, E_1)$ is a normed linear space;

b) If E_1 is complete, so is $\mathscr{L}(E, E_1)$;

c) If E_1 is complete, $A_k \in \mathscr{S}(E, E_1)$ and

$$\sum_{k=1}^{\infty} \|A_k\| < \infty,$$

then the series

$$\sum_{k=1}^{\infty} A_k$$

converges to an operator $A \in \mathscr{L}(E, E_1)$ and

$$\|A\| = \left\| \sum_{k=1}^{\infty} A_k \right\| < \sum_{k=1}^{\infty} \|A_k\|.$$

23. Inverse and Adjoint Operators

23.1. The inverse operator. Invertibility. Given two topological linear spaces E and E_1, let A be an operator from E to E_1, with domain $D_A \subset E$ and range $R_A = \{y : y = Ax, x \in D_A\}$. Then A is said to be *invertible* if the equation

$$Ax = y \tag{1}$$

has a unique solution for every $y \in R_A$. If A is invertible, we can associate the unique solution of (1) with each $y \in R_A$. This gives an operator, with domain R_A, called the *inverse* of A and denoted by A^{-1}.

THEOREM 1. *The inverse A^{-1} of a linear operator A is itself linear.*

Proof. If

$$Ax_1 = y_1, \qquad Ax_2 = y_2,$$

then

$$A^{-1}y_1 = x_1, \qquad A^{-1}y_2 = x_2,$$

and hence

$$\alpha_1 A^{-1}y_1 + \alpha_2 A^{-1}y_2 = \alpha_1 x_1 + \alpha_2 x_2. \tag{2}$$

On the other hand,

$$A(\alpha_1 x_1 + \alpha_2 x_2) = \alpha_1 y_1 + \alpha_2 y_2,$$

by the linearity of A, and hence

$$A^{-1}(\alpha_1 y_1 + \alpha_2 y_2) = \alpha_1 x_1 + \alpha_2 x_2. \tag{3}$$

Comparing (2) and (3), we get

$$A^{-1}(\alpha_1 y_1 + \alpha_2 y_2) = \alpha_1 A^{-1}y_1 + \alpha_2 A^{-1}y_2. \quad \blacksquare$$

LEMMA. *If M is an everywhere dense subset of a normed linear space E, then every nonzero element $y \in E$ is the sum of a series of the form*

$$y = y_1 + y_2 + \cdots + y_k + \cdots,$$

where $y_k \in M$ and

$$\|y_k\| < \frac{3\,\|y\|}{2^k} \qquad (k = 1, 2, \ldots).$$

Proof. Since M is everywhere dense in E, given any $y \in E$, there is an element $y_1 \in M$ such that

$$\|y - y_1\| < \frac{\|y\|}{2}.$$

By the same token, there are elements $y_2, y_3, \ldots, y_k, \ldots$ such that

$$\| y - y_1 - y_2 \| \leqslant \frac{\|y\|}{4},$$

$$\| y - y_1 - y_2 - y_3 \| \leqslant \frac{\|y\|}{8},$$

$$\cdots \cdots \cdots \cdots \cdots \cdots$$

$$\| y - y_1 - \cdots - y_k \| \leqslant \frac{\|y\|}{2^k},$$

$$\cdots \cdots \cdots \cdots \cdots \cdots$$

Then

$$\left\| y - \sum_{k=1}^{n} y_k \right\| \to 0$$

as $n \to \infty$, by the construction of the sequence $\{y_k\}$, i.e., the series

$$\sum_{k=1}^{\infty} y_k$$

converges to y. Moreover

$$\| y_1 \| = \| y_1 - y + y \| \leqslant \| y_1 - y \| + \| y \| \leqslant \frac{\|y\|}{2} + \|y\| = \frac{3\,\|y\|}{2},$$

$$\| y_2 \| = \| y_2 + y_1 - y + y - y_1 \|$$
$$\leqslant \| y - y_1 - y_2 \| + \| y - y_1 \| \leqslant \frac{\|y\|}{4} + \frac{\|y\|}{2} = \frac{3\,\|y\|}{4},$$

and in general,

$$\| y_k \| = \| y_k + y_{k-1} + \cdots + y_1 - y + y - y_1 - \cdots - y_{k-1} \|$$

$$\leqslant \| y - y_1 - \cdots - y_k \| + \| y - y_1 - \cdots - y_{k-1} \|$$

$$\leqslant \frac{\|y\|}{2^k} + \frac{\|y\|}{2^{k-1}} = \frac{3\,\|y\|}{2^k}. \quad \blacksquare$$

THEOREM 2 (**Banach**). *Let A be an invertible bounded linear operator mapping a Banach space E onto another Banach space E_1. Then the inverse operator A^{-1} is itself bounded.*

Proof. Let M_k be the subset of E_1 consisting of all $y \in E_1$ such that

$$\| A^{-1} y \| < k \, \| y \|.$$

Every element in E_1 belongs to some M_k, i.e.,

$$E_1 = \bigcup_{k=1}^{\infty} M_k.$$

By Baire's theorem (Theorem 3, p. 61), at least one of the sets M_k, say M_n, is dense in some (open) sphere $S \subset E_1$. Choosing a point $y_0 \in S \cap M_n$, we can find numbers α and β ($\alpha < \beta$) such that S contains the spherical layer

$$P = \{z : \alpha < \|z - y_0\| < \beta, z \in M_1\}.$$

Shifting P so that its center coincides with the origin, we get another spherical layer P_0. Some set M_N is dense in P_0. In fact, if $z \in P \cap M_n$, then $z - y_0 \in P_0$ and

$$\|A^{-1}(z - y_0)\| \leqslant \|A^{-1}z + \|A^{-1}y_0\| \leqslant n(\|z\| + \|y_0\|)$$
$$\leqslant n(\|z - y_0\| + 2\|y_0\|)$$
$$= n\|z - y_0\|\left(1 + \frac{2\|y_0\|}{\|z - y_0\|}\right) \leqslant n\|z - y_0\|\left(1 + \frac{2\|y_0\|}{\alpha}\right),$$

$$(4)$$

where the quantity

$$\gamma = n\left(1 + \frac{2\|y_0\|}{\alpha}\right)$$

is independent of z. Let

$$N = 1 + [\gamma]$$

(recall footnote 4, p. 8). Then, by (4), $z - y_0 \in M_N$. Hence M_N is dense in P_0, since M_n is dense in P.

Now, given any nonzero element $y \in E_1$, we can always find a number $\lambda \neq 0$ such that $\alpha < \|\lambda y\| < \beta$, i.e., such that $\lambda y \in P_0$. Since M_N is dense in P_0, there is a sequence $\{\eta_k\}$, $\eta_k \in M_N$ converging to λy. Then $\{\eta_k/\lambda\}$ converges to y. Clearly, if $\eta_k \in M_N$, then $\eta_k/\lambda \in M_N$ for any $\lambda \neq 0$. Therefore M_N is dense in $E_1 - \{0\}$ and hence in E_1 itself. It follows from the lemma that y is the sum of a series of the form

$$y = y_1 + y_2 + \cdots + y_k + \cdots,$$

where $y_k \in M_N$ and

$$\|y_k\| < \frac{3\|y\|}{2^k}.$$

Consider the series

$$\sum_{k=1}^{\infty} x_k \qquad (5)$$

with terms $x_k = A^{-1}y_k \in E$, equal to the preimages of the elements $y_k \in E_1$. Since

$$\|x_k\| = \|A^{-1}y_k\| < N\|y_k\| < N\frac{3\|y\|}{2^k},$$

the series (5) converges to an element $x \in E$, where

$$\|x\| \leqslant \sum_{k=1}^{\infty} \|x_k\| \leqslant 3N \|y\| \sum_{k=1}^{\infty} \frac{1}{2^k} = 3N \|y\|.$$

Since (5) is convergent and the operator A is continuous on E (being bounded), we can apply A term by term to (5), obtaining

$$Ax = Ax_1 + Ax_2 + \cdots + Ax_k + \cdots = y_1 + y_2 + \cdots + y_k + \cdots = y,$$

which implies

$$x = A^{-1}y.$$

Moreover,

$$\|A^{-1}y\| = \|x\| \leqslant 3N \|y\|$$

for all $y \neq 0$, and hence A^{-1} is bounded. ∎

THEOREM 3. *Let A_0 be an invertible bounded linear operator mapping a Banach space E into another Banach space E_1, and let ΔA be a bounded linear operator mapping E into E_1 such that*

$$\|\Delta A\| < \frac{1}{\|A_0^{-1}\|}. \tag{6}$$

Then the operator

$$A = A_0 + \Delta A$$

maps E onto E_1 and has a bounded inverse.

Proof. Let y be a fixed element of E_1, and consider the mapping B of the space E into itself defined by

$$Bx = A_0^{-1}y - A_0^{-1}\Delta Ax.$$

It follows from (6) that B is a contraction mapping. Hence, by Theorem 1, p. 66, B has a unique fixed point x such that

$$x = Bx = A_0^{-1}y - A_0^{-1}\Delta Ax. \tag{7}$$

But (7) implies

$$Ax = A_0x + \Delta Ax = y.$$

Clearly, if $Ax' = y$, then x' is also a fixed point of B, and hence $x' = x$. Therefore, given any $y \in E_1$, the equation $Ax = y$ has a unique solution in E, i.e., the operator A is invertible with inverse A^{-1}. Moreover, A^{-1} is bounded, by Theorem 2. ∎

THEOREM 4. *Let E be a Banach space, and let I be the identity operator on E. Suppose A is a bounded linear operator mapping E into itself, such that*

$$\|A\| < 1. \tag{8}$$

Then the operator $(I - A)^{-1}$ exists, is bounded and can be represented in the form

$$(I - A)^{-1} = \sum_{k=0}^{\infty} A^k. \tag{9}$$

Proof. The existence and boundedness of $(I - A)^{-1}$ follows from Theorem 3 (and will also emerge in the course of the proof). It follows from (8) that

$$\sum_{k=0}^{\infty} \|A^k\| \leqslant \sum_{k=0}^{\infty} \|A\|^k < \infty.$$

But then, by the completeness of E, the sum of the series

$$\sum_{k=0}^{\infty} A^k$$

is a bounded linear operator (see Problem 11c, p. 227). Given any n, we have

$$(I - A) \sum_{k=0}^{n} A^k = \sum_{k=0}^{n} A^k (I - A) = I - A^{n+1}.$$

Hence, taking the limit as $n \to \infty$ and bearing in mind that

$$\|A^{n+1}\| \leqslant \|A\|^{n+1} \to 0,$$

we get

$$(I - A) \sum_{k=0}^{\infty} A^k = I,$$

which implies (9). ∎

23.2. The adjoint operator. Given two topological linear spaces E and E_1, let A be a continuous linear operator mapping E into E_1, and let g be a continuous linear functional on E_1, i.e., an element of the conjugate space E_1^*. Suppose we apply g to the element $y = Ax$, thereby obtaining a new functional

$$f(x) = g(Ax) \qquad (x \in E). \tag{10}$$

Clearly, f is continuous and linear (why?), and hence an element of the conjugate space E^*. Thus (10) associates a functional $f \in E^*$ with each functional $g \in E_1^*$, i.e., (10) defines an operator mapping E_1^* into E^*. This operator is called the *adjoint* of A, and is denoted by A^*. Using the symmetric notation (f, x) for the functional $f(x)$, we can write (10) in the form

$$(g, Ax) = (f, x).$$

or

$$(g, Ax) = (A^*g, x). \tag{11}$$

Equation (11) can be regarded as a concise definition of the adjoint of A.

Example. As in Example 3, p. 222, suppose A is a linear operator with matrix $\|a_{ij}\|$ mapping m-space R^m into n-space R^n. Then the mapping $y = Ax$ can be written as a system of equations

$$y_i = \sum_{j=1}^{m} a_{ij} x_j \quad (i = 1, \ldots, n), \tag{12}$$

while the functional $f(x)$ can be written in the form

$$f(x) = \sum_{j=1}^{m} f_j x_j,$$

where $f_j = f(e_j)$ in terms of a basis $e_1, \ldots, e_m$ in R^m. Since

$$f(x) = g(Ax) = \sum_{i=1}^{n} g_i y_i = \sum_{i=1}^{n} \sum_{j=1}^{m} g_i a_{ij} x_j = \sum_{j=1}^{m} x_j \sum_{i=1}^{n} g_i a_{ij},$$

we find that

$$f_j = \sum_{i=1}^{n} a_{ij} g_i,$$

or

$$f_i = \sum_{j=1}^{n} a_{ji} g_i \tag{13}$$

after interchanging the roles of the indices i and j. But $f = A^*g$, and hence comparing (12) and (13), we see that the matrix of the operator A^* is $\|a_{ji}\|$, i.e., the transpose of the matrix of A.

It follows at once from the definition of the adjoint of an operator that

1) A^* is linear;
2) $(A + B)^* = A^* + B^*$;
3) $(\alpha A)^* = \alpha A^*$ for arbitrary complex α.

A somewhat less obvious property of the adjoint operator is given by

THEOREM 5. *Let A be a bounded linear operator mapping a Banach space E into another Banach space E_1, and let A^* be the adjoint of A. Then A^* is bounded and*

$$\|A^*\| = \|A\|. \tag{14}$$

Proof. By the properties of the norm of an operator, we have

$$|(A^*g, x)| = |(g, Ax)| \leqslant \|g\| \, \|A\| \, \|x\|,$$

which implies

$$\|A^*g\| \leqslant \|A\| \, \|g\|,$$

and hence

$$\|A^*\| \leqslant \|A\|. \tag{15}$$

Suppose $x \in E$, $Ax \neq 0$, and let

$$y_0 = \frac{Ax}{\|Ax\|} \in E_1,$$

so that, in particular, $\|y_0\| = 1$. Let g be the functional such that

$$g(\lambda y_0) = \lambda$$

on the set $L \subset E_1$ of all elements of the form λy_0. Then clearly $(g, y_0) = 1$, $\|g\|_{\text{on } L} = 1$. Using the Hahn-Banach theorem, we can extend g to a functional on the whole space E_1 such that $\|g\| = 1$ and

$$(g, y_0) = 1, \quad \text{i.e.,} \quad (g, Ax) = \|Ax\|.$$

Therefore

$$\|Ax\| = (g, Ax) = |(A^*g, x)| \leqslant \|A^*g\| \, \|x\| \leqslant \|A^*\| \, \|g\| \, \|x\| = \|A^*\| \, \|x\|,$$

which implies

$$\|A\| \leqslant \|A^*\|. \tag{16}$$

Comparing (15) and (16), we get (14). ∎

23.3. The adjoint operator in Hilbert space. Self-adjoint operators. Next we consider the case where A is a bounded linear operator mapping a (real or complex) Hilbert space H into itself. According to the corollary to Theorem 2, p. 188, the mapping τ assigning the linear functional

$$(\tau y)(x) = (x, y)$$

to every $y \in H$ establishes an isomorphism between H and the conjugate space H^*.[2] Let A^* be the adjoint of the operator A. Then clearly the mapping $\tilde{A}^* = \tau^{-1} A^* \tau$ is a bounded linear operator mapping H into itself, such that

$$(Ax, y) = (x, \tilde{A}^* y) \tag{17}$$

for all $x, y \in H$. Moreover $\|\tilde{A}^*\| = \|A\|$, since $\|A^*\| = \|A\|$ and the mappings τ and τ^{-1} are isometric.

We now establish the following convention: If H is a Hilbert space, then by the adjoint of an operator A mapping H into H, we mean the operator $\tilde{A}^*$ defined by (17). Note that $\tilde{A}^*$, like A, maps H into H. To keep the notation simple, we will henceforth drop the tilde, writing A^* instead of $\tilde{A}^*$. Replacing $\tilde{A}^*$ by A^* in (17), we get

$$(Ax, y) = (x, A^* y) \tag{17'}$$

for all $x, y \in H$.

---

[2] Or a "conjugate-linear isomorphism" in the case where H is complex (see Problem 6, p. 194).

Remark. It should be emphasized that this definition of $A*$ differs from the definition of the adjoint of an operator A mapping an arbitrary Banach space E into itself, in which case $A*$ is defined on the conjugate space $E*$ rather than on the space E itself. The context will always make it clear whether $A*$ is the operator defined by (11) or the operator defined by (17').

Let A be a bounded linear operator mapping a Hilbert space H into itself. Then it makes sense to ask whether or not $A = A*$, since A and $A*$ are defined on the same space. This leads to the following

DEFINITION. *A bounded linear operator A mapping a Hilbert space H into itself is said to be* **self-adjoint** *if $A = A*$, i.e., if*

$$(Ax, y) = (x, Ay)$$

for all $x, y \in H$.

Remark. Everything said above continues to hold if we replace H by the real n-space R^n or complex n-space C^n.

23.4. The spectrum of an operator. The resolvent. In the theory of linear operators and their applications, a central role is played by the notion of the "spectrum" of an operator.[3] Let A be a linear operator mapping a topological linear space E into itself. Then a number λ is called an *eigenvalue* of A if the equation

$$Ax = \lambda x$$

has at least one nonzero solution, and every such solution x is called an *eigenvector* of A (corresponding to the eigenvalue λ). Suppose E is finite-dimensional. Then the set of all eigenvalues of A is called the *spectrum* of A, and all other values of λ are said to be *regular* (*points*). In other words, λ is regular if and only if the operator $A - \lambda I$ is invertible. The operator $(A - \lambda I)^{-1}$ is then automatically bounded, like every operator on a finite-dimensional space (cf. Problem 1, p. 226). Thus there are just two possibilities in the finite-dimensional case:

1) The equation $Ax = \lambda x$ has a nonzero solution, i.e., λ is an eigenvalue of A, so that the operator $(A - \lambda I)^{-1}$ fails to exist;
2) The operator $(A - \lambda I)^{-1}$ exists and is bounded, i.e., λ is a regular point.

However, in the case where E is infinite-dimensional, there is a third possibility:

3) The operator $(A - \lambda I)^{-1}$ exists (i.e., the equation $Ax = \lambda x$ has no nonzero solutions), but is not bounded.

---

[3] In talking about the spectrum of an operator, it will always be tacitly assumed that the operator is defined on a *complex* space.

To describe this more general situation, we introduce some new terminology and make an important modification in the definition of the spectrum. Given an operator A mapping a (complex) topological linear space E into itself, the operator

$$R_\lambda = (A - \lambda I)^{-1} \tag{18}$$

is called the *resolvent* of A. The values of λ for which R_λ is *defined for all E and continuous* are said to be *regular (points)* of A, and the set of all other values of λ is called the *spectrum* of A. The eigenvalues of A still belong to the spectrum, since if $(A - \lambda I)x = 0$ for some $x \neq 0$, then (18) fails to exist. The set of all these eigenvalues is now called the *point spectrum*, and the rest of the spectrum is called the *continuous spectrum*. In other words, the continuous spectrum consists of all λ for which (18) exists but fails to be continuous. Thus there are now exactly three possibilities for any given value of λ:

1) λ is a regular point;
2) λ is an eigenvalue;
3) λ is a point of the continuous spectrum.

The possibility of an operator having a continuous spectrum is a characteristic feature of the theory of operators in infinite-dimensional spaces, distinguishing it from the finite-dimensional case.

THEOREM 6. *Let A be a linear operator mapping a Banach space E into itself. Then the set Λ of all regular points of A is open (equivalently, the complement of Λ is closed).*

Proof. If λ is regular, the operator $(A - \lambda I)^{-1}$ exists and is bounded. Hence, for sufficiently small δ, the operator $(A - (\lambda + \delta)I)^{-1}$ also exists and is bounded, by Theorem 3. In other words, the point $\lambda + \delta$ is regular for sufficiently small δ. ∎

THEOREM 7. *If A is a bounded linear operator mapping a Banach space E into itself and if $|\lambda| > \|A\|$, then λ is a regular point. In other words, the spectrum of A is contained in the disk of radius $\|A\|$ with center at the origin.*

Proof. Obviously

$$A - \lambda I = -\lambda\left(I - \frac{A}{\lambda}\right),$$

and

$$R_\lambda = (A - \lambda I)^{-1} = -\frac{1}{\lambda}\left(I - \frac{A}{\lambda}\right)^{-1}.$$

If $\|A\| < \lambda$, then $\|A/\lambda\| < 1$, and hence R_λ exists and is bounded, by Theorem 4. ∎

Example 1. In the space $C = C_{[0,1]}$, consider the operator A defined by

$$Ax(t) = \mu(t)x(t),$$

where $\mu(t)$ is a fixed function continuous on $[0, 1]$. Then

$$(A - \lambda I)x(t) = (\mu(t) - \lambda)x(t),$$

and

$$(A - \lambda I)^{-1}x(t) = \frac{1}{\mu(t) - \lambda} x(t).$$

Hence the spectrum of A consists of all λ such that $\mu(t) - \lambda$ vanishes for some t in the interval $[0, 1]$, i.e., the spectrum is the range of the function $\mu(t)$.

Example 2. Suppose $\mu(t) = t$ in the preceding example. Then the spectrum is just the interval $[0, 1]$. On the other hand, there are obviously no eigenvalues. Thus the operator A defined by

$$Ax(t) = tx(t)$$

is an example of an operator with a purely continuous spectrum.

Finally, for self-adjoint operators in a Hilbert space, we have the following analogue of a well-known result for finite-dimensional Euclidean spaces (proved in exactly the same way):

THEOREM 8. *Let A be a self-adjoint operator mapping a (complex) Hilbert space H into itself. Then all the eigenvalues of A are real, and two eigenvectors of A corresponding to distinct eigenvalues are orthogonal.*

Proof. If

$$Ax = \lambda x \qquad (x \neq 0),$$

then

$$\lambda(x, x) = (Ax, x) = (x, Ax) = (x, \lambda x) = \bar\lambda(x, x),$$

and hence $\lambda = \bar\lambda$. Moreover, if

$$Ax = \lambda x, \quad Ay = \mu y \qquad (\lambda \neq \mu),$$

then

$$\lambda(x, y) = (Ax, y) = (x, Ay) = (x, \mu y) = \bar\mu(x, y) = \mu(x, y),$$

and hence

$$(x, y) = 0,$$

i.e., the vectors x and y are orthogonal. ∎

Problem 1. Given two normed linear spaces E and E_1, a linear operator A from E to E_1, with domain D_A, is said to be *closed* if $x_n \in D_A$, $x_n \to x$,

$Ax_n \to y$ implies $x \in D_A$, $Ax = y$. Prove that every bounded operator is closed.

Problem 2. Let E and E_1 be normed linear spaces, with norms $\|\cdot\|$ and $\|\cdot\|_1$, respectively. By the *direct* (or *Cartesian*) *product* of E and E_1, denoted by $E \times E_1$, we mean the set of all ordered pairs (x, y), $x \in E$, $y \in E_1$. Prove that $E \times E_1$ is a normed linear space when equipped with the norm

$$\|(x, y)\| = \|x\| + \|y\|_1$$

(addition of elements and multiplication of elements by numbers being defined in the obvious way). By the *graph* of a linear operator A from E to E_1 we mean the subset of $E \times E_1$ equal to

$$G_A = \{(x, y) : x \in D_A, y = Ax\}.$$

Prove that

a) G_A is a linear subspace of $E \times E_1$;
b) G_A is closed if and only if the operator A is closed;
c) If E and E_1 are Banach spaces and if A is closed and defined for all $x \in E$, so that $D_A = E$, then A is bounded (this is Banach's *closed graph theorem*).

Hint. In c) apply Theorem 2 to the projection operator P carrying each ordered pair $(x, Ax) \in G_A$ into the element $x \in E$.

Problem 3. Prove that if A is an invertible continuous linear operator mapping a complete countably normed space E into another complete countably normed space E_1, then the inverse operator A^{-1} is itself continuous. State and prove the closed graph theorem for countably normed spaces.

Problem 4. Let A be a continuous linear operator mapping a Banach space E onto another Banach space E_1. Prove that there is a constant $\alpha > 0$ such if $B \in \mathscr{L}(E, E_1)$ and $\|A - B\| < \alpha$, then B also maps E onto (all of) E_1.

Problem 5. Let A be an operator mapping a Hilbert space H into itself. Then a subspace $M \subset H$ is said to be *invariant* under A if $x \in M$ implies $Ax \in M$. Prove that if M is invariant under A, then its orthogonal complement $M' = H \ominus M$ is invariant under the adjoint operator A^* (in particular, under A itself if A is self-adjoint).

Problem 6. Let A and B be bounded linear operators mapping a complex Hilbert space H into itself. Prove that

a) $(\alpha a + \beta B)^* = \bar{\alpha}A^* + \bar{\beta}B^*$;
b) $(AB)^* = B^*A^*$;
c) $(A^*)^* = A$;
d) $I^* = I$, where I is the identity operator.

Problem 7. Give an example of an operator whose spectrum consists of a single point.

Problem 8. Given a bounded linear operator A mapping a Banach space E into itself, prove that the limit

$$r = \lim_{n \to \infty} \sqrt[n]{\|A^n\|}$$

exists. Show that the spectrum of A is contained in the disk of radius r with center at the origin.

Comment. The quantity r is called the *spectral radius* of the operator A. This result contains Theorem 8 as a special case, since $\|A^n\| \leqslant \|A\|^n$.

Problem 9. Let $R_\lambda = (A - \lambda I)^{-1}$ and $R_\mu = (A - \mu I)^{-1}$ be the resolvents corresponding to the points λ and μ. Prove that $R_\lambda R_\mu = R_\mu R_\lambda$ and

$$R_\mu - R_\lambda = (\mu - \lambda) R_\mu R_\lambda. \tag{19}$$

Hint. Multiply both sides of (19) by $(A - \lambda I)(A - \mu I)$.

Comment. It follows from (19) that if λ_0 is a regular point of A, then the derivative of R_λ with respect to λ at the point λ_0, i.e., the limit

$$\lim_{\Delta\lambda \to 0} \frac{R_{\lambda_0 + \Delta\lambda} - R_{\lambda_0}}{\Delta\lambda}$$

(in the sense of convergence with respect to the operator norm) exists and equals $R_{\lambda_0}^2$.

Problem 10. Let A be a bounded self-adjoint operator mapping a complex Hilbert space H into itself. Prove that the spectrum of A is a closed bounded subset of the real line.

Problem 11. Prove that every bounded linear operator defined on a complex Banach space with at least one nonzero element has a nonempty spectrum.

24. Completely Continuous Operators

24.1. Definitions and examples. We now discuss a class of operators which closely resemble operators acting in a finite-dimensional space and at the same time are very important from the standpoint of applications:

DEFINITION. *A linear operator A mapping a Banach space E into itself is said to be **completely continuous** if it maps every bounded set into a relatively compact set.*

Remark 1. If E is finite-dimensional, then every linear operator A mapping E into E is completely continuous. In fact, A maps bounded sets into bounded sets (recall Problem 1, p. 226) and hence maps bounded sets into relatively compact sets (why?).

Remark 2. In an infinite-dimensional space, complete continuity of an operator is a stronger requirement than merely being continuous (i.e., bounded). For example, the identity operator in an infinite-dimensional space is continuous but not completely continuous (see Example 1 below).

LEMMA. *Let $x_1, x_2, \ldots$ be linearly independent vectors in a normed linear space E, and let E_n be the subspace generated by the vectors $x_1, \ldots, x_n$. Then there are vectors $y_1, y_2, \ldots$ such that $y_n \in E_n$, $\|y_n\| = 1$ and*[4]

$$\rho(E_{n-1}, y_n) = \inf_{x \in E_{n-1}} \|x - y_n\| > \tfrac{1}{2}.$$

Proof. Since the vectors $x_1, x_2, \ldots$ are linearly independent, we have $x_n \notin E_{n-1}$ and hence

$$\rho(E_{n-1}, x_n) = \alpha > 0$$

(recall Problem 5a, p. 141). Let x^* be a vector in E_{n-1} such that

$$\|x_n - x^*\| < 2\alpha.$$

Then

$$\rho(E_{n-1}, x_n - x^*) = \alpha,$$

and the vectors

$$y_1 = \frac{x_1}{\|x_1\|}, \qquad y_n = \frac{x_n - x^*}{\|x_n - x^*\|} \qquad (n = 2, 3, \ldots)$$

satisfy all the conditions of the lemma. ∎

Example 1. The identity operator I in an infinite-dimensional Banach space E is not completely continuous. In fact, we need only show that the closed unit sphere S in E (which is obviously carried into itself by I) is not compact. This follows at once from the lemma, since S contains a sequence of vectors $y_1, y_2, \ldots$ such that

$$\rho(y_{n-1}, y_n) > \tfrac{1}{2},$$

and such a sequence clearly cannot contain a convergent subsequence.

Example 2. Let A be a continuous linear operator on an infinite-dimensional Banach space E, where A is "degenerate" in the sense that it maps E into a finite-dimensional subspace of E. Then A is completely continuous,

---

[4] The quantity $\rho(E_{n-1}, y_n)$ is, of course, just the distance between the set E_{n-1} and the point y_n (cf. Problem 9, p. 54).

since it maps every bounded subset $M \subset E$ into a bounded subset of a finite-dimensional space, and hence into a relatively compact set.

Turning to the space $C_{[a,b]}$ of functions continuous on the interval $[a, b]$, we now establish conditions under which the "integral operator" A defined by

$$\psi(x) = (A\varphi)(x) = \int_a^b K(x, y)\varphi(y) \, dy \qquad (1)$$

is completely continuous.

THEOREM 1. *Suppose the kernel $K(x, y)$ is such that*

1) *$K(x, y)$ is bounded on the square $a \leqslant x \leqslant b, a \leqslant y \leqslant b$;*
2) *The discontinuities (if any) of $K(x, y)$ all lie on a finite number of curves*

$$y = f_k(x) \qquad (k = 1, \ldots, n),$$

where the functions f_k are continuous.

Then (1) is a completely continuous operator mapping $C_{[a,b]}$ into $C_{[a,b]}$.

Proof. First we note that the conditions 1) and 2) guarantee the existence of the integral (1) for every $x \in [a, b]$, so that $\psi(x)$ is defined on $[a, b]$. Let R be the square $a \leqslant x \leqslant b, a \leqslant y \leqslant b$, and let

$$M = \sup_{(x,y)\in R} |K(x, y)|. \qquad (2)$$

Moreover, let G be the set of all points $(x, y) \in R$ such that

$$|y - f_k(x)| < \frac{\varepsilon}{12Mn}$$

for at least one integer $k = 1, \ldots, n$, and let $F = R - G$. Since F is compact (why?) and $K(x, y)$ is continuous on F, given any $\varepsilon > 0$, there is a $\delta > 0$ such that

$$|K(x', y) - K(x'', y)| < \frac{\varepsilon}{3(b - a)} \qquad (3)$$

for any two points $(x', y), (x'', y) \in F$ satisfying the condition

$$|x' - x''| < \delta \qquad (4)$$

(recall Theorem 1, p. 109).

Now suppose (4) holds. Then

$$|\psi(x') - \psi(x'')| \leqslant \int_a^b |K(x', y) - K(x'', y)| \, |\varphi(y)| \, dy. \qquad (5)$$

To estimate the integral on the right, we divide the interval $a \leqslant y \leqslant b$ into the set

$$P = \bigcup_{k=1}^{n} \left\{ y: |y - f_k(x')| < \frac{\varepsilon}{12Mn} \right\} \cup \bigcup_{k=1}^{n} \left\{ y: |y - f_k(x'')| < \frac{\varepsilon}{12Mn} \right\}$$

and the complementary set $Q = [a, b] - P$. Using (2) and noting that P is a union of intervals of total length no greater than $\varepsilon/3M$, we have

$$\int_P |K(x', y) - K(x'', y)| \, |\varphi(y)| \, dy < \frac{2\varepsilon}{3} \, \|\varphi\|, \tag{6}$$

where, as usual,

$$\|\varphi\| = \sup_{a \leqslant y \leqslant b} |\varphi(y)|.$$

On the other hand, it follows from (3) and (4) that

$$\int_Q |K(x', y) - K(x'', y)| \, |\varphi(y)| \, dy < \frac{\varepsilon}{3} \, \|\varphi\|. \tag{7}$$

Comparing (5)–(7), we find that (4) implies

$$|\psi(x') - \psi(x'')| < \varepsilon \, \|\varphi\|. \tag{8}$$

In particular, ψ is continuous on $[a, b]$, so that the operator A defined by (1) actually maps the space $C_{[a,b]}$ into itself. Moreover, it follows from (8) and from the estimate.

$$\|\psi\| = \sup_{a \leqslant x \leqslant b} |\psi(x)| \leqslant \sup_{a \leqslant x \leqslant b} \int_a^b |K(x, y)| \, |\varphi(y)| \, dy \leqslant M(b - a) \, \|\varphi\|$$

that A carries any (uniformly) bounded set of functions $\Phi \subset C_{[a,b]}$ into a (uniformly) bounded equicontinuous set $\Psi \subset C_{[a,b]}$ (recall Definitions 3 and 4, p. 102). But then Ψ is relatively compact, by Arzelà's theorem (Theorem 4, p. 102), and hence A is completely continuous. ∎

Remark 1. The requirement that the discontinuities of the kernel $K(x, y)$ lie on a finite number of curves, each intersecting the lines $x = \text{const}$ in a single point, is essential. For example, let $K(x, y)$ be the function

$$K(x, y) = \begin{cases} 1 & \text{if } x < \frac{1}{2}, \\ 0 & \text{if } x > \frac{1}{2}, \end{cases}$$

defined on the square $0 \leqslant x \leqslant 1, 0 \leqslant y \leqslant 1$. Then $K(x, y)$ is discontinuous at every point of the line segment $x = \frac{1}{2}, 0 \leqslant y \leqslant 1$, and the operator (1) with this kernel maps the function $x(t) \equiv 1$ into a discontinuous function.

Remark 2. If $K(x, y) = 0$ for $y > x$, then (1) takes the form

$$\psi(x) = (A\varphi)(x) = \int_a^x K(x, y)\varphi(y) \, dy.$$

Suppose $K(x, y)$ is continuous for $y < x$. Then it follows from Theorem 1 that the operator A, called a *Volterra operator*, is completely continuous.

24.2. Basic properties of completely continuous operators. We begin with

THEOREM 2. *Given a sequence* $\{A_n\}$ *of completely continuous operators mapping a Banach space E into itself, suppose* $\{A_n\}$ *converges in norm to an operator A, i.e., suppose* $\|A - A_n\| \to 0$ *as* $n \to \infty$. *Then A is itself completely continuous.*

Proof. To prove that A is completely continuous, we need only show that the sequence $\{Ax_n\}$ contains a convergent subsequence whenever the sequence $\{x_n\}$ of elements $x_n \in E$ is bounded, i.e., such that

$$\|x_n\| \leqslant M \tag{9}$$

for some $M > 0$ and all $n = 1, 2, \ldots$ (why is A linear?). Since A_1 is completely continuous, the sequence $\{A_1 x_n\}$ contains a convergent subsequence. In other words, there is a subsequence $\{x_n^{(1)}\}$ of the sequence $\{x_n\}$ such that $\{A_1 x_n^{(1)}\}$ converges. Similarly, since A_2 is completely continuous, the sequence $\{A_2 x_n^{(1)}\}$ in turn contains a convergent subsequence. Thus there is a subsequence $\{x_n^{(2)}\}$ of the sequence $\{x_n^{(1)}\}$ such that $\{A_2 x_n^{(2)}\}$ converges. Then obviously $\{A_1 x_n^{(2)}\}$ also converges. Continuing this argument, we find a subsequence $\{x_n^{(3)}\}$ of the sequence $\{x_n^{(2)}\}$ such that $\{A_1 x_n^{(3)}\}$, $\{A_2 x_n^{(3)}\}$, $\{A_3 x_n^{(3)}\}$ all converge, and so on. Consider the "diagonal sequence"

$$x_1^{(1)}, x_2^{(2)}, \ldots, x_n^{(n)}, \ldots$$

The clearly each of the operators $A_1, A_2, \ldots, A_n, \ldots$ maps this sequence into a convergent sequence.

We now show that the sequence $\{Ax_n^{(n)}\}$ also converges, thereby completing the proof. Since the space E is complete, it is enough to show that $\{Ax_n^{(n)}\}$ is a Cauchy sequence. Clearly

$$\|Ax_n^{(n)} - Ax_{n'}^{(n')}\| \leqslant \|Ax_n^{(n)} - A_k x_n^{(n)}\| + \|A_k x_n^{(n)} - A_k x_{n'}^{(n')}\| \\ + \|A_k x_{n'}^{(n')} - Ax_{n'}^{(n')}\|. \tag{10}$$

Given any $\varepsilon > 0$, first choose k such that

$$\|A - A_k\| < \frac{\varepsilon}{3M}. \tag{11}$$

Next, using the fact that $\{A_k x_n^{(n)}\}$ converges and hence is a Cauchy sequence, choose N such that

$$\|A_k x_n^{(n)} - A_k x_{n'}^{(n')}\| < \frac{\varepsilon}{3} \tag{12}$$

for all $n, n' > N$. Then it follows from (9)–(12) that

$$\|Ax_n^{(n)} - Ax_{n'}^{(n')}\| < \frac{\varepsilon}{3} + \frac{\varepsilon}{3} + \frac{\varepsilon}{3} = \varepsilon$$

for all sufficiently large n and n', i.e., $\{Ax_n^{(n)}\}$ is a Cauchy sequence. ∎

Not only is the set of completely continuous operators closed (algebraically) under operator multiplication, but we have the following much stronger result:

THEOREM 3. *Let A be a completely continuous operator and B a bounded operator mapping a Banach space E into itself. Then the operators AB and BA are completely continuous.*

Proof. If the set $M \subset E$ is bounded, then $BM = \{y : y = Bx, x \in M\}$ is also bounded. Therefore ABM is relatively compact, and hence AB is completely continuous. Moreover, if M is bounded, then AM is relatively compact, and hence BAM is also relatively compact by the continuity of B, i.e., BA is completely continuous. ∎

COROLLARY. *A completely continuous operator A mapping a Banach space E into itself cannot have a bounded inverse if E is infinite-dimensional.*

Proof. If A^{-1} were bounded, then, by Theorem 3, the identity operator $I = A^{-1}A$ would be completely continuous. But this is impossible, by Example 1, p. 240. ∎

THEOREM 4. *Let A be a completely continuous operator mapping a Banach space E into itself. Then the adjoint operator A^* is also completely continuous.*

Proof. We must show that A^* carries every bounded subset of the conjugate space E^* into a relatively compact set. Since every bounded subset of a normed linear space is contained in some closed sphere, it is enough to show that A^* maps every closed sphere into a relatively compact set. In fact, by the linearity of A^*, we need only show that the image A^*S^* of the closed unit sphere $S^* \subset E^*$ is relatively compact.

Now suppose we regard the elements of E^* as functionals not on the whole space E but only on the compactum $[AS]$ equal to the closure of the image of the closed unit sphere under the operator A. Then the set Φ of functionals on $[AS]$ corresponding to those in S^* is uniformly bounded and equicontinuous, since $\|\varphi\| \leqslant 1$ implies

$$\sup_{x \in [AS]} |\varphi(x)| = \sup_{x \in AS} |\varphi(x)| \leqslant \|\varphi\| \sup_{x \in S} \|Ax\| \leqslant \|A\|$$

and

$$|\varphi(x') - \varphi(x'')| \leqslant \|\varphi\| \|x' - x''\| \leqslant \|x' - x''\|.$$

Hence, by Arzelà's theorem (Theorem 4, p. 102), Φ is relatively compact in the space $C_{[AS]}$ of all continuous linear functionals on $[AS]$. But the set Φ, with the metric induced by the usual metric of $C_{[AS]}$, is isometric to the set A^*S^*, with the metric induced by the norm of the space E^*. In fact, if $g_1, g_2 \in S^*$, then

$$\|A^*g_1 - A^*g_2\| = \sup_{x \in S} |(A^*g_1 - A^*g_2, x)| = \sup_{x \in S} |(g_1 - g_2, Ax)|$$

$$= \sup_{z \in AS} |(g_1 - g_2, z)| = \sup_{z \in [AS]} |(g_1 - g_2, z)| = \rho(g_1, g_2).$$

Being relatively compact, the set Φ is totally bounded, by Theorem 3, p. 101. Therefore the set A^*S^* isometric to Φ is also totally bounded, and hence relatively compact, by the same theorem. ∎

THEOREM 5. *Let A be a completely continuous operator mapping a Banach space E into itself. Then, given any $\rho > 0$, there are only finitely many linearly independent eigenvectors of A corresponding to eigenvalues of absolute value greater than ρ.*

Proof. Given nonzero eigenvalue λ of A, let E_λ be the subspace of E consisting of all eigenvectors of A corresponding to λ.[5] Then E_λ is finite-dimensional, since otherwise A would fail to be completely continuous in E_λ and hence in E itself, by virtually the same argument as in Example 1, p. 240. Therefore, to complete the proof, we need only show that if $\{\lambda_n\}$ is any sequence of distinct eigenvalues of A, then $\lambda_n \to 0$ as $n \to \infty$. This in turn will be proved once we show that there is no infinite sequence $\{\lambda_n\}$ of distinct eigenvalues of A such that the sequence $\{1/\lambda_n\}$ is bounded.

Thus, suppose there is a sequence $\{\lambda_n\}$ of distinct eigenvalues of A such that $\{1/\lambda_n\}$ is bounded, and let x_n be an eigenvector of A corresponding to the eigenvalue λ_n. Then the vectors $x_1, x_2, \ldots$ are linearly independent, by the same argument as in the case where E is finite-dimensional.[6] Let E_n be the subspace generated by $x_1, \ldots, x_n$, i.e., the set of all elements of the form

$$y = \sum_{k=1}^{n} \alpha_k x_k.$$

For every $y \in E_n$, we have

$$y - \frac{1}{\lambda_n} A y = \sum_{k=1}^{n} \alpha_k x_k - \sum_{k=1}^{n} \frac{\alpha_k \lambda_k}{\lambda_n} x_k = \sum_{k=1}^{n-1} \alpha_k \left(1 - \frac{\lambda_k}{\lambda_n}\right) x_k,$$

---

[5] Note that E_λ is *invariant* under A in the sense that $x \in E_\lambda$ implies $Ax \in E_\lambda$ (cf. Problem 5, p. 238).

[6] See e.g., G. E. Shilov, *op. cit.*, Lemma 1, p. 182.

so that

$$y - \frac{1}{\lambda_n} Ay \in E_{n-1}.$$

Let $\{y_n\}$ be a sequence such that $y_n \in E_n$, $\|y_n\| = 1$ and

$$\wp(E_{n-1}, y_n) = \inf_{x \in E_{n-1}} \|x - y_n\| > \tfrac{1}{2}$$

(such a sequence exists by the lemma on p. 240). Then $\{y_n/\lambda_n\}$ is a bounded sequence in E, since the numerical sequence $\{1/\lambda_n\}$ is bounded. But at the same time the sequence $\{A(y_n/\lambda_n)\}$ cannot contain a convergent subsequence, contrary to the complete continuity of A, since

$$\left\| A\left(\frac{y_p}{\lambda_p}\right) - A\left(\frac{y_q}{\lambda_q}\right) \right\| = \left\| y_p - \frac{1}{\lambda_p} Ay_p + A\left(\frac{y_q}{\lambda_q}\right) \right\| > \frac{1}{2}$$

for all $p > q$, since

$$y_p - \frac{1}{\lambda_p} Ay_p + A\left(\frac{y_q}{\lambda_q}\right) \in E_{p-1}.$$

This contradiction proves the theorem. ∎

24.3. Completely continuous operators in Hilbert space. Specializing to the case of completely continuous operators mapping a *Hilbert space* into itself, we have

THEOREM 6. *Let A be a linear operator mapping a Hilbert space H into itself. Then A is completely continuous if and only if*

1) *A maps every relatively compact set in the weak topology into a relatively compact set in the strong topology;*
2) *A maps every weakly convergent sequence into a strongly convergent sequence.*

Proof. To prove 1), we merely note that H is the conjugate of a separable space, since $H = H^*$, and hence, by Corollary 2′, p. 205, a subset of H is bounded if and only if it is relatively compact in the weak topology.

To prove 2), suppose A maps every weakly convergent sequence into a strongly convergent sequence, and let M be a bounded closed subset of H. Then M contains a weakly convergent sequence and hence AM contains a strongly convergent sequence, i.e., AM is relatively compact in the strong topology. It follows that A is completely continuous. Conversely, if A is completely continuous, let $\{x_n\}$ be a weakly convergent sequence with weak limit x. Then $\{Ax_n\}$ contains a strongly convergent subsequence. At the same time, $\{Ax_n\}$ converges weakly to Ax, by the continuity of A, so that $\{Ax_n\}$ cannot have more than one limit point. Therefore $\{Ax_n\}$ is a strongly convergent sequence. ∎

Let A be a self-adjoint operator in a finite-dimensional complex Euclidean space, and suppose A has matrix $\|a_{ij}\|$ (recall Example 3, p. 222). Then it will be recalled from linear algebra that $\|a_{ij}\|$ can be reduced to diagonal form with respect to a suitable orthonormal basis.[7] We now generalize this result to the case of a completely continuous self-adjoint operator in a (real or complex) Hilbert space (see Theorem 7 below), after first proving two preliminary lemmas:

LEMMA 1. *Let A be a completely continuous self-adjoint operator mapping a Hilbert space H into itself, and let $\{x_n\}$ be a sequence in H converging weakly to x. Then*

$$(Ax_n, x_n) \to (Ax, x) \tag{13}$$

as $n \to \infty$.

Proof. Clearly,

$$|(Ax_n, x_n) - (Ax, x)| \leqslant |(Ax_n, x_n) - (Ax, x_n)| + |(Ax, x_n) - (Ax, x)|.$$

But

$$|(Ax_n, x_n) - (Ax, x_n)| \leqslant \|x_n\| \, \|A(x_n - x)\|,$$

and

$$|(Ax, x_n) - (Ax, x)| = |(x, A(x_n - x))| \leqslant \|x\| \, \|A(x_n - x)\|,$$

where the numbers $\|x_n\|$, $n = 1, 2, \ldots$ are bounded, by Theorem 2, p. 196, and $\|A(x_n - x)\| \to 0$ by Theorem 6. Therefore

$$|(Ax_n, x_n) - (Ax, x)| \to 0$$

as $n \to \infty$, which is equivalent to (13). ∎

LEMMA 2. *Given a bounded linear operator A mapping a Hilbert space H into itself, let A be self-adjoint and suppose the least upper bound of the functional*

$$|Q(x)| = |(Ax, x)|$$

on the closed unit sphere $\|x\| \leqslant 1$ is achieved at the point $x = x_0$. Then

$$(x_0, y) = 0 \tag{14}$$

implies

$$(Ax_0, y) = (x_0, Ay) = 0.$$

In particular, x_0 is an eigenvector of A.

---

[7] See e.g., V. I. Smirnov, *Linear Algebra and Group Theory* (translated by R. A. Silverman), McGraw-Hill Book Co., New York (1961), Sec. 40. Dover reprint (1970).

Proof. Obviously,

$$\|x_0\| = 1. \tag{15}$$

Let

$$x = \frac{x_0 + ay}{\sqrt{1 + |a|^2 \|y\|^2}},$$

where a is an arbitrary complex number. Then $\|x\| = 1$, because of (14) and (15). Since

$$Q(x) = \frac{1}{1 + |a|^2 \|y\|^2} [Q(x_0) + 2 \operatorname{Re} \bar{a}(Ax_0, y) + |a|^2 Q(y)],$$

we have

$$Q(x) = Q(x_0) + 2 \operatorname{Re} \bar{a}(Ax_0, y) + O(|a|^2) \tag{16}$$

for small $|a|$. But it is clear from (16) that if $(Ax_0, y) \neq 0$, then a can be chosen to make $|Q(x)| > |Q(x_0)|$, contrary to the assumption that the least upper bound of $|Q(x)|$ on the closed unit sphere is achieved at the point $x = x_0$. Therefore $(Ax_0, y) = 0$ as asserted, i.e., A is orthogonal to every vector orthogonal to x_0. It follows that Ax_0 and x_0 are proportional (why?), so that x_0 is an eigenvector of A. ∎

THEOREM 7 (*Hilbert-Schmidt*). *Let A be a completely continuous self-adjoint operator mapping a Hilbert space H into itself. Then there is an orthonormal system $\varphi_1, \varphi_2, \ldots$ of eigenvectors of A, with corresponding nonzero eigenvalues $\lambda_1, \lambda_2, \ldots$, such that every element $x \in H$ has a unique representation of the form*[8]

$$x = \sum c_n \varphi_n + x', \tag{17}$$

where x' satisfies the condition $Ax' = 0$. Moreover

$$Ax = \sum \lambda_n c_n \varphi_n, \tag{18}$$

and

$$\lim_{n \to \infty} \lambda_n = 0$$

in the case where there are infinitely many nonzero eigenvalues.

Proof. Let

$$M_1 = \sup_{\substack{\|x\| \leq 1 \\ x \in H}} |(Ax, x)|,$$

and let $\{x_n\}$ be a sequence of elements of H such that $\|x_n\| = 1$ and

$$|(Ax_n, x_n)| \to M_1$$

as $n \to \infty$. Since the closed unit sphere in H is weakly compact (recall

---

[8] As will appear in the course of the proof, the sums in (17) and (18) may be finite or infinite, and x' may vanish.

Corollary 2′, p. 205), we can find a subsequence of $\{x_n\}$ which converges weakly to an element $y \in H$, where clearly $\|y\| \leqslant 1$. By Lemma 1,

$$|(Ay, y)| = M_1,$$

and hence, by Lemma 2, y is an eigenvector of A. Moreover $\|y\| = 1$, since if $\|y\| < 1$, then choosing

$$y' = \frac{y}{\|y\|},$$

we would have $\|y'\| = 1$ and

$$|(Ay', y')| > M_1,$$

contrary to the meaning of M_1. We choose y as our first eigenvector φ_1. Let λ_1 be the corresponding eigenvalue, so that

$$A\varphi_1 = \lambda_1\varphi_1.$$

Then

$$|\lambda_1| = |(A\varphi_1, \varphi_1)| = M_1.$$

Next let E_1 be the subspace of H consisting of all vectors of the form $\alpha\varphi_1$, and let $E_1' = H \ominus E_1$ be the orthogonal complement of E_1. Clearly E_1' is again a Hilbert space, mapped into itself by the operator A (this follows from Problem 5, p. 238 and the fact that A is self-adjoint). Let

$$M_2 = \sup_{\substack{\|x\| \leqslant 1 \\ x \in E_1'}} |(Ax, x)|. \tag{19}$$

Then, by the same argument as before, we can find an eigenvector φ_2 of A such that $\varphi_2 \in E_1'$, $\|\varphi_2\| = 1$. Let λ_2 be the corresponding eigenvalue, so that

$$A\varphi_2 = \lambda_2\varphi_2.$$

Then

$$|\lambda_2| = |(A\varphi_2, \varphi_2)| = M_2,$$

and hence

$$|\lambda_1| \geqslant |\lambda_2|,$$

since $H \supset E_1'$ implies

$$M_1 = \sup_{\substack{\|x\| \leqslant 1 \\ x \in H}} |(Ax, x)| \geqslant \sup_{\substack{\|x\| \leqslant 1 \\ x \in E_1'}} |(Ax, x)| = M_2.$$

By its very construction, φ_2 is orthogonal to φ_1.

To construct further eigenvectors of A, we argue inductively, replacing (19) by

$$M_{n+1} = \sup_{\substack{\|x\| \leqslant 1 \\ x \in E_n'}} |(Ax, x)| \qquad (n = 1, 2, \ldots),$$

where $E'_n = H \ominus E_n$ is the orthogonal complement of the subspace E_n generated by the previously constructed eigenvectors $\varphi_1, \varphi_2, \ldots, \varphi_n$. Then E'_n is again a Hilbert space mapped into itself by A, and there is an eigenvector $\varphi_{n+1} \in E'_n$ of unit norm, with corresponding eigenvalue λ_{n+1} satisfying the inequality

$$|\lambda_n| \geqslant |\lambda_{n+1}| \qquad (n = 1, 2, \ldots).$$

In this way, we construct an orthonormal system $\{\varphi_n\}$ of eigenvectors of A.

There are now just two possibilities, which we examine in turn:

Case 1. Suppose the construction of the sequence $\{\varphi_n\}$ terminates after a finite number of steps, i.e., suppose there is a positive integer n_0 such that $(Ax, x) \equiv 0$ on E'_{n_0}. Then it follows from Lemma 2 that A maps the whole space E'_{n_0} into the zero vector. According to Theorem 14, p. 158, every element $x \in H$ has a unique representation of the form

$$x = h + x',$$

where $h \in E_{n_0}$, $x' \in E'_{n_0}$, and hence of the form

$$x = \sum c_n \varphi_n + x',$$

where the sum is finite (consisting of n_0 terms) and $Ax' = 0$. Obviously we have

$$Ax = \sum \lambda_n c_n \varphi_n,$$

thereby completing the proof in this case.

Case 2. Suppose the construction of the sequence $\{\varphi_n\}$ never terminates, i.e., suppose $(Ax, x) \not\equiv 0$ on E'_n for all $n = 1, 2, \ldots$. We then have infinitely many nonzero eigenvalues $\lambda_1, \lambda_2, \ldots, \lambda_n, \ldots$. Clearly $\lambda_n \to 0$ as $n \to \infty$. In fact, the sequence $\{\varphi_n\}$ converges weakly to zero, like any sequence of orthonormal vectors (why?), and hence the sequence $\{A\varphi_n\}$ converges to zero in norm, so that $\|A\varphi_n\| \to 0$ and hence $\|\lambda_n \varphi_n\| = |\lambda_n| \to 0$. Let E_∞ be the subspace of H generated by all the eigenvectors $\varphi_1, \varphi_2, \ldots, \varphi_n, \ldots$, i.e., the set of all linear combinations of the form

$$\sum_{n=1}^{\infty} c_n \varphi_n,$$

and let

$$E'_\infty = H \ominus E_\infty = \bigcap_{n=1}^{\infty} E'_n.$$

If $E'_\infty = \{0\}$, then $H = E_\infty$ and x obviously has a representation of the form (17) with $x' = 0$ (so that $Ax' = 0$ trivially). If $E'_\infty \neq 0$, let x be any nonzero element of E'_∞. Then

$$|(Ax, x)| \leqslant |\lambda_n| \, \|x\|^2$$

for all $n = 1, 2, \ldots$, and hence $(Ax, x) \equiv 0$ on E'_∞. It follows from Lemma 2 that A maps the whole space E'_∞ into the zero vector. The rest of the proof is the same as in Case 1, where (18) follows from (17) by the continuity of A. ∎

COROLLARY. *Let A be a completely continuous self-adjoint operator mapping a Hilbert space H into itself. Then there is an orthonormal system $\{\psi_n\}$ of eigenvectors of A such that every element $x \in H$ has a unique representation of the form*

$$x = \sum_{n=1}^{\infty} c_n \psi_n.$$

Moreover

$$Ax = \sum_{n=1}^{\infty} \lambda_n c_n \psi_n,$$

where $\lambda_1, \lambda_2, \ldots$ are the eigenvalues corresponding to $\psi_1, \psi_2, \ldots$.

Proof. Noting that every element of E'_{n_0} or E'_∞ is an eigenvector of A corresponding to the eigenvalue $\lambda = 0$, let $\{\psi_n\}$ consist of the orthonormal system $\{\varphi_n\}$ constructed in the proof of Theorem 7, together with an arbitrary orthonormal basis in E'_{n_0} or E'_∞. ∎

Problem 1. Prove that the projection operator of Example 4, p. 222 is completely continuous if and only if the subspace H_1 is finite-dimensional.

Problem 2. Prove that the operator A mapping the point

$$x = (x_1, x_2, \ldots, x_n, \ldots) \in l_2$$

into the point

$$Ax = \left(x_1, \frac{x_2}{2}, \ldots, \frac{x_n}{2^{n-1}}, \ldots\right) \in l_2$$

is completely continuous. More generally, suppose

$$Ax = (a_1 x_1, a_2 x_2, \ldots, a_n x_n, \ldots).$$

Under what conditions on the sequence $\{a_n\}$ is A completely continuous?

Hint. Since every bounded set in l_2 is contained in some closed sphere, it is enough to show that the images of spheres are relatively compact. In fact, by the linearity of A, it need only be shown that the image of the unit sphere is compact. In this regard, recall Example 5, p. 98.

Problem 3. Let A be the integral operator on $C_{[-1,1]}$ defined by

$$\psi(x) = (A\varphi)(x) = \int_{-1}^{x} \varphi(y) \, dy.$$

Prove that A maps the closed unit sphere in $C_{[-1,1]}$ into a noncompact set. Reconcile this with Theorem 1.

Hint. Let

$$\varphi_n(x) = \begin{cases} 0 & \text{if } -1 \leqslant x \leqslant 0, \\ nx & \text{if } 0 < x \leqslant \dfrac{1}{n}, \\ 1 & \text{if } \dfrac{1}{n} < x \leqslant 1. \end{cases}$$

Then $\varphi_n \in C_{[-1,1]}$, $\|\varphi_n\| = 1$ for all n, and

$$\psi_n(x) = (A\varphi_n)(x) = \begin{cases} 0 & \text{if } -1 \leqslant x \leqslant 0, \\ \dfrac{1}{2}nx^2 & \text{if } 0 < x \leqslant \dfrac{1}{n}, \\ x - \dfrac{1}{2n} & \text{if } \dfrac{1}{n} < x \leqslant 1. \end{cases}$$

The sequence $\{\psi_n\}$ converges in $C_{[-1,1]}$ to the function

$$\psi(x) = \begin{cases} 0 & \text{if } -1 \leqslant x \leqslant 0, \\ x & \text{if } 0 < x \leqslant 1, \end{cases}$$

which, having a discontinuous derivative, cannot be the image under A of any function in $C_{[-1,1]}$.

Problem 4. Let A be a completely continuous operator mapping a reflexive Banach space E (e.g. a Hilbert space) into itself. Prove that A maps the closed unit sphere in E into a compact set. Reconcile this with the preceding problem.

Hint. Use Theorem 6, p. 205.

Problem 5. Prove that

a) A linear combination of completely continuous operators is itself a completely continuous operator;

b) The set $\mathscr{C}(E, E)$ of all completely continuous operators mapping a Banach space E into itself is a closed subspace of the linear space $\mathscr{L}(E, E)$ of all bounded linear operators mapping E into E.

Problem 6. Let $\mathscr{C}(E, E)$ and $\mathscr{L}(E, E)$ be the same as in the preceding problem. Prove that besides being a linear space, $\mathscr{L}(E, E)$ is also a ring when equipped with the usual operations of addition and multiplication of operators. Prove that $\mathscr{C}(E, E)$ is a two-sided ideal in $\mathscr{L}(E, E)$.

Comment. By a *two-sided ideal* in a ring $\mathscr{R}$ is meant a subring $\mathscr{A} \subset \mathscr{R}$ such that $a \in \mathscr{A}$, $r \in \mathscr{R}$ implies $ar \in \mathscr{A}$, $ra \in \mathscr{A}$.

Problem 7. Let Φ and A^*S^* be the same as in the proof of Theorem 4. Show that Φ is closed and hence compact. Deduce from this that A^*S^* is compact, even though as shown in Problem 3, the image of the closed unit sphere under a completely continuous operator need not be compact.

Problem 8. Discuss the connection between Theorem 4 and the theory of Sec. 20.4, in particular Corollary 1', p. 204.

Problem 9. Let A be a bounded linear operator mapping a Banach space E into itself. Show that if A^* is completely continuous, then so is A.

Problem 10. Prove that a linear operator A mapping a Hilbert space H into itself is completely continuous if and only if its adjoint (in the sense of Sec. 23.3) is completely continuous.

Problem 11. Give an example of a completely continuous operator A mapping a Hilbert space H into itself, such that A has no eigenvectors. Reconcile this with Theorem 7.

Hint. Let A be the operator in l_2 such that

$$Ax = A(x_1, x_2, x_3, \ldots, x_n, \ldots) = \left(0, x_1, \frac{x_2}{2}, \ldots, \frac{x_{n-1}}{n-1}, \ldots\right).$$

Then $Ax = \lambda x$ implies

$$\lambda x_1 = 0, \lambda x_2 = x_1, \lambda x_3 = \frac{x_2}{2}, \ldots, \lambda x_n = \frac{x_{n-1}}{n-1}, \ldots,$$

and hence $x = 0$.

Comment. This situation differs from the finite-dimensional case, where every linear operator (self-adjoint or not) has at least one eigenvector.

7

MEASURE

The concept of the measure $\mu(E)$ of a set E is a natural generalization of such concepts as

1) The length $l(\Delta)$ of a line segment Δ;
2) The area $A(F)$ of a plane figure F;
3) The volume $V(G)$ of a space figure G;
4) The increment $\varphi(b) - \varphi(a)$ of a nondecreasing function $\varphi(t)$ over a half-open interval $[a, b)$;
5) The integral of a nonnegative function over a set on the line or over a region in the plane or in space.

Although the notion of measure first arose in the theory of functions of a real variable, it was subsequently used extensively in functional analysis, probability theory, the theory of dynamical systems, and other branches of mathematics. In Sec. 25 we discuss the measure of plane sets, starting from the notion of the area of a rectangle. Measure in general will then be studied in Secs. 26 and 27. The reader will easily confirm that the considerations in Sec. 25 are of a general nature and carry over to the case of the more abstract theory without essential changes.

25. Measure in the Plane

25.1. Measure of elementary sets. Consider the system $\mathscr{S}$ of sets in the xy-plane, each defined by one of the inequalities

$$a \leqslant x \leqslant b, \qquad a < x \leqslant b, \qquad a \leqslant x < b, \qquad a < x < b$$

and one of the inequalities

$$c \leqslant y \leqslant d, \quad c < y \leqslant d, \quad c \leqslant y < d, \quad c < y < d,$$

where a, b, c and d are arbitrary real numbers. The sets in $\mathscr{S}$ will be called *rectangles*. The closed rectangle defined by the inequalities

$$a \leqslant x \leqslant b, \quad c \leqslant y \leqslant d$$

is a rectangle in the usual sense (including its boundary) if $a < b$ and $c < d$, a line segment (including its end points) if $a = b$ and $c < d$ or if $a < b$ and $c = d$, a point if $a = b$, $c = d$, or even the empty set if $a > b$ or $c > d$. The open rectangle

$$a < x < b, \quad c < y < d$$

is either a rectangle in the usual sense (without its boundary) if $a < b$ and $c < d$ or the empty set if $a > b$ or $c > d$. Each of the rectangles of the remaining types will be called *half-open* and is an ordinary rectangle minus one, two or three sides, a line segment minus one or two end points, or possibly the empty set.

In keeping with the concept of area familiar from elementary geometry, we now define the *measure* of each set in $\mathscr{S}$ as follows:

1) The measure of the empty set equals 0;
2) The measure of the nonempty rectangle (closed, open or half-open) specified by the numbers a, b, c, and d equals

$$(b - a)(d - c).$$

Thus with each rectangle $P \in \mathscr{S}$ we associate a number $m(P)$, called its *measure*, where clearly

1) $m(P)$ is real and nonnegative;
2) $m(P)$ is *additive* in the sense that if

$$P = \bigcup_{k=1}^{n} P_k, \quad P_k \cap P_l = \varnothing$$

then

$$m(P) = \sum_{k=1}^{n} m(P_k).$$

Our problem is to define the concept of measure for sets more general than rectangles, while preserving these two properties. The first step in this direction is to define measure for elementary sets, where by an *elementary set* we mean any set which can be represented in at least one way as a union of a finite number of pairwise disjoint rectangles. First we prove

THEOREM 1. *The union, intersection, difference and symmetric difference of two elementary sets are again elementary sets.*

Proof. If

$$A = \bigcup_k P_k, \qquad B = \bigcup_l Q_l$$

are two elementary sets, then clearly

$$A \cap B = \bigcup_{k,l} (P_k \cap Q_l)$$

is also an elementary set, since each $P_k \cap Q_l$ is obviously either a rectangle or the empty set. Moreover, it is easy to see that the difference of two rectangles is an elementary set. Hence, subtracting an elementary set from a rectangle gives another elementary set (as an intersection of elementary sets). Suppose A and B are elementary sets, and let P be a rectangle containing both of them (such a rectangle obviously exists). It follows from what has just been proved that

$$A \cup B = P - [(P - A) \cap (P - B)]$$

is an elementary set. It is then an easy consequence of the formulas

$$A - B = A \cap (P - B),$$

$$A \triangle B = (A \cup B) - (A \cap B)$$

that the difference and symmetric difference of two elementary sets is again an elementary set. ∎

Remark. In other words, the system of all elementary sets is a *ring* $\mathscr{R}$, as defined on p. 31.

We now define measure for elementary sets:

DEFINITION 1. *Given an elementary set A, suppose*

$$A = \bigcup_k P_k,$$

*where the P_k are pairwise disjoint rectangles. Then by the **measure** of A, denoted by $\tilde{m}(A)$, is meant the number*

$$\tilde{m}(A) = \sum_k m(P_k), \qquad (1)$$

where $m(P_k)$ is the measure of the rectangle P.

Remark. Clearly, $\tilde{m}(A)$ is nonnegative and additive. Moreover, in defining $\tilde{m}(A)$, we have tacitly relied on the fact that the sum (1) does not depend on how A is represented as a union of sets. To verify this, suppose

$$A = \bigcup_k P_k = \bigcup_l Q_l,$$

where P_k and Q_l are rectangles such that

$$P_i \cap P_j = \varnothing, \qquad Q_i \cap Q_j = \varnothing \qquad (i \neq j).$$

Since the intersection $P_k \cap Q_l$ of two rectangles is itself a rectangle, it follows from the additivity of the measure of rectangles that

$$\sum_k m(P_k) = \sum_{k,l} m(P_k \cap Q_l) = \sum_l m(Q_l).$$

THEOREM 2. *If A is an elementary set and $\{A_n\}$ is a finite or countable system of elementary sets such that*

$$A \subset \bigcup_n A_n,$$

then

$$\tilde{m}(A) \leqslant \sum_n \tilde{m}(A_n). \tag{2}$$

Proof. Given any $\varepsilon > 0$, there is a closed elementary set $\bar{A}$ contained in A and satisfying the condition

$$\tilde{m}(\bar{A}) \geqslant \tilde{m}(A) - \frac{\varepsilon}{2}.$$

In fact, to get $\bar{A}$ we need only replace each of the k rectangles P_j making up A by a closed rectangle contained in P_j of area no less than

$$m(P_j) - \frac{\varepsilon}{2k}.$$

Moreover, for each A_n there is clearly an open elementary set $\hat{A}_n$ containing A_n and satisfying the condition

$$\tilde{m}(\hat{A}_n) \leqslant \tilde{m}(A_n) + \frac{\varepsilon}{2^{n+1}}.$$

Obviously,

$$\bar{A} \subset \bigcup_n \hat{A}_n.$$

Hence, by the Heine-Borel theorem (recall p. 92), there is a finite system $\hat{A}_{n_1}, \ldots, \hat{A}_{n_s}$ covering $\bar{A}$, where

$$\tilde{m}(\bar{A}) \leqslant \sum_{i=1}^{s} \tilde{m}(\hat{A}_{n_i}),$$

since otherwise $\bar{A}$ would be covered by a finite number of rectangles of total area less than $\tilde{m}(\bar{A})$, which is impossible. Therefore

$$\tilde{m}(A) \leqslant \tilde{m}(\bar{A}) + \frac{\varepsilon}{2} \leqslant \sum_{i=1}^{s} \tilde{m}(\hat{A}_{n_i}) + \frac{\varepsilon}{2} \leqslant \sum_n \tilde{m}(A_n) + \frac{\varepsilon}{2}$$

$$\leqslant \sum_n \tilde{m}(A_n) + \sum_n \frac{\varepsilon}{2^{n+1}} + \frac{\varepsilon}{2} = \sum_n \tilde{m}(A_n) + \varepsilon,$$

which implies (2), since $\varepsilon > 0$ is arbitrary. ∎

25.2. Lebesgue measure of plane sets. Elementary sets are, of course, far from being the most general plane sets considered in geometry and analysis. Thus we naturally arrive at the problem of extending the concept of measure (while preserving its basic properties) to sets more general than finite unions of rectangles with sides parallel to the coordinate axes. This problem is solved in a definitive way by Lebesgue's theory of measure, in which we consider countably *infinite* unions of rectangles, as well as finite unions. To avoid sets of "infinite measure," we restrict our discussion to subsets of the closed unit square E, defined by the inequalities

$$0 \leqslant x \leqslant 1, \qquad 0 \leqslant y \leqslant 1$$

(this restriction is dropped in Remarks 2 and 3, p. 267).

DEFINITION 2. *By the **outer measure** of a set $A \subset E$ is meant the number*

$$\mu^*(A) = \inf_{A \subset \bigcup_k P_k} \sum_k m(P_k),$$

where the greatest lower bound is taken over all coverings of A by a finite or countable system of rectangles P_k.

DEFINITION 3. *By the **inner measure** of a set $A \subset E$ is meant the number*

$$\mu_*(A) = 1 - \mu^*(E - A).$$

THEOREM 3. *The inequality*

$$\mu_*(A) \leqslant \mu^*(A)$$

holds for any set $A \subset E$.

Proof. Suppose

$$\mu_*(A) > \mu^*(A),$$

i.e.,

$$\mu^*(A) + \mu^*(E - A) < 1.$$

Then, by the definition of a greatest lower bound, there are systems of rectangles $\{P_j\}$ and $\{Q_k\}$ covering A and $E - A$, respectively, such that

$$\sum_j m(P_j) + \sum_k m(Q_k) < 1.$$

Let $\{R_l\}$ denote the union of the systems $\{P_j\}$ and $\{Q_k\}$. Then

$$E \subset \bigcup_l R_l,$$

while

$$m(E) > \sum_l m(R_l),$$

contrary to Theorem 2. ∎

DEFINITION 4. *A set A is said to be* (*Lebesgue*) *measurable if*

$$\mu_*(A) = \mu^*(A),$$

i.e., if its inner and outer measures coincide.

DEFINITION 5. *If a set A is measurable, the number $\mu(A)$ equal to the common value of $\mu_*(A)$ and $\mu^*(A)$ is called the* (*Lebesgue*) *measure of A.*

For outer measure, we have the following analogue of Theorem 2:

THEOREM 4. *If A is any set and $\{A_n\}$ is a finite or countable system of sets such that*

$$A \subset \bigcup_n A_n,$$

then

$$\mu^*(A) \leqslant \sum_n \mu^*(A_n). \tag{2'}$$

Proof. Given any $\varepsilon > 0$, for each A_n there is a finite or countable system of rectangles $\{P_{nk}\}$ such that

$$A_n \subset \bigcup_r P_{nk}$$

and

$$\sum_k m(P_{nk}) \leqslant \mu^*(A_n) + \frac{\varepsilon}{2^n},$$

by the definition of outer measure. Then

$$A \subset \bigcup_n \bigcup_k P_{nk}$$

and

$$\mu^*(A) \leqslant \sum_n \sum_k m(P_{nk}) \leqslant \sum_n \mu^*(A_n) + \varepsilon,$$

which implies (2'), since $\varepsilon > 0$ is arbitrary. ∎

COROLLARY. *If A is any measurable set and $\{A_n\}$ is a finite or countable system of measurable sets such that*

$$A \subset \bigcup_n A_n,$$

then

$$\mu(A) \leqslant \sum_n \mu(A_n). \tag{2''}$$

Proof. Merely replace μ^* by μ in (2'). ∎

Next we show that the Lebesgue measure of an elementary set coincides with its measure as previously defined:

THEOREM 5. *Every elementary set $A \subset E$ is measurable, with Lebesgue measure $\mu(A)$ equal to the measure $\tilde{m}(A)$ introduced in Definition 1.*

Proof. Suppose A is the union of the pairwise disjoint rectangles $P_1, \ldots, P_k$. Then

$$\tilde{m}(A) = \sum_{j=1}^{k} m(P_j),$$

by Definition 1. Therefore, since the rectangles $P_1, \ldots, P_k$ obviously cover A,

$$\mu^*(A) \leqslant \sum_j m(P_j) = \tilde{m}(A), \tag{3}$$

by Definition 2. Moreover, if $\{Q_j\}$ is any finite or countable system of rectangles covering A, we have

$$\tilde{m}(A) \leqslant \sum_j m(Q_j)$$

by Theorem 2, and hence

$$\tilde{m}(A) \leqslant \mu^*(A), \tag{4}$$

by Definition 2 again. Comparing (3) and (4), we get

$$\tilde{m}(A) = \mu^*(A).$$

Now $E - A$ is also an elementary set, and hence

$$\tilde{m}(E - A) = \mu^*(E - A).$$

But

$$\tilde{m}(E - A) = 1 - \tilde{m}(A),$$

while

$$\mu^*(E - A) = 1 - \mu_*(A).$$

It follows that

$$\tilde{m}(A) = \mu_*(A),$$

and hence

$$\tilde{m}(A) = \mu_*(A) = \mu^*(A). \quad \blacksquare$$

COROLLARY. *Theorem 2 is a special case of Theorem 4.*

Proof. Merely replace μ^* by $\tilde{m}$ in (2′) or μ by $\tilde{m}$ in (2″). $\blacksquare$

LEMMA. *The inequality*

$$|\mu^*(A) - \mu^*(B)| \leqslant \mu^*(A \vartriangle B) \tag{5}$$

holds for any two sets A and B.

Proof. Since

$$A \subset B \cup (A \vartriangle B)$$

it follows from Theorem 4 that

$$\mu^*(A) \leqslant \mu^*(B) + \mu^*(A \vartriangle B). \tag{6}$$

This implies (5) if $\mu^*(A) \geqslant \mu^*(B)$. If $\mu^*(A) \leqslant \mu^*(B)$, we deduce (5) from the inequality

$$\mu^*(B) \leqslant \mu^*(A) + \mu^*(A \,\triangle\, B)$$

obtained by interchanging the roles of A and B in (6). ∎

THEOREM 6. *A set A is measurable if and only if, given any $\varepsilon > 0$, there is an elementary set B such that*

$$\mu^*(A \,\triangle\, B) < \varepsilon. \tag{7}$$

Proof. Suppose that given any $\varepsilon > 0$, there is an elementary set B such that (7) holds. Then, by the lemma,

$$|\mu^*(A) - \mu^*(B)| = |\mu^*(A) - \tilde{m}(B)| < \varepsilon, \tag{8}$$

and similarly

$$|\mu^*(E - A) - \tilde{m}(E - B)| < \varepsilon, \tag{9}$$

since

$$(E - A) \,\triangle\, (E - B) = A \,\triangle\, B.$$

Bearing in mind that

$$\tilde{m}(B) + \tilde{m}(E - B) = \tilde{m}(E) = 1,$$

we deduce from (8) and (9) that

$$|\mu^*(A) - \mu^*(E - A) - 1| < 2\varepsilon,$$

and hence that

$$\mu^*(A) + \mu^*(E - A) = 1, \tag{10}$$

since $\varepsilon > 0$ is arbitrary. But then $\mu_*(A) = \mu^*(A)$, so that A is measurable.

Conversely, suppose A is measurable, i.e., suppose (10) holds. Then, given any $\varepsilon > 0$, there are systems of rectangles $\{B_n\}$ and $\{C_n\}$ covering A and $E - A$, respectively, such that

$$\sum_n m(B_n) \leqslant \mu^*(A) + \frac{\varepsilon}{3}, \tag{11}$$

$$\sum_n m(C_n) \leqslant \mu^*(E - A) + \frac{\varepsilon}{3}. \tag{12}$$

Moreover, since $\sum_n m(B_n) < \infty$, there is an N such that

$$\sum_{n > N} m(B_n) < \frac{\varepsilon}{3}.$$

We now show that (7) holds for the elementary set

$$B = \bigcup_{n=1}^{N} B_n.$$

Clearly, the set

$$P = \bigcup_{n>N} B_n$$

contains $A - B$, while the set

$$Q = \bigcup_{n} (B \cap C_n)$$

contains $B - A$, and hence

$$A \triangle B \subset P \cup Q. \tag{13}$$

Moreover,

$$\mu^*(P) \leqslant \sum_{n>N} m(B_n) < \frac{\varepsilon}{3}. \tag{14}$$

To estimate $\mu^*(Q)$, we note that

$$\left(\bigcup_{n} B_n\right) \cup \left(\bigcup_{n} (C_n - B)\right) = E,$$

and hence

$$\sum_{n} m(B_n) + \sum_{n} \tilde{m}(C_n - B) \geqslant 1. \tag{15}$$

But (11) and (12) imply

$$\sum_{n} m(B_n) + \sum_{n} m(C_n) \leqslant \mu^*(A) + \mu^*(E - A) + \frac{2\varepsilon}{3} = 1 + \frac{2\varepsilon}{3}. \tag{16}$$

Subtracting (15) from (16), we get

$$\sum_{n} m(C_n) - \sum_{n} \tilde{m}(C_n - B) = \sum_{n} \tilde{m}(C_n \cap B) < \frac{2\varepsilon}{3},$$

i.e.,

$$\mu^*(Q) < \frac{2\varepsilon}{3}. \tag{17}$$

Finally, comparing (13), (16) and (17), we find that

$$\mu^*(A \triangle B) \leqslant \mu^*(P \cup Q) \leqslant \mu^*(P) + \mu^*(Q) < \varepsilon. \quad \blacksquare$$

THEOREM 7. *The union and intersection of a finite number of measurable sets are again measurable sets.*

Proof. It is enough to prove the theorem for two sets. Thus suppose A_1 and A_2 are measurable sets. Then, by Theorem 6, there are elementary sets B_1 and B_2 such that

$$\mu^*(A_1 \triangle B_1) < \frac{\varepsilon}{2}, \qquad \mu^*(A_2 \triangle B_2) < \frac{\varepsilon}{2}.$$

Since

$$(A_1 \cup A_2) \triangle (B_1 \cup B_2) \subset (A_1 \triangle B_1) \cup (A_2 \triangle B_2),$$

we have

$$\mu^*[(A_1 \cup A_2) \triangle (B_1 \cup B_2)] \leqslant \mu^*(A_1 \triangle B_1) + \mu^*(A_2 \triangle B_2) < \varepsilon.$$

But $B_1 \cup B_2$ is an elementary set, and hence $A_1 \cup A_2$ is measurable, by Theorem 6 again. Moreover, a set A is measurable if and only if

$$\mu^*(A) + \mu^*(E - A) = 1,$$

and hence if A is measurable, so is $E - A$. Therefore the measurability of $A_1 \cap A_2$ follows from that of $A_1 \cup A_2$ and the formula

$$A_1 \cap A_2 = E - [(E - A_1) \cup (E - A_2)]. \quad \blacksquare$$

COROLLARY. *The difference and symmetric difference of two measurable sets are again measurable sets.*

Proof. An immediate consequence of Theorem 7 and the formulas

$$A_1 - A_2 = A_1 \cap (E - A_2),$$
$$A_1 \triangle A_2 = (A_1 - A_2) \cup (A_2 - A_1). \quad \blacksquare$$

THEOREM 8. *If $A_1, \ldots, A_N$ are pairwise disjoint measurable sets, then*

$$\mu\left(\bigcup_{n=1}^{N} A_n\right) = \sum_{n=1}^{N} \mu(A_n).$$

Proof. As in the proof of Theorem 7, we need only consider the case $n = 2$. By Theorem 6, given any $\varepsilon > 0$, there are elementary sets B_1 and B_2 such that

$$\mu^*(A_1 \triangle B_1) < \varepsilon, \qquad \mu^*(A_2 \triangle B_2) < \varepsilon. \qquad (18)$$

Let

$$A = A_1 \cup A_2, \qquad B = B_1 \cup B_2.$$

Then A is measurable, by Theorem 7. Since A_1 and A_2 are disjoint, we have

$$B_1 \cap B_2 \subset (A_1 \triangle B_1) \cup (A_2 \triangle B_2),$$

and hence

$$\tilde{m}(B_1 \cap B_2) < 2\varepsilon. \qquad (19)$$

Moreover, it follows from (18) and the lemma on p. 260 that

$$|\tilde{m}(B_1) - \mu^*(A_1)| < \varepsilon, \qquad |\tilde{m}(B_2) - \mu^*(A_2)| < \varepsilon. \qquad (20)$$

Since measure is additive on elementary sets, it follows from (19) and (20) that

$$\tilde{m}(B) = \tilde{m}(B_1) + \tilde{m}(B_2) - \tilde{m}(B_1 \cap B_2) \geqslant \mu^*(A_1) + \mu^*(A_2) - 4\varepsilon.$$

Noting also that
$$A \triangle B \subset (A_1 \triangle B_1) \cup (A_2 \triangle B_2),$$
we have
$$\mu^*(A) \geqslant \tilde{m}(B) - \mu^*(A \triangle B) \geqslant \tilde{m}(B) - 2\varepsilon > \mu^*(A_1) + \mu^*(A_2) - 6\varepsilon.$$
Therefore
$$\mu^*(A) > \mu^*(A_1) + \mu^*(A_2), \tag{21}$$
since $\varepsilon > 0$ can be made arbitrarily small. On the other hand, it follows from $A = A_1 \cup A_2$ and Theorem 4 that
$$\mu^*(A) \leqslant \mu^*(A_1) + \mu^*(A_2). \tag{22}$$
Comparing (21) and (22), we get
$$\mu^*(A) = \mu^*(A_1) + \mu^*(A_2),$$
where μ^* can be replaced by μ, since A_1, A_2, and A are measurable. ∎

THEOREM 9. *The union and intersection of a countable number of measurable sets are again measurable sets.*

Proof. Given a countable system of measurable sets $\{A_n\}$, let
$$A = \bigcup_{n=1}^{\infty} A_n,$$
and let
$$A_1' = A_1, \quad A_n' = A_n - \bigcup_{k=1}^{n-1} A_k \quad (n = 2, 3, \ldots).$$
Then the sets A_n' are pairwise disjoint, and
$$A = \bigcup_{n=1}^{\infty} A_n'.$$
By Theorem 7 and its corollary, the sets A_n' are all measurable. Moreover, by Theorems 4 and 8,
$$\sum_{n=1}^{N} \mu(A_n') = \mu\left(\bigcup_{n=1}^{N} A_n' \right) \leqslant \mu^*(A)$$
for every $N = 1, 2, \ldots$. Therefore the series
$$\sum_{n=1}^{\infty} \mu(A_n')$$
converges, and hence, given any $\varepsilon > 0$, there is an integer $\nu > 0$ such that
$$\sum_{n > \nu} \mu(A_n') < \frac{\varepsilon}{2}. \tag{23}$$

Since the set

$$C = \bigcup_{n=1}^{v} A'_n$$

is measurable, being the union of a finite number of measurable sets, there is an elementary set B such that

$$\mu^*(C \triangle B) < \frac{\varepsilon}{2}. \tag{24}$$

Moreover, since

$$A \triangle B \subset (C \triangle B) \cup \left(\bigcup_{n>v} A'_n \right),$$

it follows from (23) and (24) that

$$\mu^*(A \triangle B) < \varepsilon.$$

Therefore A is measurable, by Theorem 6. Finally, since complements of measurable sets are themselves measurable, the intersection

$$\bigcap_{n=1}^{\infty} A_n = E - \bigcup_{n=1}^{\infty} (E - A_n)$$

is measurable. ∎

Theorem 9 generalizes Theorem 7 to the case of a *countable* number of measurable sets. The corresponding generalization of Theorem 8 is given by

THEOREM 10. *If $A_1, A_2, \ldots, A_n, \ldots$ are pairwise disjoint measurable sets, then*

$$\mu\left(\bigcup_{n=1}^{\infty} A_n \right) = \sum_{n=1}^{\infty} \mu(A_n). \tag{25}$$

Proof. Let

$$A = \bigcup_{n=1}^{\infty} A_n.$$

Then, since

$$\bigcup_{n=1}^{N} A_n \subset A$$

for every $N = 1, 2, \ldots$, it follows from Theorem 8 and the corollary to Theorem 4 that

$$\sum_{n=1}^{N} \mu(A_n) = \mu\left(\bigcup_{n=1}^{N} A_n \right) \leqslant \mu(A).$$

Taking the limit as $N \to \infty$, we get

$$\sum_{n=1}^{\infty} \mu(A_n) \leqslant \mu(A). \tag{26}$$

On the other hand, since obviously

$$A \subset \bigcup_{n=1}^{\infty} A_n,$$

it follows from the same corollary that

$$\mu(A) \leqslant \sum_{n=1}^{\infty} \mu(A_n). \tag{27}$$

Comparing (26) and (27), we get

$$\mu(A) = \sum_{n=1}^{\infty} \mu(A_n),$$

or equivalently (25). ∎

The key property of the measure μ expressed by (25) is described by saying that μ is *countably additive* or *σ-additive*.

THEOREM 11. *Let* $\{A_n\}$ *be a sequence of measurable sets which is* **decreasing** *in the sense that*

$$A_1 \supset A_2 \supset \cdots \supset A_n \supset \cdots.$$

Then

$$\lim_{n \to \infty} \mu(A_n) = \mu(A), \tag{28}$$

where

$$A = \bigcap_{n=1}^{\infty} A_n.$$

Proof. We need only consider the case $A = \varnothing$, to which the general case reduces if A_n is replaced by $A_n - A$. Clearly

$$A_1 = (A_1 - A_2) \cup (A_2 - A_3) \cup \cdots,$$

and

$$A_n = (A_n - A_{n+1}) \cup (A_{n+1} - A_{n+2}) \cup \cdots.$$

Therefore, by the σ-additivity of μ,

$$\mu(A_1) = \sum_{k=1}^{\infty} \mu(A_k - A_{k+1}) \tag{29}$$

and

$$\mu(A_n) = \sum_{k=n}^{\infty} \mu(A_k - A_{k+1}). \tag{30}$$

Since the series (29) converges, its remainder (30) approaches 0 as $n \to \infty$. It follows that

$$\lim_{n \to \infty} \mu(A_n) = 0 = \mu(\varnothing). \quad \blacksquare$$

COROLLARY. *Let* $\{A_n\}$ *be a sequence of measurable sets which is in-creasing in the sense that*

$$A_1 \subset A_2 \subset \cdots \subset A_n \subset \cdots.$$

Then

$$\lim_{n \to \infty} \mu(A_n) = \mu(A), \qquad (28')$$

where

$$A = \bigcup_{n=1}^{\infty} A_n.$$

Proof. Apply Theorem 11 to the complements of the sets A_n. ∎

The property of the measure μ expressed by (28) and (28') is described by saying that μ is *continuous*.

Remark 1. To recapitulate, starting from a measure m defined on the class $\mathscr{S}_m$ of all rectangles (with sides parallel to the coordinate axes), we have succeeded in extending m first to a measure $\tilde{m}$ defined on the larger class $\mathscr{S}_{\tilde{m}}$ of all elementary sets and then to a Lebesgue measure μ defined on the still larger class $\mathscr{S}_{\mu}$ of all measurable sets. The class $\mathscr{S}_{\mu}$ is closed under the operations of taking countable unions and intersections. Moreover, the measure μ is σ-additive on $\mathscr{S}_{\mu}$.

Remark 2. So far we have required all our sets to be subsets of the closed unit square

$$E = \{(x, y) : 0 \leqslant x \leqslant 1, 0 \leqslant y \leqslant 1\}.$$

It is easy to get rid of this restriction. For example, representing the whole plane as the union of the squares

$$E_{mn} = \{(x, y) : m \leqslant x \leqslant m + 1, n \leqslant y \leqslant n + 1\},$$

where m and n are arbitrary integers, we say that a plane set A is *measurable* if its intersection $A_{mn} = A \cap E_{mn}$ with every square E_{mn} is measurable as previously defined and if the series

$$\sum_{m,n} \mu(A_{mn})$$

converges. The *measure* of A is then defined as

$$\mu(A) = \sum_{m,n} \mu(A_{mn}). \qquad (31)$$

All the properties of measure proved above carry over to this more general case in a straightforward way (give the details).

Remark 3. We might go still further, calling a set A measurable with "infinite measure" if every A_{mn} is measurable and if the series (31) diverges.

Alternatively, we can regard the whole plane as the union of the squares

$$E_n = \{(x, y): -n \leqslant x \leqslant n, -n \leqslant y \leqslant n\},$$

calling a plane set measurable, with (possibly infinite) measure

$$\mu(A) = \lim_{n \to \infty} \mu(A_n) \tag{32}$$

if its intersection $A_n = A \cap E_n$ with every square E_n is measurable as previously defined. As an exercise, prove the consistency of (31) and (32).

Problem 1. Let E be the closed unit square. Prove that

a) Every open subset of E is measurable;
b) Every closed subset of E is measurable;
c) Every set obtained from open and closed subsets of E by forming no more than a countable number of unions, intersections and complements is measurable.

Comment. There are measurable subsets of E which are not of the type c).

Problem 2. Construct a theory of Lebesgue measure for sets on the line, starting from intervals (closed, open and half-open) instead of rectangles. Do the same for

a) Sets on the circumference of a circle;
b) Three-dimensional sets;
c) Sets in R^n.

Problem 3. Prove that the set of all rational points on the line is measurable, with measure zero.

Problem 4. Prove that the Cantor set constructed in Example 4, p. 52 is measurable, with measure zero.

Problem 5. Prove that every set of positive measure in the interval $[0, 1]$ contains a pair of points whose distance apart is a rational number.

Problem 6. Show that the power of the set of all measurable subsets of the interval $[0, 1]$ is greater than the power of the continuum.

Problem 7. Let C be a circle of circumference 1, and let α be an irrational number. Let all points of C which can be obtained from each other by rotating C through an angle $n\alpha\pi$ (where n is any integer, positive, negative or zero) be assigned to the same class. (Clearly, each such class contains countably many points.) Let Φ_0 be any set containing one point from each class. Prove that Φ_0 is nonmeasurable.

Hint. Let Φ_n be the set obtained by rotating Φ_0 through the angle $n\alpha\pi$. Then

$$C = \bigcup_{n=-\infty}^{\infty} \Phi_n,$$

and

$$\Phi_m \cap \Phi_n = \varnothing \qquad (m \neq n).$$

If Φ_0 were measurable, the congruent sets Φ_n would also be measurable. This would imply

$$\sum_{n=-\infty}^{\infty} \mu(\Phi_n) = 1, \tag{33}$$

by the σ-additivity of μ. But congruent sets must have the same measure, i.e., if Φ_0 were measurable, then

$$\mu(\Phi_n) = \mu(\Phi_0),$$

which contradicts (33).

26. General Measure Theory

26.1. Measure on a semiring. In Sec. 25 we constructed a theory of measure of plane sets, starting from a measure (area) m defined on the class $\mathscr{S}_m$ of all rectangles (with sides parallel to the coordinate axes) and then extending m to a Lebesgue measure μ defined on the much larger class $\mathscr{S}_\mu$ of all measurable sets. The explicit formula for the area of a rectangle played no role in this construction. In fact, a moment's thought shows that we only used the following properties of the set function m:

1) The domain of definition $\mathscr{S}_m$ of m, i.e., the class of all rectangles, is a semiring;[1]
2) m is real and nonnegative;
3) m is additive in the sense that if P is a rectangle such that

$$P = \bigcup_{k=1}^{n} P_k,$$

where $P_1, \ldots, P_n$ are pairwise disjoint rectangles, then

$$m(P) = \sum_{k=1}^{n} m(P_k).$$

As will be shown in this section and the next, the construction given in Sec. 25 for the case of plane sets can be carried out in an abstract setting, whose very generality greatly enhances its range of applicability.

---

[1] We now draw freely from the material in Sec. 4, on systems of sets.

Guided by the above properties of m, we introduce

DEFINITION 1. *A set function $\mu(A)$ is called a* **measure** *if*

1) *The domain of definition $\mathscr{S}_\mu$ of μ is a semiring;*
2) *μ is real and nonnegative;*
3) *μ is additive in the sense that if A is a set in $\mathscr{S}_\mu$ such that*

$$A = \bigcup_{k=1}^{n} A_k,$$

where $A_1, \ldots, A_n$ are pairwise disjoint sets in $\mathscr{S}_\mu$, then

$$\mu(A) = \sum_{k=1}^{n} \mu(A_k).$$

Remark. It follows from $\varnothing = \varnothing \cup \varnothing$ that

$$\mu(\varnothing) = 2\mu(\varnothing),$$

and hence

$$\mu(\varnothing) = 0.$$

THEOREM 1. *Let μ be a measure on a semiring $\mathscr{S}_\mu$ and suppose the sets $A, A_1, \ldots, A_n$, where $A_1, \ldots, A_n$ are disjoint subsets of A, all belong to $\mathscr{S}_\mu$. Then*

$$\sum_{k=1}^{n} \mu(A_k) \leqslant \mu(A).$$

Proof. By Lemma 1, p. 33, there is a finite expansion

$$A = \bigcup_{k=1}^{s} A_k \quad (s \geqslant n)$$

with $A_1, \ldots, A_n$ as its first n terms, where

$$A_k \in \mathscr{S}_\mu, \qquad A_k \cap A_l = \varnothing \qquad (k \neq l)$$

for all $k, l = 1, 2, \ldots$. Hence

$$\sum_{k=1}^{n} \mu(A_k) \leqslant \sum_{k=1}^{s} \mu(A_k) = \mu(A),$$

since μ is nonnegative and additive. ∎

THEOREM 2. *Let μ be a measure on a semiring $\mathscr{S}_\mu$, and suppose the sets $A, A_1, \ldots, A_n$ all belong to $\mathscr{S}_\mu$ and satisfy the condition*

$$A \subset \bigcup_{k=1}^{n} A_k.$$

Then

$$\mu(A) \leqslant \sum_{k=1}^{n} \mu(A_k).$$

Proof. According to Lemma 2, p. 33, there is a finite system of pairwise disjoint sets $B_1, \ldots, B_t$ belonging to $\mathscr{S}_\mu$ such that each of the sets $A, A_1, \ldots, A_n$ has a finite expansion

$$A = \bigcup_{s \in M_0} B_s, \qquad A_k = \bigcup_{s \in M_k} B_s \qquad (k = 1, \ldots, n)$$

with respect to certain of the sets B_s, where each index $s \in M_0$ belongs to at least one of the sets M_k (recall footnote 16, p. 33). Hence each term in the sum

$$\sum_{s \in M_0} \mu(B_s)$$

appears at least once in the double sum

$$\sum_{k=1}^{n} \sum_{s \in M_k} \mu(B_s)$$

It follows that

$$\mu(A) = \sum_{s \in M_0} \mu(B_s) \leqslant \sum_{k=1}^{n} \sum_{s \in M_k} \mu(B_s) = \sum_{k=1}^{n} \mu(A_k). \quad \blacksquare$$

COROLLARY. *If $A \subset A'$, then $\mu(A) \leqslant \mu(A')$.*

Proof. Choose $n = 1$. $\quad \blacksquare$

It will be recalled that the first step in constructing Lebesgue measure of plane sets was to extend measure from rectangles to elementary sets, i.e., to finite unions of disjoint rectangles. We now consider the abstract analogue of this process:

DEFINITION 2. *A measure μ is called an **extension** of a measure m if $\mathscr{S}_m \subset \mathscr{S}_\mu$ and $\mu(A) = m(A)$ for every $A \in \mathscr{S}_m$.*

THEOREM 3. *Any measure m defined on a semiring $\mathscr{S}_m$ has a unique extension μ defined on the ring $\mathscr{R}(\mathscr{S}_m)$, i.e., the minimal ring generated by $\mathscr{S}_m$.*

Proof. By Theorem 3, p. 34, every set $A \in \mathscr{R}(\mathscr{S}_m)$ has a finite expansion

$$A = \bigcup_{k=1}^{n} B_k, \tag{1}$$

where the sets $B_1, \ldots, B_n$ are pairwise disjoint and belong to $\mathscr{S}_m$. Let

$$\mu(A) = \sum_{k=1}^{n} m(B_k). \tag{2}$$

Then μ is obviously real, nonnegative and additive. Moreover, the quantity $\mu(A)$ defined by (2) is independent of the expansion (1). In fact, suppose A has another expansion of the form

$$A = \bigcup_{l=1}^{s} C_l, \tag{1'}$$

where the sets $C_1, \ldots, C_l$ are pairwise disjoint and belong to $\mathscr{S}_m$. Then, since the intersections $B_k \cap C_l$ all belong to $\mathscr{S}_m$, it follows from the additivity of the measure m that

$$\sum_{k=1}^{n} m(B_n) = \sum_{k=1}^{n} \sum_{l=1}^{s} m(B_k \cap C_l) = \sum_{l=1}^{s} m(C_l),$$

and hence

$$\sum_{l=1}^{s} m(C_l) = \mu(A),$$

as asserted. This proves the existence of the extension μ. To prove the uniqueness of μ, suppose m has another extension μ', and let A be the set (1). Then, by the additivity of μ',

$$\mu'(A) = \sum_{k=1}^{n} \mu'(B_k) = \sum_{k=1}^{n} m(B_k) = \mu(A).$$

Hence, since every set $A \in \mathscr{R}(\mathscr{S}_m)$ has a representation of the form (1), the extensions μ and μ' coincide. ∎

Remark. As already noted, the proof of Theorem 3 is a repetition in abstract language of the extension of measure from the semiring of rectangles to the minimal ring generated by this semiring, i.e., the class of elementary sets.

26.2. Countably additive measures. Many problems in analysis involve unions of countably many sets, as well as unions of only finitely many sets. Correspondingly, the (finite) additivity imposed on measures in Definition 1 turns out to be inadequate, and it is natural to introduce a stronger kind of additivity:

DEFINITION 2. *A measure μ with domain of definition $\mathscr{S}_\mu$ is said to be* **countably additive** *or* **σ-additive** *if*

$$\mu(A) = \sum_{n=1}^{\infty} \mu(A_n)$$

for all sets $A, A_1, \ldots, A_n, \ldots \in \mathscr{S}_\mu$ satisfying the conditions

$$A = \bigcup_{n=1}^{\infty} A_n, \quad A_i \cap A_j = \varnothing \qquad (i \neq j).$$

Example. According to Theorem 10, p. 265, Lebesgue measure in the plane is σ-additive.

THEOREM 4. *Suppose a σ-additive measure m on a semiring $\mathscr{S}_m$ is extended to a measure μ on the ring $\mathscr{R}(\mathscr{S}_m)$. Then μ is also σ-additive.*

Proof. Suppose

$$A \in \mathscr{R}(\mathscr{S}_m), \quad B_n \in \mathscr{R}(\mathscr{S}_m) \quad (n = 1, 2, \ldots)$$

and

$$A = \bigcup_{n=1}^{\infty} B_n,$$

where

$$B_k \cap B_l = \varnothing \quad (k \neq l).$$

Then, by Theorem 3, p. 34, there exist finite expansions

$$A = \bigcup_j A_j, \quad B_n = \bigcup_i B_{ni},$$

where

$$A_k \cap A_l = \varnothing, \quad B_{nk} \cap B_{nl} = \varnothing \quad (k \neq l).$$

Let

$$C_{nij} = B_{ni} \cap A_j.$$

Then the sets C_{nij} are pairwise disjoint and

$$A_j = \bigcup_n \bigcup_i C_{nij},$$

$$B_{ni} = \bigcup_j C_{nij}.$$

Therefore

$$m(A_j) = \sum_n \sum_i m(C_{nij}), \tag{3}$$

$$m(B_{ni}) = \sum_j m(C_{nij}), \tag{4}$$

since m is σ-additive on $\mathscr{S}_m$, and moreover

$$\mu(A) = \sum_j m(A_j), \tag{5}$$

$$\mu(B_n) = \sum_i m(B_{ni}), \tag{6}$$

by the definition of the measure μ. Comparing (3)–(6), we find that

$$\mu(A) = \sum_j m(A_j) = \sum_j \sum_n \sum_i m(C_{nij}) = \sum_n \sum_i m(B_{ni}) = \sum_n \mu(B_n)$$

(the sums over i and j are finite, while those over n are convergent). ∎

Next we generalize Theorems 1 and 2 to the case of σ-additive measures:

THEOREM 1'. *Let* μ *be a σ-additive measure on a semiring* $\mathscr{S}_\mu$, *and suppose the sets* $A, A_1, \ldots, A_k, \ldots$, *where* $A_1, \ldots, A_k, \ldots$ *are pairwise disjoint subsets of* A, *all belong to* $\mathscr{S}_\mu$. *Then*

$$\sum_{k=1}^{\infty} \mu(A_k) \leqslant \mu(A). \tag{7}$$

Proof. By Theorem 1,

$$\sum_{k=1}^{n} \mu(A_k) \leqslant \mu(A)$$

for all $n = 1, 2, \ldots$. Taking the limit as $n \to \infty$, we get (7). ∎

THEOREM 2'. *Let* μ *be a σ-additive measure on a semiring* $\mathscr{S}_\mu$, *and suppose the sets* $A, A_1, \ldots, A_k, \ldots$ *all belong to* $\mathscr{S}_\mu$ *and satisfy the condition*

$$A \subset \bigcup_{k=1}^{\infty} A_k.$$

Then

$$\mu(A) \leqslant \sum_{k=1}^{\infty} \mu(A_k). \tag{8}$$

Proof. By Theorem 4, we can assume that μ is defined on the ring $\mathscr{R}(\mathscr{S}_\mu)$, instead of just on the semiring $\mathscr{S}_\mu$. In fact, if μ is σ-additive, so is its extension on $\mathscr{R}(\mathscr{S}_\mu)$, which we continue to denote by μ, and the validity of (8) on $\mathscr{R}(\mathscr{S}_\mu)$ obviously implies its validity on $\mathscr{S}_\mu$. The sets

$$B_n = (A \cap A_n) - \bigcup_{k=1}^{n-1} A_k$$

belong to $\mathscr{R}(\mathscr{S}_\mu)$ and clearly satisfy the conditions

$$A = \bigcup_{n=1}^{\infty} B_n, \quad B_n \subset A_n, \quad B_k \cap B_l = \varnothing \qquad (k \neq l).$$

Therefore

$$m(A) = \sum_{n=1}^{\infty} m(B_n) \leqslant \sum_{n=1}^{\infty} m(A_n). \quad ∎$$

Problem 1. Let $X = \{x_1, x_2, \ldots\}$ be any countable set, and let $p_1, p_2, \ldots$ be positive numbers such that

$$\sum_{n=1}^{\infty} p_n = 1.$$

On the set $\mathscr{S}_\mu$ of all subsets of X, define a measure μ by the formula

$$\mu(A) = \sum_{x_n \in A} p_n \qquad (A \subset X),$$

where the sum is over all n such that $x_n \in A$. Prove that μ is a σ-additive measure, with $\mu(X) = 1$.

Comment. This kind of measure arises quite naturally in many problems of probability theory.

Problem 2. Let X be the set of all rational points in the closed unit interval $[0, 1]$, and let $\mathscr{S}_\mu$ be the set of all intersections of the set X with arbitrary closed, open and half-open subintervals of $[0, 1]$, including the degenerate closed intervals consisting of a single point. Prove that $\mathscr{S}_\mu$ is a semiring. Define a measure μ on $\mathscr{S}_\mu$ by the formula

$$\mu(A_{ab}) = b - a,$$

where A_{ab} is the intersection of X with any of the intervals $[a, b]$, (a, b), $(a, b]$, $[a, b)$. Prove that μ is additive, but not σ-additive.

Hint. Although $\mu(X) = 1$, X is a countable union of single-element sets, each of measure zero.

Problem 3. Let μ be a measure which is additive, but not σ-additive. Prove that

a) Theorem 1' continues to hold for μ;
b) Theorem 2' fails to hold for μ.

Hint. Use Problem 2.

Problem 4. Given a measure μ on a semiring $\mathscr{S}_\mu$, suppose

$$\mu(A) \leqslant \sum_{k=1}^{\infty} \mu(A_k)$$

whenever the sets $A, A_1, \ldots, A_k, \ldots$ all belong to $\mathscr{S}_\mu$ and satisfy the condition

$$A \subset \bigcup_{k=1}^{\infty} A_k.$$

Prove that μ is σ-additive.

Comment. It is often easier to verify that μ has this property than to prove the σ-additivity of μ directly.

27. Extensions of Measures

Any measure m defined on a semiring $\mathscr{S}_m$ can be extended to a measure defined on the ring $\mathscr{R}(\mathscr{S}_m)$, i.e., the minimal ring generated by $\mathscr{S}_m$. However, if m is σ-additive, we can extend m to a measure defined on a much larger class of sets than $\mathscr{R}(\mathscr{S}_m)$. This is done by the abstract analogue of the procedure used in Sec. 25.2 to construct Lebesgue measure in the plane.

Assuming that $\mathscr{S}_m$ has a unit,[2] we begin with the analogues of Definitions 2–5, pp. 259–260.

DEFINITION 1. *Let m be a σ-additive measure on a semiring $\mathscr{S}_m$ with a unit E. Then by the* **outer measure** *of a set $A \subset E$ is meant the number*

$$\mu^*(A) = \inf_{A \subset \bigcup_k B_k} \sum_k m(B_k),$$

where the greatest lower bound is taken over all coverings of A by a finite or countable system of sets $B_k \in \mathscr{S}_m$.

DEFINITION 2. *By the* **inner measure** *of a set $A \subset E$ is meant the number*

$$\mu_*(A) = m(E) - \mu^*(E - A).$$

Remark. By the exact analogue of Theorem 3, p. 258, it follows that

$$\mu_*(A) \leqslant \mu^*(A).$$

DEFINTION 3. *A set A is said to be* **(Lebesgue) measurable** *if*

$$\mu_*(A) = \mu^*(A),$$

i.e., if its inner and outer measures coincide.

DEFINITION 4. *If a set A is measurable, the number $\mu(A)$ equal to the common value of $\mu_*(A)$ and $\mu^*(A)$ is called the* **Lebesgue measure** *of A.*[3]

Remark. Clearly, a set $A \subset E$ is measurable if and only if

$$\mu^*(A) + \mu^*(E - A) = m(E). \tag{1}$$

In particular, it follows from (1) that if A is measurable, so is $E - A$.

THEOREM 1. *If A is any set and $\{A_n\}$ is any finite or countable system of sets such that*

$$A \subset \bigcup_n A_n,$$

then

$$\mu^*(A) \leqslant \sum_n \mu^*(A_n).$$

Proof. Exactly analogous to that of Theorem 4, p. 259. ∎

---

[2] The case where $\mathscr{S}_m$ fails to have a unit will be discussed later (after Theorem 7).

[3] It turns out, of course, that μ is a measure as defined in Sec. 26.1 (see Theorem 5, where the additivity of μ is proved). In particular, this justifies the use of the notation $\mathscr{S}_\mu$ for the system of all measurable sets.

THEOREM 2. *Every set $A \in \mathscr{R}(\mathscr{S}_m)$ is measurable, with Lebesgue measure equal to $\tilde{m}(A)$, where $\tilde{m}$ is the extension of m from the semiring $\mathscr{S}_m$ to the ring $\mathscr{R}(\mathscr{S}_m)$.*

Proof. Exactly analogous to that of Theorem 5, p. 259. ∎

THEOREM 3. *A set A is measurable if and only if, given any $\varepsilon > 0$, there is a set $B \in \mathscr{R}(\mathscr{S}_m)$ such that*

$$\mu^*(A \triangle B) < \varepsilon.$$

Proof. Exactly analogous to that of Theorem 6, p. 261. ∎

THEOREM 4. *The system $\mathscr{S}_\mu$ of all measurable sets is a ring.*

Proof. Exactly analogous to that of Theorem 7, p. 262 and its corollary. ∎

Remark. Obviously E is the unit of $\mathscr{S}_\mu$, so that $\mathscr{S}_\mu$ is an algebra of sets (see p. 31).

THEOREM 5. *The set function $\mu(A)$ is additive on $\mathscr{S}_\mu$.*

Proof. Exactly analogous to that of Theorem 8, p. 263. ∎

THEOREM 6. *The set function $\mu(A)$ is σ-additive on $\mathscr{S}_\mu$.*

Proof. Exactly analogous to that of Theorem 10, p. 265. ∎

Remark. Thus μ is a σ-additive measure of the system $\mathscr{S}_\mu$ of all measurable sets. This measure is called the *Lebesgue extension* of the original measure m.

THEOREM 7. *The system $\mathscr{S}_\mu$ of all measurable sets is a Borel algebra with unit E.*

Proof. Recall from p. 35 that a Borel algebra is closed under the operations of taking countable unions and intersections. The proof is the exact analogue of that of Theorem 9, p. 264. ∎

It is interesting to note that an arbitrary measurable set can be approximated to within a set of measure zero by a set of a very special kind:

THEOREM 8. *Given any set $A \in \mathscr{S}_\mu$, there are sets*

$$B_{nk} \in \mathscr{R}(\mathscr{S}_m) \qquad (B_{n1} \subset B_{n2} \subset \cdots \subset B_{nk} \subset \cdots)$$

and corresponding sets

$$B_n = \bigcup_k B_{nk} \in \mathscr{S}_\mu \qquad (B_1 \supset B_2 \supset \cdots \supset B_n \supset \cdots)$$

such that

$$A \subset B = \bigcap_n B_n,$$

$$\mu(A) = \mu(B).$$

Proof. Given any n, we can cover A by a union

$$C_n = \bigcup_r \Delta_{nr}$$

of sets $\Delta_{nr} \in \mathscr{S}_m$ such that

$$\mu(C_n) < \mu(A) + \frac{1}{n}.$$

Let

$$B_n = \bigcap_{k=1}^n C_k,$$

so that, in particular, $B_1 \supset B_2 \supset \cdots \supset B_n \supset \cdots$. Then it is easy to see that

$$B_n = \bigcup_s \delta_{ns},$$

where $\delta_{ns} \in \mathscr{S}_m$. Next let

$$B_{nk} = \bigcup_{s=1}^k \delta_{ns},$$

so that, in particular,

$$B_n = \bigcup_k B_{nk}.$$

Then obviously $B_{nk} \in \mathscr{R}(\mathscr{S}_m)$ and $B_{n1} \subset B_{n2} \subset \cdots \subset B_{nk} \subset \cdots$. Moreover

$$A \subset B = \bigcap_n B_n,$$

since B is an intersection of sets containing A. It follows that

$$\mu(A) \leqslant \mu(B). \tag{2}$$

On the other hand, $B \subset B_n \subset C_n$ for every n, and therefore

$$\mu(B) \leqslant \mu(B_n) \leqslant \mu(C_n) < \mu(A) + \frac{1}{n}.$$

Taking the limit as $n \to \infty$, we get

$$\mu(B) \leqslant \mu(A),$$

which, together with (2), implies $\mu(A) = \mu(B)$. ∎

Our construction of the Lebesgue extension of a measure m defined on a semiring $\mathscr{S}_m$ must be modified somewhat if $\mathscr{S}_m$ fails to have a unit. We continue to use Definition 1 to define the outer measure μ^*, but μ^* is now defined only on the system $\mathscr{S}_{\mu*}$ of all sets with coverings

$$\bigcup_k B_k \qquad (B_k \in \mathscr{S}_m)$$

such that

$$\sum_k m(B_k) < \infty.$$

Since Definition 2 is meaningless in the absence of a unit, we now define measurable sets by using the property figuring in Theorem 3:

> DEFINITION 3′. *A set A is said to be* **(Lebesgue) measurable** *if, given any* ε > 0, *there is a set* $B \in \mathscr{R}(\mathscr{S}_m)$ *such that* $\mu^*(A \triangle B) < \varepsilon$.

> DEFINITION 4′. *If a set A is measurable, the number* $\mu(A)$ *equal to its outer measure* $\mu^*(A)$ *is called the* **(Lebesgue) measure of A.**

Remark. Note that Definitions 3′ and 4′ are equivalent to Definitions 3 and 4 if $\mathscr{S}_m$ has a unit.

In the case where $\mathscr{S}_m$ has no unit, Theorems 4–6 continue to hold, since the proofs of Theorems 5 and 6 do not require $\mathscr{S}_m$ to have a unit, while the proof of Theorem 4 can easily be freed of this requirement (see Problem 4). However, Theorem 7 now takes a new form (see Problem 5). As before, the σ-additive measure μ on the system $\mathscr{S}_\mu$ of all measurable sets is called the *Lebesgue extension* of the original measure m.

Remark. There is an interesting analogy between the construction of the Lebesgue extension of a measure m defined on a semiring $\mathscr{S}_m$ and the process of completing a metric space. Let $\tilde{m}$ be the extension of m from the semiring $\mathscr{S}_m$ to the ring $\mathscr{R}(\mathscr{S}_m)$, and suppose we regard $\tilde{m}(A \triangle B)$ as the distance between the elements $A, B \in \mathscr{R}(\mathscr{S}_m)$. Then $\mathscr{R}(\mathscr{S}_m)$ becomes a metric space (in general, incomplete), whose completion, according to Theorem 3, is just the system $\mathscr{S}_\mu$ of all Lebesgue-measurable sets. However, note that from a metric point of view, two sets $A, B \in \mathscr{S}_\mu$ are indistinguishable if $\mu(A \triangle B) = 0$.

Problem 1. Let m be a σ-additive measure on a semiring $\mathscr{S}_m$ with a unit E, let μ be the Lebesgue extension of m, and let $\tilde{\mu}$ be an arbitrary σ-additive extension of m. Prove that $\tilde{\mu}(A) = \mu(A)$ for every measurable set A on which $\tilde{\mu}$ is defined.

Hint. First show that $\mu_*(A) \leqslant \tilde{\mu}(A) \leqslant \mu^*(A)$.

Problem 2. Let m be the same as in the preceding problem, and let $\tilde{m}$ be the extension of m to a measure defined on $\mathscr{R}(\mathscr{S}_m)$. Prove that the outer measure of a set $A \subseteq E$ is given by

$$\mu^*(A) = \inf_{\substack{A \subset \bigcup_k B_k}} \sum_k \tilde{m}(B_k),$$

where the greatest lower bound is taken over all coverings of A by a finite or countable system of sets $B_k \in \mathcal{R}(\mathcal{S}_m)$.

Problem 3. State and prove the analogues of Theorem 11, p. 266 and its corollary for an arbitrary σ-additive measure μ defined on a Borel algebra $\mathcal{S}_\mu$ with unit E.

Problem 4. Give a proof of Theorem 7 valid in the case where $\mathcal{S}_\mu$ fails to have a unit.

Hint. Suppose $A_1, A_2 \in \mathcal{S}_\mu$. Then $A_1 \cup A_2 \in \mathcal{S}_\mu$, by the same proof as before (cf. p. 262). Moreover, there are sets $B_1, B_2 \in \mathcal{R}(\mathcal{S}_m)$ such that

$$\mu^*(A_1 \bigtriangleup B_1) < \frac{\varepsilon}{2}, \qquad \mu^*(A_2 \bigtriangleup B_2) < \frac{\varepsilon}{2}.$$

But

$$(A_1 - A_2) \bigtriangleup (B_1 - B_2) \subset (A_1 \bigtriangleup B_1) \cup (A_2 \bigtriangleup B_2),$$

and hence $\mu^*(A \bigtriangleup B) < \varepsilon$ where $B = B_1 - B_2 \in \mathcal{R}(\mathcal{S}_m)$. Therefore $A_1 - A_2 \in \mathcal{S}_\mu$. To prove that $A_1 \cap A_2$ and $A_1 \bigtriangleup A_2$ belong to $\mathcal{S}_\mu$, use the formulas

$$A_1 \cap A_2 = A_1 - (A_1 - A_2),$$
$$A_1 \bigtriangleup A_2 = (A_1 - A_2) \cup (A_2 - A_1).$$

Problem 5. Given a measure m on a semiring $\mathcal{S}_m$ with no unit, let μ be the Lebesgue extension of m and $\mathcal{S}_\mu$ the corresponding system of all measurable sets. Prove that

a) $\mathcal{S}_\mu$ is a δ-ring (see p. 35);
b) The set

$$A = \bigcup_k A_k \qquad (A_k \in \mathcal{S}_\mu)$$

belongs to $\mathcal{S}_\mu$ if and only if there is a constant $C > 0$ such that

$$\mu\left(\bigcup_{k=1}^n A_k\right) \leqslant C \tag{3}$$

for all $n = 1, 2, \ldots$

Comment. The necessity of the condition (3) is obvious, since our measures are always finite.

Problem 6. Let μ and $\mathcal{S}_\mu$ be the same as in the preceding problem. Prove that the system of all sets $B \in \mathcal{S}_\mu$ which are subsets of a fixed set $A \in \mathcal{S}_\mu$ is a Borel algebra with unit A.

Problem 7. A measure μ is said to be *complete* if every subset of a set of measure zero is measurable, i.e., if $A' \subset A$, $\mu(A) = 0$ implies $A' \in \mathcal{S}_\mu$. (If $A' \in \mathcal{S}_\mu$, then obviously $\mu(A') = 0$.) Prove that the Lebesgue extension of any measure m is complete.

Hint. If $A' \subset A$ and $\mu(A) = 0$, then $\mu^*(A') = 0$. But $\varnothing \in \mathscr{R}(\mathscr{S}_m)$ and $\mu^*(A' \triangle \varnothing) = \mu^*(A') = 0$.

Problem 8. Let $\tilde{m}$ be a measure defined on a ring $\mathscr{R}$. For example, $\tilde{m}$ might be the extension of a measure m originally defined on a semiring $\mathscr{S}_m$ to a measure defined on the minimal ring $\mathscr{R} = \mathscr{R}(\mathscr{S}_m)$ generated by $\mathscr{S}_m$. Then a set A is said to be *Jordan measurable* if, given any $\varepsilon > 0$, there are sets $A', A'' \in \mathscr{R}$ such that

$$A' \subset A \subset A'', \qquad \tilde{m}(A'' - A') < \varepsilon.$$

Prove that the system $\mathscr{R}^*$ of all Jordan-measurable sets is a ring containing $\mathscr{R}$.

Problem 9. Let $\tilde{m}$, $\mathscr{R}$ and $\mathscr{R}^*$ be the same as in the preceding problem, and let $\mathscr{A}$ be the system of all sets A such that there is a set $B \in \mathscr{R}$ containing A. Given any set $A \in \mathscr{A}$, let

$$\bar{\mu}(A) = \inf_{\substack{B \supset A \\ B \in \mathscr{R}}} \tilde{m}(B),$$

$$\underline{\mu}(A) = \sup_{\substack{B \subset A \\ B \in \mathscr{R}}} \tilde{m}(B)$$

(since $\varnothing \subset A$, A always contains a set in $\mathscr{R}$). Prove that

a) $\underline{\mu}(A) \leqslant \bar{\mu}(A)$;
b) The ring $\mathscr{R}^*$ coincides with the system of all sets $A \in \mathscr{A}$ for which $\underline{\mu}(A) = \bar{\mu}(A)$;
c) If

$$A \subset \bigcup_{k=1}^{n} A_k,$$

where $A, A_1, \ldots, A_n$ all belong to $\mathscr{A}$, then

$$\bar{\mu}(A) \leqslant \sum_{k=1}^{n} \bar{\mu}(A_k);$$

d) If $A_1, \ldots, A_n$ are pairwise disjoint sets contained in a set A, then

$$\underline{\mu}(A) \geqslant \sum_{k=1}^{n} \underline{\mu}(A_k).$$

By the *Jordan measure* of a set $A \in \mathscr{R}^*$, we mean the number $\mu(A)$ equal to the common value of $\underline{\mu}(A)$ and $\bar{\mu}(A)$. Prove that μ is a measure on $\mathscr{R}^* = \mathscr{S}_\mu$.

Comment. The measure μ is called the *Jordan extension* of the measure $\tilde{m}$. If $\tilde{m}$ is itself an extension of a measure m originally defined on a semiring $\mathscr{S}_m$, we write $\mathscr{R}^* = \mathscr{R}^*(\mathscr{S}_m)$ and call μ the Jordan extension of the measure m, as well as of the "intermediate" measure $\tilde{m}$.

Problem 10. Given two measures $\tilde{m}_1$ and $\tilde{m}_2$ defined on rings $\mathscr{R}_1$ and $\mathscr{R}_2$, let μ_1 and μ_2 be their Jordan extensions onto the larger rings $\mathscr{R}_1^* = \mathscr{S}_{\mu_1}$ and $\mathscr{R}_2^* = \mathscr{S}_{\mu_2}$. Prove that μ_1 and μ_2 coincide if and only if

$$\mathscr{R}_1 \subset \mathscr{S}_{\mu_2}, \quad \tilde{m}_1(A) = \mu_2(A) \text{ for all } A \in \mathscr{R}_1,$$
$$\mathscr{R}_2 \subset \mathscr{S}_{\mu_1}, \quad \tilde{m}_2(A) = \mu_1(A) \text{ for all } A \in \mathscr{R}_2.$$

Problem 11. Let $\tilde{m}$ be the measure defined in Sec. 25.1 on the ring $\mathscr{R}$ of all elementary sets (i.e., all finite unions of disjoint rectangles with sides parallel to the coordinate axes), and let μ be the Jordan extension of $\tilde{m}$. Prove that μ does not depend on the particular choice of the underlying rectangular coordinate system. In other words, prove that μ (as well as the corresponding ring $\mathscr{R}^* = \mathscr{S}_\mu$) does not change if all the sets in $\mathscr{R}$ are subjected to the same shift and rigid rotation.

Problem 12. We say that a set A is a *set of uniqueness* for a measure m if

1) There is an extension of m defined on A;
2) If μ_1 and μ_2 are two such extensions, then $\mu_1(A) = \mu_2(A)$.

Prove that the system of sets of uniqueness of a measure m defined on a semiring $\mathscr{S}_m$ coincides with the ring $\mathscr{R}^* = \mathscr{R}^*(\mathscr{S}_m)$ of sets which are Jordan measurable (with respect to m). In other words, prove that the Jordan extension of a measure m originally defined on a semiring $\mathscr{S}_m$ is the unique extension of m to a measure defined on $\mathscr{R}^* = \mathscr{R}^*(\mathscr{S}_m)$, but that the extension of m to a larger system is no longer unique.

Problem 13. Prove that if a set A is Jordan measurable, then

a) A is Lebesgue measurable;
b) The Jordan and Lebesgue measures of A coincide.

Prove that every Jordan extension of a σ-additive measure is σ-additive.

Problem 14. Give an example of a set which is Lebesgue measurable, but not Jordan measurable.

Problem 15. We say that a set A is a *set of σ-uniqueness* for a σ-additive measure m if

1) There is a σ-additive extension of m defined on A;
2) If μ_1 and μ_2 are two such extensions, then $\mu_1(A) = \mu_2(A)$.

Prove that the system of sets of σ-uniqueness of a σ-additive measure m defined on a semiring $\mathscr{S}_m$ coincides with the system of sets which are Lebesgue measurable (with respect to m).

Hint. To show that every Lebesgue-measurable set A is a set of σ-uniqueness for m, choose any $\varepsilon > 0$. Then there is a set $B \in \mathscr{R} = \mathscr{R}(\mathscr{S}_m)$

such that $\mu^*(A \triangle B) < \varepsilon$. If μ is any extension of m defined on A (and on $\mathcal{R}$), then $\mu(B) = \tilde{m}(B)$, where $\tilde{m}$ is the unique extension of m onto $\mathcal{R}$. Moreover, $\mu(A \triangle B) \leqslant \mu^*(A \triangle B) < \varepsilon$, and hence $|\mu(A) - \tilde{m}(B)| < \varepsilon$. Therefore $|\mu_1(A) - \mu_2(A)| < 2\varepsilon$ if μ_1 and μ_2 are two σ-additive extensions of m defined on A (and on $\mathcal{R}$). Hence $\mu_1(A) = \mu_2(A)$, by the arbitrariness of ε.

Problem 16. Let m be a σ-additive measure defined on a semiring $\mathcal{S}_m$, and let $\mathcal{L}$ be the domain of the Lebesgue extension of m. Let m' be a σ-additive extension of m to a semiring $\mathcal{S}_{m'}$ such that

$$\mathcal{S}_m \subset \mathcal{S}_{m'} \subset \mathcal{L},$$

and let $\mathcal{L}'$ be the domain of the Lebesgue extension of m'. Prove that $\mathcal{L}' = \mathcal{L}$.

8

INTEGRATION

28. Measurable Functions

28.1. Basic properties of measurable functions. Given any two sets X and Y, let $\mathscr{S}$ be a system of subsets of X and $\mathscr{S}'$ a system of subsets of Y. Then an abstract function $y = f(x)$ defined on X and taking values in Y is said to be $(\mathscr{S}, \mathscr{S}')$-*measurable* if $A \in \mathscr{S}'$ implies $f^{-1}(A) \in \mathscr{S}$.

Example. Let X and Y both be the real line R^1, so that $y = f(x)$ is a "function of a real variable." Moreover, let $\mathscr{S}$ and $\mathscr{S}'$ both be the system of all open (or closed) subsets of R^1. Then our definition of measurability reduces to that of continuity (recall Sec. 9.6). On the other hand, if we choose both $\mathscr{S}$ and $\mathscr{S}'$ to be the system $\mathscr{B}^1$ of all Borel sets on the real line (recall p. 36), our definition becomes that of a *Borel-measurable* (or simply *B-measurable*) function.

In what follows, we will be primarily concerned with the notion of real functions measurable with respect to some underlying measure μ, this being the case of greatest interest from the standpoint of integration theory. More exactly, let X be any set and Y the real line R^1, with $\mathscr{S} = \mathscr{S}_\mu$ the domain of definition of some σ-additive measure μ and $\mathscr{S}'$ the system $\mathscr{B}^1$ of all Borel sets $B \subset R^1$. For simplicity, we assume that $\mathscr{S}_\mu$ has a unit equal to X itself. Moreover, since any σ-additive measure can be extended onto a Borel algebra (by Theorem 7, p. 277), we might as well assume from the outset that $\mathscr{S}_\mu$ is a Borel algebra. These considerations suggest

DEFINITION 1. *Given a σ-additive measure μ defined on a Borel algebra $\mathscr{S}_\mu$ of subsets of a set X, where X is the unit of $\mathscr{S}_\mu$, let $y = f(x)$ be a real* *function defined on X, and let $\mathscr{B}^1$ be the set of all Borel sets on the real line. Then the function f is said to be **μ-measurable** (on X) if $f^{-1}(A) \in \mathscr{S}_\mu$ for every $A \in \mathscr{B}^1$, or equivalently if $f^{-1}(\mathscr{B}^1) \subset \mathscr{S}_\mu$.*

THEOREM 1. *A function f is μ-measurable if and only if the set $\{x : f(x) < c\}$ is μ-measurable (i.e., belongs to $\mathscr{S}_\mu$) for every real c.*

Proof. If f is μ-measurable, then obviously so is $\{x : f(x) < c\}$, since $(-\infty, c)$ is a Borel set. Conversely, let Σ be the system of all semi-infinite intervals $(-\infty, c)$, and suppose $f^{-1}(\Sigma) \subset \mathscr{S}_\mu$. Since $\mathscr{B}(\Sigma)$, the Borel closure of Σ (see p. 36), coincides with the system $\mathscr{B}^1$ of all Borel sets on the line (why?), we have

$$f^{-1}(\mathscr{B}^1) = f^{-1}(\mathscr{B}(\Sigma)) = \mathscr{B}(f^{-1}(\Sigma)) \subset \mathscr{B}(\mathscr{S}_\mu)$$

(recall Problem 3e, p. 36). But $\mathscr{B}(\mathscr{S}_\mu) = \mathscr{S}_\mu$, since $\mathscr{S}_\mu$ is a Borel algebra, and hence

$$f^{-1}(\mathscr{B}^1) \subset \mathscr{S}_\mu. \quad \blacksquare$$

THEOREM 2. *Let $\{f_n\}$ be a sequence of μ-measurable functions on X, and let f be a function on X such that*

$$f(x) = \lim_{n \to \infty} f_n(x)$$

for every $x \in X$. Then f is itself μ-measurable.

Proof. First we verify that

$$\{x : f(x) < c\} = \bigcup_k \bigcup_n \bigcap_{m > n} \left\{ x : f_m(x) < c - \frac{1}{k} \right\}. \tag{1}$$

In fact, if $f(x) < c$, there is an integer $k > 0$ such that

$$f(x) < c - \frac{2}{k},$$

and then for this k, there is an integer $n > 0$ so large that

$$f_m(x) < c - \frac{1}{k} \tag{2}$$

for all $m > n$. Therefore every x belonging to the left-hand side of (1) also belongs to the right-hand side. Conversely, if x belongs to the right-hand side of (1), there is a k such that (2) holds for all sufficiently large m. But then $f(x) < c$, i.e., x belongs to the left-hand side of (1).

Now, since the functions f_m are μ-measurable, the sets

$$\left\{ x : f_m(x) < c - \frac{1}{k} \right\}$$

all belong to $\mathscr{S}_\mu$, and hence so does the right-hand side of (1), since $\mathscr{S}_\mu$ is a Borel algebra. Therefore $\{x : f(x) < c\} \in \mathscr{S}_\mu$. But then f is μ-measurable, by Theorem 1. ∎

THEOREM 3. *A B-measurable function of a μ-measurable function is itself μ-measurable.*

Proof. Let $f(x) = \varphi[\psi(x)]$, where φ is B-measurable and ψ is μ-measurable. If $A \subset R^1$ is any B-measurable set, then its preimage $A' = \varphi^{-1}(A)$ is B-measurable, and hence the preimage $A'' = \psi^{-1}(A')$ is μ-measurable. But $A'' = f^{-1}(A)$, and hence f is μ-measurable. ∎

COROLLARY. *A continuous function of a μ-measurable function is itself μ-measurable.*

Proof. A continuous function is clearly B-measurable. ∎

28.2. Simple functions. Algebraic operations on measurable functions.
A function f is said to be *simple* if it is μ-measurable and takes no more than countably many distinct values. This notion clearly depends on the choice of the measure μ.

The structure of simple functions is clarified by

THEOREM 4. *A function f taking no more than countably many distinct values $y_1, y_2, \ldots$ is μ-measurable if and only if the sets*

$$A_n = \{x : f(x) = y_n\} \qquad (n = 1, 2, \ldots)$$

are μ-measurable.

Proof. Since each single-element set $\{y_n\}$ is a Borel set, the set A_n, being the preimage of $\{y_n\}$, is measurable if f is measurable.[1] Conversely, suppose the sets A_n are all measurable. Then the preimage $f^{-1}(B)$ of any Borel set $B \subset R^1$ is measurable, being a union

$$\bigcup_{y_n \in B} A_n$$

of no more than countably many measurable sets A_n. But then f is measurable. ∎

The relation between measurable functions and simple functions is shown by

THEOREM 5. *A function f is μ-measurable if and only if it can be represented as the limit of a uniformly convergent sequence of simple functions.*

---

[1] For simplicity, we often say "measurable" instead of "μ-measurable," omitting explicit reference to the underlying measure μ.

Proof. If f is the (uniform) limit of a convergent sequence of simple functions, then f is μ-measurable by Theorem 2, since simple functions are μ-measurable by definition. Conversely, given any μ-measurable function f, let

$$f_n(x) = \frac{m}{n} \quad \text{if} \quad \frac{m}{n} \leqslant f(x) < \frac{m+1}{n},$$

where m and n are positive integers. Then the functions f_n are simple and moreover converge uniformly to f as $n \to \infty$, since

$$|f(x) - f_n(x)| < \frac{1}{n}. \quad \blacksquare$$

The next few theorems show that the class of measurable functions is closed under the usual algebraic operations.

THEOREM 6. *If f and g are measurable, then so is $f + g$.*

Proof. First let f and g be simple functions, taking value $y_1, y_2, \ldots$ and $z_1, z_2, \ldots$, respectively. Then the sum $h = f + g$ can only take the values $c_{ij} = y_i + z_j$, where each such value is taken on a set of the form

$$\{x : h(x) = c_{ij}\} = \bigcup_{y_i + z_j = c_{ij}} (\{x : f(x) = y_i\} \cap \{x : g(x) = z_j\}). \quad (3)$$

There are no more than countably many values w of the function $h = f + g$, and moreover each set $\{x : h(x) = c_{ij}\}$ is measurable, since the right-hand side of (3) is clearly measurable. Therefore $h = f + g$ is a simple function.

Now let f and g be arbitrary measurable functions, and let $\{f_n\}$ and $\{g_n\}$ be sequences of simple functions converging uniformly to f and g, respectively, as in the proof of Theorem 5. Then the sequence of simple functions $\{f_n + g_n\}$ converges uniformly to $f + g$, and hence $f + g$ is measurable, by Theorem 5. $\blacksquare$

THEOREM 7. *If f is measurable, then so is cf, where c is an arbitrary constant.*

Proof. Obviously, the product of a simple function and a constant is again simple. But if $\{f_n\}$ is a sequence of simple functions converging uniformly to f, then $\{cf_n\}$ converges uniformly to cf, and hence cf is measurable, by Theorem 5. $\blacksquare$

THEOREM 8. *If f and g are measurable, then so is $f - g$.*

Proof. An immediate consequence of Theorems 6 and 7. $\blacksquare$

THEOREM 9. *If f and g are measurable, then so is fg.*

Proof. Clearly,

$$fg = \frac{1}{4} [(f + g)^2 - (f - g)^2].$$

But the expression on the right is a measurable function, by Theorems 6–8 and the fact that the square of a measurable function is measurable (this follows from the corollary to Theorem 3). ∎

THEOREM 10. *If f is measurable, then so is* $1/f$, *provided f does not vanish.*

Proof. We have

if $c > 0$,
$$\left\{ x: \frac{1}{f(x)} < c \right\} = \left\{ x: f(x) > \frac{1}{c} \right\} \cup \{ x: f(x) < 0 \}$$

if $c < 0$, and
$$\left\{ x: \frac{1}{f(x)} < c \right\} = \left\{ x: \frac{1}{c} < f(x) < 0 \right\}$$

$$\left\{ x: \frac{1}{f(x)} < c \right\} = \{ x: f(x) < c \}$$

if $c = 0$. But in each case the set on the right is measurable. ∎

COROLLARY. *If f and g are measurable, then so is* f/g, *provided g does not vanish.*

Proof. An immediate consequence of Theorems 9 and 10. ∎

28.3. Equivalent functions. The values of a function can often be neglected on a set of measure zero. This suggests

DEFINITION 2. *Two functions f and g defined on the same set are said to be* **equivalent** (*with respect to a measure* μ) *if*

$$\mu\{ x: f(x) \neq g(x) \} = 0.$$

A property is said to hold *almost everywhere* (on E) if it holds at all points (of E) except possibly on a set of measure zero. Thus two functions f and g are said to be equivalent (written $f \sim g$) if they coincide almost everywhere.

THEOREM 11. *Given two functions f and g continuous on an interval E, suppose f and g are equivalent* (*with respect to Lebesgue measure* μ *on the line*). *Then f and g coincide.*

Proof. Suppose $f(x_0) \neq g(x_0)$ at some point $x_0 \in E$, so that $f(x_0) - g(x_0) \neq 0$. Since $f - g$ is continuous, there is a neighborhood of x_0 (possibly one-sided) in which $f - g$ is nonzero. This neighborhood has positive measure, and hence

$$\mu\{x:f(x) \neq g(x)\} > 0,$$

i.e., f and g cannot be equivalent, contrary to hypothesis. ∎

Remark. Thus two continuous functions cannot be equivalent if they differ at even a single point. However, discontinuous functions can obviously be equivalent without being identical. For example, the *Dirichlet function*

$$f(x) = \begin{cases} 1 & \text{if } x \text{ is rational,} \\ 0 & \text{if } x \text{ is irrational} \end{cases}$$

is equivalent to the function $g(x) \equiv 0$ (recall Problem 3, p. 268).

THEOREM 12. *A function f equivalent to a measurable function g is itself measurable.*

Proof. It follows from Definition 2 that the sets $\{x:f(x) < c\}$ and $\{x:g(x) < c\}$ can differ only by a set of measure zero. Hence if the second set is measurable, so is the first set. The proof is now an immediate consequence of Theorem 1. ∎

28.4. Convergence almost everywhere. Since the behavior of measurable functions on sets of measure zero is often unimportant, it is natural to introduce the following generalization of the ordinary notion of convergence of a sequence of functions:

DEFINITION 3. *A sequence of functions $\{f_n(x)\}$ defined on a space X is said to **converge almost everywhere** to a function $f(x)$ if*

$$\lim_{n \to \infty} f_n(x) = f(x) \qquad (4)$$

for almost all $x \in X$, i.e., if the set of points for which (4) fails to hold is of measure zero.

Example. The sequence $\{f_n(x)\} = \{(-x)^n\}$ defined on $[0, 1]$ converges almost everywhere to the function $f(x) \equiv 0$, in fact everywhere except at the point $x = 1$.

Theorem 2 now has the following generalization:

THEOREM 2′. *Let $\{f_n\}$ be a sequence of μ-measurable functions on X, and let f be a function on X such that*

$$f(x) = \lim_{n \to \infty} f_n(x) \qquad (5)$$

almost everywhere on X. Then f is itself μ-measurable, provided μ is complete.[2]

Proof. If A is the set on which (5) holds, then $\mu(X - A) = 0$. The function f is measurable on A, by Theorem 2, and also on $X - A$, since every function is measurable on a set of measure zero if μ is complete (why?). Hence f is measurable on the whole set $X = A \cup (X - A)$. ∎

28.5. Egorov's theorem. The following important theorem shows the relation between the concepts of convergence almost everywhere and uniform convergence:

THEOREM 12 (*Egorov*). *Let $\{f_n\}$ be a sequence of measurable functions converging almost everywhere on a measurable set E to a function f. Then, given any $\delta > 0$, there exists a measurable set $E_\delta \subset E$ such that*

1) $\mu(E_\delta) > \mu(E) - \delta$;
2) $\{f_n\}$ *converges uniformly to f on E_δ.*

Proof. The function f is measurable, by Theorem 2'. Let

$$E_n^m = \bigcap_{i > n} \left\{ x \colon |f_i(x) - f(x)| < \frac{1}{m} \right\}. \tag{6}$$

Thus, for fixed m and n, E_n^m is the set of all points x such that

$$|f_i(x) - f(x)| < \frac{1}{m}$$

holds for all $i > n$. Moreover, let

$$E^m = \bigcup_{n=1}^{\infty} E_n^m.$$

It follows from (6) that

$$E_1^m \subset E_2^m \subset \cdots \subset E_n^m \subset \cdots,$$

and hence, by the corollary to Theorem 11, p. 267,[3] given any m and any $\delta > 0$, there is an $n_0(m)$ such that

$$\mu(E^m - E_{n_0(m)}^m) < \frac{\delta}{2^m}. \tag{7}$$

Let

$$E_\delta = \bigcap_{m=1}^{\infty} E_{n_0(m)}^m.$$

---

[2] See Problem 7, p. 280.
[3] See also Problem 3, p. 280.

Then E_δ satisfies the two conditions of the theorem. The fact that the sequence $\{f_n\}$ is uniformly convergent on E_δ is almost obvious, since if $x \in E_\delta$, then, given any $m = 1, 2, \ldots,$

$$|f_i(x) - f(x)| < \frac{1}{m}$$

for every $i > n_0(m)$.

To verify condition 2), we now estimate the measure of the set $E - E_\delta$, noting first that $\mu(E - E^m) = 0$ for every m. In fact, if $x_0 \in E - E^m$, then there are arbitrarily large values of i such that

$$|f_i(x_0) - f(x_0)| > \frac{1}{m},$$

which means that the sequence $\{f_n\}$ cannot converge to f at the point x_0. Therefore $\mu(E - E^m) = 0$, as asserted, since $\{f_n\}$ converges to f almost everywhere, by hypothesis. It follows from (7) that

$$\mu(E - E^m_{n_0(m)}) = \mu(E^m - E^m_{n_0(m)}) < \frac{\delta}{2^m}.$$

Therefore

$$\mu(E - E_0) = \mu\left(E - \bigcap_{m=1}^{\infty} E^m_{n_0(m)}\right) = \mu\left(\bigcup_{m=1}^{\infty} (E - E^m_{n_0(m)})\right)$$

$$< \sum_{m=1}^{\infty} \mu(E - E^m_{n_0(m)}) < \sum_{m=1}^{\infty} \frac{\delta}{2^m} = \delta,$$

and hence $\mu(E_\delta) > \mu(E) - \delta$. ∎

Problem 1. Prove that the Dirichlet function

$$f(x) = \begin{cases} 1 & \text{if } x \text{ is rational,} \\ 0 & \text{if } x \text{ is irrational} \end{cases}$$

is measurable on every interval $[a, b]$.

Problem 2. Do the same for the function

$$f(x) = \begin{cases} \dfrac{1}{q} & \text{if } x = \dfrac{p}{q} \text{ is rational,} \\ 0 & \text{if } x \text{ is irrational.} \end{cases}$$

Problem 3. Suppose $f(x)$ is measurable on $[a, b]$. Is $g(x) = e^{f(x)}$ measurable on $[a, b]$?

Problem 4. Prove that if f is measurable, then so is $|f|$.

Problem 5. Let $\{f_n\}$ be a sequence of measurable functions converging almost everywhere to a function f. Prove that $\{f_n\}$ converges almost everywhere to a function g if and only if f and g are equivalent.

Problem 6. A sequence $\{f_n\}$ of μ-measurable functions is said to *converge in measure* to a function f if

$$\lim_{n \to \infty} \mu\{x : |f_n(x) - f(x)| \geqslant \delta\} = 0$$

for every $\delta > 0$. Prove that if a sequence $\{f_n\}$ of measurable functions converges to f almost everywhere, then it converges to f in measure.

Hint. Let A be the set (of measure zero) on which $\{f_n\}$ fails to converge to f, and let

$$E_k(\delta) = \{x : |f_k(x) - f(x)| \geqslant \delta\},$$

$$R_n(\delta) = \bigcup_{k=n}^{\infty} E_k(\delta), \tag{8}$$

$$M = \bigcap_{n=1}^{\infty} R_n(\delta).$$

Then the sets (8) are all measurable (why?), and $\mu(R_n(\delta)) \to \mu(M)$ as $n \to \infty$, since $R_1(\delta) \supset R_2(\delta) \supset \cdots$. Prove that $M \subset A$ and hence that $\mu(M) = 0$ (as always, we assume that μ is complete). It follows that $\mu(R_n(\delta)) \to 0$ as $n \to \infty$. Now use the fact that $E_n(\delta) \subset R_n(\delta)$.

Problem 7. Let $\{f_n\}$ be a sequence of measurable functions converging in measure to a function f. Prove that $\{f_n\}$ converges in measure to a function g if and only if f and g are equivalent.

Problem 8. Given any positive integer k, consider the function

$$f_i^{(k)}(x) = \begin{cases} 1 & \text{if } \dfrac{i-1}{k} < x \leqslant \dfrac{i}{k}, \\ 0 & \text{otherwise,} \end{cases}$$

defined on the half-open interval $(0, 1]$. Show that the sequence

$$f_1^{(1)}, f_1^{(2)}, f_2^{(2)}, \ldots, f_1^{(k)}, f_2^{(k)}, \ldots, f_k^{(k)}, \ldots$$

converges in measure to zero, but does not converge at any point whatsoever.

Comment. Thus the converse of the proposition in Problem 6 is false. Instead we have the weaker proposition considered in the next problem.

Problem 9. Prove that if a sequence $\{f_n\}$ of functions converges to f in measure, then it contains a subsequence $\{f_{n_k}\}$ converging to f almost everywhere.

Hint. Let $\{\delta_n\}$ be a sequence of positive numbers such that

$$\lim_{n \to \infty} \delta_n = 0,$$

and let $\{\varepsilon_n\}$ be a sequence of positive numbers such that

$$\sum_{n=1}^{\infty} \varepsilon_n < \infty.$$

Let $\{n_k\}$ be a sequence of positive integers such that $n_k > n_{k-1}$ and

$$\mu\{x : |f_{n_k}(x) - f(x)| \geqslant \delta_k\} < \varepsilon_k \qquad (k = 1, 2, \ldots).$$

Moreover, let

$$R_i = \bigcup_{k=i}^{\infty} \{x : |f_{n_k}(x) - f(x)| \geqslant \delta_k\}, \qquad Q = \bigcap_{i=1}^{\infty} R_i.$$

Then $\mu(R_i) \to \mu(Q)$ as $i \to \infty$, since $R_1 \supset R_2 \supset \cdots$. On the other hand,

$$\mu(R_i) < \sum_{k=1}^{\infty} \varepsilon_k,$$

and hence $\mu(R_i) \to 0$, so that $\mu(Q) = 0$. Now show that $\{f_{n_k}\}$ converges to f on $E - Q$.

Problem 10. Prove that a function f defined on a closed interval $[a, b]$ is μ-measurable if and only if, given any $\varepsilon > 0$, there is a continuous function φ on $[a, b]$ such that $\mu\{x : f(x) \neq \varphi(x)\} < \varepsilon$.

Hint. Use Egorov's theorem.

Comment. This result, known as *Luzin's theorem*, shows that a measurable function "can be made continuous by altering it on a set of arbitrarily small measure."

29. The Lebesgue Integral

The concept of the Riemann integral, familiar from calculus, applies only to functions which are either continuous or else do not have "too many" points of discontinuity. Hence we cannot form the Riemann integral of a general measurable function f. In fact, f may be discontinuous everywhere, or it may even be meaningless to talk about the continuity of f in the case where f is defined on an abstract set. For such functions, there is another fully developed notion of the integral, due to Lebesgue, which is more flexible that the notion of the Riemann integral.

Let f be a function defined on a closed interval $[a, b]$ of the x-axis. Then to form the Riemann integral of f, we divide $[a, b]$ into many subintervals, thereby grouping together neighboring points of the x-axis. On the other hand, as we will see below, the Lebesgue integral is formed by grouping together points of the x-axis at which the function f takes neighboring values. In other words, the key idea of the theory of Lebesgue integration is to partition the *range* of the function f rather than its domain. This immediately makes it possible to extend the notion of integral to a very large class of functions.

Another advantage of the Lebesgue integral is that it is constructed in exactly the same way for functions defined on an abstract "measure space" (an arbitrary set X equipped with a measure) as for functions defined on the real line. This is to be contrasted with the situation for the Riemann integral, which is first introduced for functions of a single real variable and then extended, with suitable modifications, to the case of functions of several real variables, but fails to make any sense at all for functions defined on an abstract measure space.

In what follows, unless the contrary is explicitly stated, we will consider a σ-additive measure μ defined on a Borel algebra of subsets of a set X, with X as the unit. We will assume that all sets under consideration are μ-measurable, and that all functions under consideration are defined and μ-measurable on X.

29.1. Definition and basic properties of the Lebesgue integral. Let f be a simple function, i.e., a μ-measurable function taking no more than countably many distinct values

$$y_1, y_2, \ldots, y_n, \ldots \tag{1}$$

Then by the (*Lebesgue*) *integral* of f over the set A, denoted by

$$\int_A f(x)\, d\mu,$$

we mean the quantity

$$\sum_n y_n \mu(A_n) \tag{2}$$

where

$$A_n = \{x : x \in A, f(x) = y_n\},$$

provided the series (2) is *absolutely* convergent. If the Lebesgue integral of f exists, we say that f is *integrable* or *summable* (with respect to the measure μ) on the set A.

Example. Obviously,

$$\int_A 1 \cdot d\mu = \int_A d\mu = \mu(A).$$

We now get rid of the restriction that the numbers (1) be distinct:

LEMMA. *Given a simple function* f *defined on a set* A, *suppose* A *is a union*

$$A = \bigcup_k B_k$$

of pairwise disjoint sets B_k such that f takes only one value c_k on B_k. Then f is integrable on A if and only if the series

$$\sum_k c_k \mu(B_k) \tag{3}$$

is absolutely convergent, in which case

$$\int_A f(x)\,d\mu = \sum_k c_k \mu(B_k).$$

Proof. Each set

$$A_n = \{x : x \in A, f(x) = y_n\}$$

is the union of the sets B_k for which $c_k = y_n$. Therefore[4]

$$\sum_n y_n \mu(A_n) = \sum_n y_n \sum_{c_k = y_n} \mu(B_k) = \sum_k c_k \mu(B_k).$$

Moreover, since μ is nonnegative, we have

$$\sum_n |y_n|\,\mu(A_n) = \sum_n |y_n| \sum_{c_k = y_n} \mu(B_k) = \sum_k |c_k|\,\mu(B_k),$$

so that the series (2) is absolutely convergent if and only if the series (3) is absolutely convergent. ∎

THEOREM 1. *Let f and g be simple functions integrable on a set A, and let k be any constant. Then $f + g$ and kf are integrable over A, and*

$$\int_A [f(x) + g(x)]\,d\mu = \int_A f(x)\,d\mu + \int_A g(x)\,d\mu, \tag{4}$$

$$\int_A [kf(x)]\,d\mu = k \int_A f(x)\,d\mu. \tag{5}$$

Proof. Suppose f takes distinct values y_i on sets $F_i \subset A$, while g takes distinct values z_j on sets $G_j \subset A$, where $i, j = 1, 2, \ldots$. Then

$$\int_A f(x)\,d\mu = \sum_i y_i \mu(F_i), \tag{6}$$

$$\int_A g(x)\,d\mu = \sum_j z_j \mu(G_j). \tag{7}$$

Clearly, $f + g$ takes the values $c_{ij} = y_i + z_j$ (not necessarily distinct) on the pairwise disjoint sets $B_{ij} = F_i \cap G_j$. It follows from

$$\mu(F_i) = \sum_j \mu(F_i \cap G_j), \qquad \mu(G_j) = \sum_i \mu(F_i \cap G_j)$$

---

[4] The notation $\displaystyle\sum_{c_k = y_n}$ calls for the sum over all k such that $c_k = y_n$.

and the absolute convergence of the series (6) and (7) that the series

$$\sum_i \sum_j c_{ij}\mu(B_{ij}) = \sum_i \sum_j (y_i + z_j)\mu(F_i \cap G_j)$$

is absolutely convergent. Hence, by the lemma, $f + g$ is integrable on A and

$$\int_A [f(x) + g(x)] \, d\mu = \sum_i \sum_j (y_i + z_j)\mu(F_i \cap G_j)$$
$$= \sum_i y_i\mu(F_i) + \sum_j z_j\mu(G_j). \qquad (8)$$

Comparing (6)–(8), we get (4). The proof of (5) is trivial. ∎

THEOREM 2. *Let f be a bounded simple function on A, where $|f(x)| \leqslant M$ if $x \in A$. Then f is integrable on A and*

$$\left| \int_A f(x) \, d\mu \right| \leqslant M\mu(A).$$

Proof. If f takes values y_n on sets $A_n \subset A$ ($n = 1, 2, \ldots$), then

$$\left| \int_A f(x) \, d\mu \right| = \left| \sum_n y_n\mu(A_n) \right| \leqslant \sum_n |y_n| \, \mu(A_n) \leqslant M \sum_n \mu(A_n) = M\mu(A),$$

where we have incidentally proved the integrability of f on A (how?). ∎

Next we remove the restriction that f be a simple function:

DEFINITION. *A measurable function f is said to be **integrable** (or **summable**) on a set A if there exists a sequence $\{f_n\}$ of integrable simple functions converging uniformly to f on A. The limit*

$$\lim_{n \to \infty} \int_A f_n(x) \, d\mu \qquad (9)$$

*is then called the (**Lebesgue**) **integral** of f over the set A, denoted by*

$$\int_A f(x) \, d\mu.$$

This definition relies tacitly on the following conditions being met:

1) The limit (9) exists (and is finite) for any uniformly convergent sequence of integrable simple functions on A;
2) For any given f, this limit is independent of the choice of the sequence $\{f_n\}$;
3) For simple functions, the definitions of integrability and of the integral reduce to those given on p. 294.

All these conditions are indeed satisfied. Condition 1) is an immediate consequence of the estimate

$$\left| \int_A f_m(x)\, d\mu - \int_A f_n(x)\, d\mu \right| = \left| \int_A [f_m(x) - f_n(x)]\, d\mu \right|$$
$$= \mu(A) \sup_{x \in A} |f_m(x) - f_n(x)|,$$

implied by Theorem 1 and 2. To prove 2), suppose the sequences $\{f_n\}$ and $\{f_n^*\}$ both converge uniformly to f, but

$$\lim_{n \to \infty} \int_A f_n(x)\, d\mu \neq \lim_{n \to \infty} \int_A f_n^*(x)\, d\mu.$$

Let $\{\varphi_n\}$ be the sequence

$$f_1, f_1^*, f_2, f_2^*, \ldots, f_n, f_n^*, \ldots$$

Then $\{\varphi_n\}$ converges uniformly to f, but

$$\lim_{n \to \infty} \int_A \varphi_n(x)\, d\mu$$

fails to exist, contrary to condition 1). Finally, to prove 3), if f is simple, we need only consider the trivial sequence $\{f_n\}$ with general term $f_n = f$.

THEOREM 1'. *Theorem 1 continues to hold if f and g are arbitrary measurable functions integrable on A.*

Proof. An immediate consequence of Theorem 1, after taking suitable uniform limits of integrable simple functions. ∎

THEOREM 3. *If φ is nonnegative and integrable on A and if $|f(x)| \leqslant \varphi(x)$ almost everywhere on A, then f is also integrable on A and*

$$\left| \int_A f(x)\, d\mu \right| \leqslant \int_A \varphi(x)\, d\mu. \tag{10}$$

Proof. If f and φ are simple functions, then, by subtracting a set of measure zero from A, we get a set A' which can be represented as a finite or countable union

$$A' = \bigcup_n A_n$$

of subsets $A_n \subset A'$ such that

$$f(x) = a_n, \qquad \varphi(x) = b_n$$

for all $x \in A_n$ and

$$|a_n| \leqslant b_n \qquad (n = 1, 2, \ldots).$$

Since φ is integrable on A, we have

$$\sum_n |a_n|\, \mu(A_n) \leqslant \sum_n b_n \mu(A_n) = \int_{A'} \varphi(x)\, d\mu = \int_A \varphi(x)\, d\mu \tag{11}$$

(see Problem 3b). Therefore f is also integrable on A, and

$$\left| \int_A f(x)\, d\mu \right| = \left| \int_{A'} f(x)\, d\mu \right| = \left| \sum_n a_n \mu(A_n) \right| \leqslant \sum_n |a_n| \mu(A_n). \quad (12)$$

Comparing (11) and (12), we get (10).

In the case where f and φ are arbitrary measurable functions, let $\{f_n\}$ and $\{\varphi_n\}$ be sequences of simple functions converging uniformly to f and φ, respectively, constructed in the same way as in the proof of Theorem 5, p. 286. Then clearly

$$|f_n(x)| < \varphi_n(x) \qquad (n = 1, 2, \ldots)$$

on A'. Moreover each φ_n is integrable, since φ is integrable by hypothesis. It follows that each f_n and hence f itself is integrable, where

$$\int_A |f_n(x)|\, d\mu \leqslant \int_A \varphi_n(x)\, d\mu.$$

Taking the limit as $n \to \infty$, we again get (10). ∎

COROLLARY. *If f is bounded and measurable on A, then f is integrable on A.*

Proof. Choose $\varphi(x) = M$, where

$$M = \sup_{x \in A} |f(x)|.$$

29.2. Some key theorems. We now prove some important properties of the Lebesgue integral, regarded as a set function

$$F(A) = \int_A f(x)\, d\mu \quad (13)$$

defined on a system of measurable sets (with the integrand f held fixed).

THEOREM 4. *Let*

$$A = \bigcup_n A_n$$

be a finite or countable union of pairwise disjoint sets A_n, and suppose f is integrable on A. Then f is integrable on each A_n and

$$\int_A f(x)\, d\mu = \sum_n \int_{A_n} f(x)\, d\mu, \quad (14)$$

where the series on the right is absolutely convergent.

Proof. First let f be a simple function, taking the values $y_1, y_2, \ldots$, and let

$$B_k = \{x : x \in A, f(x) = y_k\}, \qquad B_{nk} = \{x : x \in A_n, f(x) = y_k\}.$$

Then

$$\int_A f(x)\, d\mu = \sum_k y_k \mu(B_k) = \sum_k y_k \sum_n \mu(B_{nk})$$

$$= \sum_n \sum_k y_k \mu(B_{nk}) = \sum_n \int_{A_n} f(x)\, d\mu. \qquad (15)$$

Since f is integrable on A, the series $\sum_k y_k \mu(B_k)$ converges absolutely, and hence so do the other series in (15). (Here we use the nonnegativity of the measure μ.) In particular, f is integrable on each set A_n.

Next let f be an arbitrary measurable function integrable on A. Then, given any $\varepsilon > 0$, there is a simple function g integrable on A such that

$$|f(x) - g(x)| < \varepsilon \qquad (x \in A). \qquad (16)$$

For g we have

$$\int_A g(x)\, d\mu = \sum_n \int_{A_n} g(x)\, d\mu, \qquad (17)$$

as just shown, where g is integrable on each A_n and the series converges absolutely. Hence, by (16), f is also integrable on each A_n and

$$\sum_n \left| \int_{A_n} f(x)\, d\mu - \int_{A_n} g(x)\, d\mu \right| < \sum_n \varepsilon\mu(A_n) = \varepsilon\mu(A),$$

$$\left| \int_A f(x)\, d\mu - \int_A g(x)\, d\mu \right| < \varepsilon\mu(A),$$

which, together with (17), implies the absolute convergence of the series

$$\sum_n \int_{A_n} f(x)\, d\mu$$

and the estimate

$$\left| \int_A f(x)\, d\mu - \sum_n \int_{A_n} f(x)\, d\mu \right| < 2\varepsilon\mu(A). \qquad (18)$$

But (18) implies (14), since $\varepsilon > 0$ is arbitrary. ∎

COROLLARY. *If f is integrable on A, then f is integrable on every measurable subset $A' \subset A$.*

Proof. Think of A as the union of the disjoint sets A' and $A - A'$. ∎

Remark. A succinct way of expressing the property (14) is to say that the set function (13) is σ-additive.

THEOREM 5 (*Chebyshev's inequality*). *If f is nonnegative and integrable on A, then*

$$\mu\{x : x \in A, f(x) > c\} < \frac{1}{c} \int_A f(x)\, d\mu.$$

Proof. If
$$A' = \{x : x \in A, f(x) \geqslant c\},$$
then
$$\int_A f(x)\, d\mu = \int_{A'} f(x)\, d\mu + \int_{A-A'} f(x)\, d\mu \geqslant \int_{A'} f(x)\, d\mu > c\mu(A')$$
(see Problem 4a). ∎

COROLLARY. *If*
$$\int_A |f(x)|\, d\mu = 0,$$
then $f(x) = 0$ almost everywhere.

Proof. By Chebyshev's inequality,
$$\mu\left\{x : x \in A, |f(x)| > \frac{1}{n}\right\} \leqslant n\int_A |f(x)|\, d\mu = 0$$
for all $n = 1, 2, \ldots$. Therefore
$$\mu\{x : x \in A, f(x) \neq 0\} \leqslant \sum_{n=1}^{\infty} \mu\left\{x : x \in A, |f(x)| > \frac{1}{n}\right\} = 0. ∎$$

THEOREM 6. *If f is integrable on a set A, then, given any $\varepsilon > 0$, there is a $\delta > 0$ such that*
$$\left| \int_E f(x)\, d\mu \right| < \varepsilon$$
for every measurable set $E \subset A$ of measure less than δ.

Proof. The proof is immediate if f is bounded, since then
$$\left| \int_E f(x)\, d\mu \right| \leqslant \int_E |f(x)|\, d\mu \leqslant \sup_{x \in E} |f(x)|\, \mu(E)$$
(see Problem 4c). In the general case, let
$$A_n = \{x : x \in A, n \leqslant |f(x)| < n + 1\},$$
$$B_N = \bigcup_{n=0}^{N} A_n,$$
$$C_N = A - B_N.$$
Then, by Theorem 4,
$$\int_A |f(x)|\, d\mu = \sum_{n=0}^{\infty} \int_{A_n} |f(x)|\, d\mu.$$
Let N be such that
$$\sum_{n=N+1}^{\infty} \int_{A_n} |f(x)|\, d\mu = \int_{C_N} |f(x)|\, d\mu < \frac{\varepsilon}{2}.$$

and let

$$0 < \delta < \frac{\varepsilon}{2(N+1)}.$$

Then $\mu(E) < \delta$ implies

$$\left| \int_E f(x)\, d\mu \right| = \int_E |f(x)|\, d\mu = \int_{E \cap B_N} |f(x)|\, d\mu + \int_{E \cap C_N} |f(x)|\, d\mu$$

$$\leqslant (N+1)\mu(E) + \int_{C_N} |f(x)|\, d\mu < \frac{\varepsilon}{2} + \frac{\varepsilon}{2} = \varepsilon. \quad \blacksquare$$

Remark. The property figuring in Theorem 6 is expressed by saying that the set function (13) is *absolutely continuous* with respect to the measure μ.

Problem 1. Prove that the Dirichlet function

$$f(x) = \begin{cases} 1 & \text{if } x \text{ is rational,} \\ 0 & \text{if } x \text{ is irrational} \end{cases}$$

fails to have a Riemann integral over any interval $[a, b]$. Prove that the Lebesgue integral of f over any measurable set A exists and equals zero.

Problem 2. Find the Lebesgue integral of the function

$$f(x) = \begin{cases} \dfrac{1}{q} & \text{if } x = \dfrac{p}{q} \text{ is rational,} \\ 1 & \text{if } x \text{ is irrational} \end{cases}$$

over the interval $[a, b]$.

Problem 3. Prove that

a) If f is integrable on a set Z of measure zero, then

$$\int_Z f(x)\, d\mu = 0;$$

b) If f is integrable on A, then

$$\int_{A'} f(x)\, d\mu = \int_A f(x)\, d\mu$$

for every subset $A' \subset A$ such that $\mu(A - A') = 0$.

Comment. We can regard a) as a limiting case of Theorem 6.

Problem 4. Prove that

a) If f is nonnegative and integrable on A, then

$$\int_A f(x)\, d\mu \geqslant 0;$$

b) If f and g are integrable on A and $f(x) \leqslant g(x)$ almost everywhere, then

$$\int_A f(x) \, d\mu \leqslant \int_A g(x) \, d\mu;$$

c) If f is integrable on A and $m \leqslant f(x) \leqslant M$ almost everywhere, then

$$m\mu(A) \leqslant \int_A f(x) \, d\mu \leqslant M\mu(A).$$

Problem 5. Prove that the existence of either of the integrals

$$\int_A f(x) \, d\mu, \qquad \int_A |f(x)| \, d\mu$$

implies the existence of the other.

Problem 6. Let

$$A = \bigcup_n A_n$$

be a finite or countable union of pairwise disjoint sets A_n, and suppose f is integrable on each A_n and satisfies the condition

$$\sum_n \int_{A_n} |f(x)| \, d\mu < \infty. \tag{19}$$

Prove that f is integrable on A.

Hint. If f is simple, with values $y_1, y_2, \ldots$, let the sets B_k and B_{nk} be the same as in the proof of Theorem 4. Then

$$\int_{A_n} |f(x)| \, d\mu = \int_k |y_k| \, \mu(B_{nk}).$$

The absolute convergence of (19) implies the convergence of

$$\sum_n \sum_k |y_k| \, \mu(B_{nk}) = \sum_k |y_k| \sum_n \mu(B_{nk}) = \sum_k |y_k| \, \mu(B_k),$$

and hence the integrability of f on A. In the general case, let g be a simple function approximating f, and show that (19) implies the convergence

$$\sum_n \int_{A_n} |g(x)| \, d\mu,$$

so that g, and hence f, is integrable on A.

Comment. This is essentially the converse of Theorem 4.

Problem 7. Let μ be a σ-additive measure defined on a Borel algebra $\mathscr{S}_\mu$ of subsets of a given set X, and let f be nonnegative and integrable on X (with respect to μ). Prove that the set function

$$F(A) = \int_A f(x) \, d\mu$$

is itself a σ-additive measure on $\mathscr{S}_\mu$, with the property that $F(A) = 0$ whenever $\mu(A) = 0$.

Problem 8. Suppose f is integrable on sets $A_1, A_2, \ldots, A_n, \ldots$ such that

$$A_1 \supset A_2 \supset \cdots \supset A_n \supset \cdots,$$

and let

$$A = \bigcap_{n=1}^{\infty} A_n.$$

Does

$$\int_{A_n} f(x) \, d\mu$$

converge to

$$\int_A f(x) \, d\mu \, ?$$

30. Further Properties of the Lebesgue Integral

30.1. Passage to the limit in Lebesgue integrals. The problem of taking limits behind the integral sign, or equivalently of integrating a convergent series term by term, is often encountered in analysis. In the classical theory of integration, it is proved that a sufficient condition for taking such a limit is that the series (or sequence) in question be uniformly convergent. We now examine the corresponding theorems for Lebesgue integrals, which constitute a rather far-reaching generalization of their classical counterparts.

THEOREM 1 (*Lebesgue's bounded convergence theorem*). *Let $\{f_n\}$ be a sequence of functions converging to a limit f on A, and suppose*

$$|f_n(x)| \leqslant \varphi(x) \qquad (x \in A, n = 1, 2, \ldots),$$

where φ is integrable on A. Then f is integrable on A and

$$\lim_{n \to \infty} \int_A f_n(x) \, d\mu = \int_A f(x) \, d\mu.$$

Proof. Clearly $|f(x)| \leqslant \varphi(x)$, and hence f is integrable, by Theorem 3, p. 297. Let

$$A_k = \{x : k - 1 \leqslant \varphi(x) < k\},$$
$$B_m = \bigcup_{k \geqslant m} A_k = \{x : \varphi(x) \geqslant m\}.$$

By Theorem 4, p. 298,

$$\int_A \varphi(x) \, d\mu = \sum_k \int_{A_k} \varphi(x) \, d\mu, \qquad (1)$$

where the series on the right is absolutely convergent. By the same token,

$$\int_{B_m} \varphi(x)\, d\mu = \sum_{k \geqslant m} \int_{A_k} \varphi(x)\, d\mu.$$

Given any $\varepsilon > 0$, there is an integer m such that

$$\int_{B_m} \varphi(x)\, d\mu < \frac{\varepsilon}{5},$$

since the series (1) converges. Moreover, $\varphi(x) < m$ on $A - B_m$. By Egorov's theorem (Theorem 12, p. 290), $A - B_m$ can be represented in the form

$$A - B_m = C \cup D,$$

where $\{f_n\}$ converges uniformly to f on C and

$$\mu(D) < \frac{\varepsilon}{5m}.$$

Let N be such that

$$|f_n(x) - f(x)| < \frac{\varepsilon}{5\mu(C)}$$

on C if $n > N$. Then

$$\int_A [f_n(x) - f(x)]\, d\mu = \int_{B_m} f_n(x)\, d\mu - \int_{B_m} f(x)\, d\mu + \int_D f_n(x)\, d\mu$$
$$- \int_D f(x)\, d\mu + \int_C [f_n(x) - f(x)]\, d\mu,$$

and hence

$$\left| \int_A f_n(x) - \int_A f(x)\, d\mu \right| = \left| \int_A [f_n(x) - f(x)]\, d\mu \right|$$
$$\leqslant \int_{B_m} |f_n(x)|\, d\mu + \int_{B_m} |f(x)|\, d\mu + \int_D |f_n(x)|\, d\mu$$
$$+ \int_D |f(x)|\, d\mu + \int_C |f_n(x) - f(x)|\, d\mu$$
$$< \frac{\varepsilon}{5} + \frac{\varepsilon}{5} + \frac{\varepsilon}{5m} m + \frac{\varepsilon}{5m} + \frac{\varepsilon}{5\mu(C)} \mu(C) = \varepsilon,$$

which implies (1), since $\varepsilon > 0$ is arbitrary. ∎

COROLLARY. *If $|f_n(x)| \leqslant M$ and $f_n \to f$, then*

$$\lim_{n \to \infty} \int_A f_n(x)\, d\mu = \int_A f(x)\, d\mu.$$

Proof. Choose $\varphi(x) = M$, noting that every constant is integrable on A. ∎

Remark. The values taken by a function on a set of measure zero have no effect on its integral. Hence in Theorem 1 we need only assume that $\{f_n\}$ converges to f almost everywhere and that the inequality $|f_n(x)| \leqslant \varphi(x)$ holds almost everywhere.

THEOREM 2 (*Levi*). *Suppose*

$$f_1(x) \leqslant f_2(x) \leqslant \cdots \leqslant f_n(x) \leqslant \cdots$$

on a set A, where the functions f_n are all integrable and

$$\int_A f_n(x)\, d\mu \leqslant M \qquad (n = 1, 2, \ldots) \tag{2}$$

for some constant M. Then the limit

$$f(x) = \lim_{n \to \infty} f_n(x)$$

exists (and is finite) almost everywhere on A.[5] *Moreover, f is integrable and*

$$\lim_{n \to \infty} \int_A f_n(x)\, d\mu = \int_A f(x)\, d\mu.$$

Proof. It can be assumed that $f_1(x) \geqslant 0$, since otherwise we need only replace the f_n by $f_n - f_1$. Let

$$\Omega = \{x : x \in A, f_n(x) \to \infty\}.$$

Then clearly

$$\Omega = \bigcap_r \bigcup_n \Omega_n^{(r)},$$

where

$$\Omega_n^{(r)} = \{x : x \in A, f_n(x) \geqslant r\}.$$

It follows from (2) and Chebyshev's inequality (Theorem 5, p. 299) that

$$\mu(\Omega_n^{(r)}) \leqslant \frac{M}{r}.$$

Moreover

$$\mu\left(\bigcup_n \Omega_n^{(r)}\right) \leqslant \frac{M}{r},$$

since

$$\Omega_1^{(r)} \subset \Omega_2^{(r)} \subset \cdots \subset \Omega_n^{(r)} \subset \cdots.$$

But

$$\Omega \subset \bigcup_n \Omega_n^{(r)}$$

---

[5] The function f can be defined in an arbitrary way on the set E where the limit (2) fails to exist, for example, by setting $f(x) = 0$ on E.

for any r, and hence

$$\mu(\Omega) \leqslant \frac{M}{r}.$$

Since r can be arbitrarily large, this implies

$$\mu(\Omega) = 0,$$

thereby showing that the sequence $\{f_n(x)\}$ has a finite limit $f(x)$ for almost all $x \in A$.

Now let

$$A_r = \{x : r - 1 \leqslant f(x) < r\},$$

and let φ be the simple function such that

$$\varphi(x) = r \text{ if } x \in A_r \qquad (r = 1, 2, \ldots).$$

Moreover, let

$$B_s = \bigcup_{r=1}^{s} A_r.$$

Since the functions f_n and f are bounded on B_s and since

$$\varphi(x) \leqslant f(x) + 1,$$

we have

$$\int_{B_s} \varphi(x) \, d\mu \leqslant \int_{B_s} f(x) \, d\mu + \mu(A)$$

$$= \lim_{n \to \infty} \int_{B_s} f_n(x) \, d\mu + \mu(A) \leqslant M + \mu(A),$$

where we use the corollary to Theorem 1. But

$$\int_{B_s} \varphi(x) \, d\mu = \sum_{r=1}^{s} r\mu(A_r),$$

and hence

$$\sum_{r=1}^{s} r\mu(A_r) < M + \mu(A)$$

for all $s = 1, 2, \ldots$. Therefore

$$\sum_{r=1}^{\infty} r\mu(A_r) < \infty,$$

i.e., φ is integrable on A, with integral

$$\int_A \varphi(x) \, d\mu = \sum_{r=1}^{\infty} r\mu(A_r).$$

Since $f_n(x) \leqslant \varphi(x)$, the validity of (3) is now an immediate consequence of Lebesgue's bounded convergence theorem (Theorem 1). ∎

COROLLARY. *If $\varphi_k(x) \geqslant 0$ and*

$$\sum_{k=1}^{\infty} \int_A \varphi_k(x) \, d\mu < \infty,$$

then the series

$$\sum_{k=1}^{\infty} \varphi_k(x)$$

converges almost everywhere on A and

$$\sum_{k=1}^{\infty} \int_A \varphi_k(x) \, d\mu = \int_A \left(\sum_{k=1}^{\infty} \varphi_k(x) \right) \, d\mu.$$

Proof. Apply Theorem 2 to the functions

$$f_n(x) = \sum_{k=1}^{n} \varphi_k(x).$$

THEOREM 3 (*Fatou*). *Let $\{f_n\}$ be a sequence of nonnegative functions integrable on a set A, such that*

$$\int_A f_n(x) \, d\mu \leqslant M \qquad (n = 1, 2, \ldots).$$

Suppose $\{f_n\}$ converges almost everywhere on A to a function f. Then f is integrable on A and

$$\int_A f(x) \, d\mu \leqslant M.$$

Proof. Let

$$\varphi_n(x) = \inf_{k \geqslant n} f_k(x).$$

Then φ_n is measurable, since

$$\{x : \varphi_n(x) < c\} = \bigcup_{k \geqslant n} \{x : f_k(x) < c\}.$$

Moreover

$$0 \leqslant \varphi_n(x) \leqslant f_n(x),$$

and hence φ_n is integrable, by Theorem 3, p. 297, with

$$\int_A \varphi_n(x) \, d\mu \leqslant \int_A f_n(x) \, d\mu \leqslant M \qquad (n = 1, 2, \ldots).$$

Clearly

$$\varphi_1(x) \leqslant \varphi_2(x) \leqslant \cdots \leqslant \varphi_n(x) \leqslant \cdots,$$

and

$$\lim_{n \to \infty} \varphi_n(x) = f(x)$$

almost everywhere. Applying Theorem 2 to the sequence $\{\varphi_n\}$, we find that f is integrable and

$$\int_A f(x) \, d\mu = \lim_{n \to \infty} \int_A \varphi_n(x) \, d\mu \leqslant M. \quad \blacksquare$$

30.2. The Lebesgue integral over a set of infinite measure. So far all our measures have been finite (except for Remark 3, p. 267), and hence everything said about the Lebesgue integral and its properties has been tacitly understood to apply only to the case of functions defined on sets of finite measure. However, one often deals with functions defined on a set X of infinite measure, for example, the real line equipped with ordinary Lebesgue measure. We will confine ourselves to the case of greatest practical interest, where X can be represented as a union

$$X = \bigcup_n X_n, \quad \mu(X_n) < \infty \tag{3}$$

of countably many sets X_n, each of finite measure with respect to some σ-additive measure μ defined on a σ-ring of subsets of X (the sets of finite measure). Such a measure is called σ-*finite*. For example, Lebesgue measure on the line, in the plane, or more generally in n-space is σ-finite. For simplicity, and without loss of generality (why?), we will assume that the sequence $\{X_n\}$ is increasing, i.e., that

$$X_1 \subset X_2 \subset \cdots \subset X_n \subset \cdots. \tag{4}$$

A sequence $\{X_n\}$ satisfying the conditions (3) and (4) will be called *exhaustive*. For example, the sequence $\{E_n\}$ in Remark 3, p. 267 is an exhaustive sequence (with respect to ordinary Lebesgue measure), whose union is the whole plane.

Now let f be a measurable function on X.[6] Then f is said to be *integrable* (or *summable*) on X if it is integrable on every measurable subset $A \subset X$ and if the limit

$$\lim_{n \to \infty} \int_{X_n} f(x) \, d\mu \tag{5}$$

exists (and is finite) for every exhaustive sequence $\{X_n\}$. The limit (5) is then called the (*Lebesgue*) *integral* of f over the set X, denoted by

$$\int_X f(x) \, d\mu.$$

Remark 1. The limit (5) is independent of the choice of the exhaustive sequence $\{X_n\}$. In fact, suppose

$$\lim_{n \to \infty} \int_{X_n} f(x) \, d\mu \neq \lim_{n \to \infty} \int_{X_n^*} f(x) \, d\mu,$$

---

[6] A real function $y = f(x)$ is now said to be measurable if the set $f^{-1}(A) \cap X_n$ is measurable for every X_n and every Borel set A (this being the obvious slight generalization of Definition 1, p. 284).

where $\{X_n^*\}$ is another exhaustive sequence. Define a new sequence $\{\Omega_n\}$ such that

$$\Omega_1 = X_1,$$

$$\Omega_{2k} \text{ is any set of } \{X_n^*\} \text{ containing } \Omega_{2k-1},$$

$$\Omega_{2k+1} \text{ is any set of } \{X_n\} \text{ containing } \Omega_{2k}$$

(why do such sets exist?). Then $\{\Omega_n\}$ is exhaustive, but

$$\lim_{n \to \infty} \int_{\Omega_n} f(x) \, d\mu$$

fails to exist, contrary to hypothesis.

Remark 2. The integral of a simple function is defined in the same way as on p. 294. It is clear that a necessary (but not sufficient) condition for integrability of a simple function f is that f take every nonzero value on a set of finite measure.

30.3. The Lebesgue integral vs. the Riemann integral.

Finally we examine the relation between the Lebesgue integral and the Riemann integral, restricting ourselves to the case of ordinary Lebesgue measure on the line:

THEOREM 4. *If the Riemann integral*

$$I = \int_a^b f(x) \, dx$$

exists, then f is Lebesgue integrable on $[a, b]$ and

$$\int_{[a,b]} f(x) \, d\mu = I. \tag{6}$$

Proof. Introducing the points of subdivision

$$x_k = a + \frac{k}{2^n}(b - a) \qquad (k = 1, \ldots, 2^n),$$

we partition $[a, b]$ into 2^n subintervals. Let

$$\Delta_n = \frac{b - a}{2^n} \sum_{k=1}^{2^n} M_{nk},$$

$$\delta_n = \frac{b - a}{2^n} \sum_{k=1}^{2^n} m_{nk}$$

be the corresponding Darboux sums, where M_{nk} is the least upper bound and m_{nk} the greatest lower bound on f on the subinterval $x_{k-1} \leqslant x \leqslant x_k$. By the definition of the Riemann integral,

$$I = \lim_{n \to \infty} \Delta_n = \lim_{n \to \infty} \delta_n.$$

Consider the functions

$$\bar{f}_n(x) = M_{nk} \qquad \text{if } x_{k-1} \leqslant x < x_k,$$

$$\underline{f}_n(x) = m_{nk} \qquad \text{if } x_{k-1} \leqslant x < x_k,$$

$$\bar{f}_n(b) = \underline{f}_n(b) = f(b).$$

Then clearly

$$\int_{[a,b]} \bar{f}_n(x)\, d\mu = \Delta_n, \qquad \int_{[a,b]} \underline{f}_n(x)\, d\mu = \delta_n. \tag{7}$$

Moreover,

$$\bar{f}_1(x) > \bar{f}_2(x) > \cdots > \bar{f}_n(x) > \cdots > f(x),$$

$$\underline{f}_1(x) \leqslant \underline{f}_2(x) \leqslant \cdots \leqslant \underline{f}_n(x) \leqslant \cdots \leqslant f(x),$$

and hence

$$\lim_{n \to \infty} \bar{f}_n(x) = \bar{f}(x) > f(x),$$

$$\lim_{n \to \infty} \underline{f}_n(x) = \underline{f}(x) < f(x).$$

Using (7) and Theorem 2, we find that

$$\int_{[a,b]} \bar{f}(x)\, d\mu = \lim_{n \to \infty} \int_{[a,b]} \bar{f}_n(x)\, d\mu = \lim_{n \to \infty} \Delta_n = I$$

$$= \lim_{n \to \infty} \delta_n = \lim_{n \to \infty} \int_{[a,b]} \underline{f}_n(x)\, d\mu = \int_{[a,b]} \underline{f}(x)\, d\mu \tag{8}$$

(see also Problem 2). Therefore

$$\int_{[a,b]} |\bar{f}(x) - \underline{f}(x)|\, d\mu = \int_{[a,b]} \{\bar{f}(x) - \underline{f}(x)\}\, d\mu = 0,$$

and hence

$$\bar{f}(x) - \underline{f}(x) = 0$$

almost everywhere, by the corollary on p. 300. In other words,

$$\bar{f}(x) = f(x) = \underline{f}(x) \tag{9}$$

almost everywhere. Comparing (8) and (9), we get (6). ∎

Problem 1. Prove that

$$\lim_{n \to \infty} \int_A f_n(x) g(x)\, d\mu = \int_A f(x)\, d\mu(x)$$

if the sequence $\{f_n\}$ satisfies the conditions of Theorem 1 (as stated more generally in the remark on p. 305) and if g is *essentially bounded* on A in the sense that there is a constant $M > 0$ such that $|g(x)| < M$ almost everywhere on A.

Comment. If g is essentially bounded on A, then the quantity

$$\operatorname*{ess\,sup}_{x \in A} |g(x)| = \inf_{\substack{Z \subset A \\ \mu(Z)=0}} \left\{ \sup_{x \in A-Z} |g(x)| \right\},$$

called the *essential supremum* of g on A, is finite.

Problem 2. Prove that Theorem 2 remains valid if

$$f_1(x) \geqslant f_2(x) \geqslant \cdots \geqslant f_n(x) \geqslant \cdots$$

and if (2) is replaced by the condition

$$\int_A f_n(x) \, d\mu \geqslant M \qquad (n = 1, 2, \ldots).$$

Problem 3. Consider the system $\mathscr{S}$ of all subsets of the real line containing only finitely many points, and let the measure $\mu(A)$ of a set $A \in \mathscr{S}$ be defined as the number of points in A. Prove that

a) $\mathscr{S}$ is a ring without a unit;
b) μ is not σ-finite.

Problem 4. Why do we talk about a σ-ring rather than a σ-algebra on p. 308?

Problem 5. Prove that if a function f vanishes outside a set of finite measure, then its Lebesgue integral as defined on p. 308 coincides with its Lebesgue integral as previously defined.

Problem 6. Show that the analogue of the definition on p. 296 cannot be used to define the Lebesgue integral in the case where A is of infinite measure.

Hint. Give an example of a uniformly convergent sequence $\{f_n\}$ of integrable simple functions such that

$$\lim_{n \to \infty} \int_A f_n(x) \, d\mu$$

fails to exist.

Problem 7. Which of the theorems of Sec. 29 continue to hold for integrals over sets of infinite measure?

Hint. The corollary on p. 298 fails if A is of infinite measure.

Problem 8. Verify that Theorems 1–3 of Sec. 30.1 continue to hold for integrals over sets of infinite measure.

Problem 9. Given a nonnegative function f, suppose the Riemann integral

$$\int_{a+\varepsilon}^{b} f(x) \, dx$$

exists for every $\varepsilon > 0$ and approaches a finite limit as $\varepsilon \to 0+$, so that the improper Riemann integral

$$\int_a^b f(x)\,dx = \lim_{\varepsilon \to 0+} \int_{a+\varepsilon}^b f(x)\,dx \qquad (10)$$

exists. Prove that f is Lebesgue integrable on $[a, b]$ and

$$\int_{[a,b]} f(x)\,d\mu = \int_a^b f(x)\,dx.$$

Comment. On the other hand, if f is of variable sign and if

$$\lim_{\varepsilon \to +0} \int_{a+\varepsilon}^b |f(x)|\,dx = \infty,$$

then the Lebesgue integral of f over $[a, b]$ fails to exist, even if the improper Riemann integral (10) exists. In fact, by Problem 5, p. 302, summability of f would imply that of $|f|$.

Problem 10. Prove that the integral

$$\int_0^1 \frac{1}{x} \sin \frac{1}{x}\,dx$$

exists as an improper Riemann integral, but not as a Lebesgue integral.

Problem 11. Suppose f is Riemann integrable over an infinite interval (such an integral can exist only in the improper sense). Prove that f is Lebesgue integrable over the same interval if and only if the improper integral converges absolutely.

Comment. For example, the function

$$f(x) = \frac{\sin x}{x}$$

is not Lebesgue integrable over $(-\infty, \infty)$, since

$$\int_{-\infty}^\infty \left| \frac{\sin x}{x} \right|\,dx = \infty.$$

On the other hand, f has an improper Riemann integral equal to

$$\int_{-\infty}^\infty \frac{\sin x}{x} = \pi.$$

9

DIFFERENTIATION

Let f be a summable function defined on a space X, equipped with a σ-additive measure μ. Then the (Lebesgue) integral

$$\int_E f(x)\, d\mu \tag{1}$$

exists for every measurable $E \subset X$, thereby defining a set function on the system $\mathscr{S}_\mu$ of all measurable subsets of X. If X is the real line, equipped with ordinary Lebesgue measure μ, and if $E = [a, b]$ is a closed interval, we write (1) simply as

$$\int_a^b f(x)\, dx,$$

or equivalently as

$$\int_a^b f(t)\, dt \tag{2}$$

in terms of the new dummy variable of integration t (here we anticipate subsequent notational convenience). Then (2) is clearly a function of the lower limit of integration a and the upper limit of integration b. Suppose we fix a, but leave b variable, indicating this by replacing b by the symbol x. Then (2) reduces to the "indefinite Lebesgue integral"

$$\int_a^x f(t)\, dt,$$

with its upper limit of integration variable.

Now let f be continuous, and let F have a continuous derivative. Then it will be recalled from elementary calculus that the connection between

313

the operations of differentiation and integration is expressed by the familiar formulas

$$\frac{d}{dx} \int_a^x f(t)\, dt = f(x), \tag{3}$$

$$\int_a^x F'(t)\, dt = F(x) - F(a). \tag{4}$$

This immediately suggests two questions:

1) Does (3) continue to hold for an arbitrary summable function f?
2) What is the largest class of functions for which (4) holds?

These questions will be answered in Secs. 31–33. The study of the general set function (1) will be resumed in Sec. 34.

31. Differentiation of the Indefinite Lebesgue Integral

31.1. Basic properties of monotonic functions. We begin our study of the indefinite Lebesgue integral

$$F(x) = \int_a^x f(t)\, dt \tag{1}$$

as a function of its upper limit by making the following obvious but important observation. If f is nonnegative, then (1) is a nondecreasing function. Moreover, since every summable function $f(t)$ is the difference

$$f(t) = f_+(t) - f_-(t)$$

of two nonnegative summable functions (which?), the integral (1) is the difference between two nondecreasing functions. Hence, the study of the Lebesgue integral as a function of its upper limit is closely related to the study of monotonic functions. Monotonic functions are interesting in their own right, and have a number of simple and important properties which we now discuss. Here all functions will be regarded as defined on some fixed interval $[a, b]$ unless the contrary is explicitly stated.

DEFINITION 1, *A function f is said to be **nondecreasing** if $x_1 < x_2$ implies $f(x_1) < f(x_2)$ and **nonincreasing** if $x_1 < x_2$ implies $f(x_1) > f(x_2)$. By a **monotonic** function is meant a function which is either nondecreasing or nonincreasing.*

DEFINITION 2. *Given any function f, the limit*

$$\lim_{\substack{\varepsilon \to 0 \\ \varepsilon > 0}} f(x_0 + \varepsilon)$$

*(provided it exists) is called the **right-hand limit** of f at the point x_0,* denoted by

$$f(x_0 + 0).$$

Similarly, the limit

$$\lim_{\substack{\varepsilon \to 0 \\ \varepsilon > 0}} f(x_0 - \varepsilon)$$

*is called the **left-hand limit** of f at x_0, denoted by*

$$f(x_0 - 0).$$

Remark. If

$$f(x_0 + 0) = f(x_0 - 0),$$

then clearly f is either continuous at x_0 or has a removable discontinuity at x_0.

DEFINITION 3. *A function f is said to be **continuous from the right** at x_0 if*

$$f(x_0) = f(x_0 + 0),$$

*and **continuous from the left** at x_0 if*

$$f(x_0) = f(x_0 - 0).$$

DEFINITION 4. *By a **discontinuity point of the first kind** of a function f is meant a point x_0 at which the limits $f(x_0 + 0)$ and $f(x_0 - 0)$ exist but are unequal. The difference*

$$f(x_0 + 0) - f(x_0 - 0)$$

*is then called the **jump** of f at x_0.*

Example. Given no more than countably many points

$$x_1, x_2, \ldots, x_n, \ldots$$

in the interval $[a, b]$, let

$$h_1, h_2, \ldots, h_n, \ldots$$

be corresponding positive numbers such that

$$\sum_n h_n < \infty.$$

Then the function

$$f(x) = \sum_{x_n < x} h_n, \tag{2}$$

where the sum is over all n such that $x_n < x$, is obviously nondecreasing. A monotonic function of this particularly simple type is called a *jump function*. A jump function such that

$$x_1 < x_2 < \cdots < x_n < \cdots,$$

is called a *step function*. For an example of a jump function which is not a step function, see Problem 1.

We now establish the basic properties of monotonic functions. To be explicit, we will talk about nondecreasing functions, but clearly everything carries over automatically to the case of nonincreasing functions.

THEOREM 1. *Every nondecreasing function f on* $[a, b]$ *is measurable and bounded, and hence summable.*[1]

Proof. Since $f(x) \leqslant f(b)$ for all $x \in [a, b]$, f is obviously bounded. Consider the set

$$E_c = \{x : f(x) < c\}.$$

If E_c is empty, then E_c is (trivially) measurable. If E_c is nonempty, let d be the least upper bound of all $x \in E_c$. Then E_c is either the closed interval $[c, d]$, if $d \in E_c$, or the half-open interval $[a, d)$ if $d \notin E_c$. In either case, E_c is measurable. ∎

THEOREM 2. *Every discontinuity point of a nondecreasing function is of the first kind.*

Proof. Let x be any point of $[a\ b]$, and let $\{x_n\}$ be any sequence such that $x_n < x_0, x_n \to x_0$. Then $\{f(x_n)\}$ is a nondecreasing sequence bounded from above, e.g., by the number $f(x_0)$. Therefore $\lim_{n \to \infty} f(x_n)$ exists for any such sequence, i.e., $f(x_0 - 0)$ exists. The existence of $f(x_0 + 0)$ is proved in the same way. ∎

Obviously, a nondecreasing function need not be continuous. However, we have

THEOREM 3. *A nondecreasing function can have no more than countably many points of discontinuity.*

Proof. The sum of the jumps of f on the interval $[a, b]$ cannot exceed $f(b) - f(a)$. Let J_n be the set of all jumps greater than $1/n$, and let J be the set of all jumps regardless of size. Then obviously

$$J = \bigcup_{n=1}^{\infty} J_n,$$

where each J_n is a finite set. Hence J has no more than countably many elements. ∎

THEOREM 4. *The jump function* (2) *is continuous from the left. Moreover, all the discontinuity points of f are of the first kind, with the jump at x_n equal to h_n.*

---

[1] See the corollary on p. 298.

Proof. Clearly,

$$f(x - 0) = \lim_{\substack{\varepsilon \to 0 \\ \varepsilon > 0}} f(x - \varepsilon) = \lim_{\substack{\varepsilon \to 0 \\ \varepsilon > 0}} \sum_{x_n < x - \varepsilon} h_n.$$

But if $x_n < x$, then $x_n < x - \varepsilon$ for sufficiently small $\varepsilon > 0$. Therefore

$$\lim_{\substack{\varepsilon \to 0 \\ \varepsilon > 0}} \sum_{x_n < x - \varepsilon} h_n = f(x),$$

and hence

$$f(x - 0) = f(x).$$

If x coincides with one of the points x_n, say with x_{n_0}, then

$$f(x_{n_0} + 0) = \lim_{\varepsilon \to 0} f(x_{n_0} + \varepsilon) = \lim_{\varepsilon \to 0} \sum_{x_n < x_{n_0} + \varepsilon} h_n = \sum_{x_n \leqslant x_{n_0}} h_n,$$

which implies

$$f(x_{n_0} + 0) - f(x_{n_0} - 0) = h_{n_0}. \quad \blacksquare$$

THEOREM 5. *If f is continuous from the left and nondecreasing, then f is the sum of a continuous nondecreasing function φ and a jump function ψ.*

Proof. If $x_1, x_2, \ldots$ are the discontinuity points of f, with corresponding jumps $h_1, h_2, \ldots$, let

$$\psi(x) = \sum_{x_n < x} h_n,$$

$$\varphi(x) = f(x) - \psi(x).$$

Then

$$\varphi(x'') - \varphi(x') = [f(x'') - f(x')] - [\psi(x'') - \psi(x')],$$

where the expression on the right is the difference between the total increment of f on the interval $[x', x'']$ and the sum of its jumps on $[x', x'']$, i.e., $\varphi(x'') - \varphi(x')$ is the measure of the set of values taken by f at its continuity points in $[x', x'']$. This quantity is clearly nonnegative, and hence φ is nondecreasing. Moreover, given any point $x \in [a, b]$, we have

$$\varphi(x - 0) = \lim_{\substack{\varepsilon \to 0 \\ \varepsilon > 0}} f(x - \varepsilon) - \lim_{\substack{\varepsilon \to 0 \\ \varepsilon > 0}} \psi(x - \varepsilon) = f(x - 0) - \sum_{x_n < x} h_n,$$

$$\varphi(x + 0) = \lim_{\substack{\varepsilon \to 0 \\ \varepsilon > 0}} f(x + \varepsilon) - \lim_{\substack{\varepsilon \to 0 \\ \varepsilon > 0}} \psi(x + \varepsilon) = f(x + 0) - \sum_{x_n \leqslant x} h_n,$$

and hence

$$\varphi(x + 0) - \varphi(x - 0) = f(x + 0) - f(x - 0) - h = 0,$$

where h is the jump of ψ at x. It follows that φ is continuous at every point $x \in [a, b]$. $\quad \blacksquare$

31.2. Differentiation of a monotonic function. The key result of this section (see Theorem 6 below) will be to show that *a monotonic function f defined on an interval* [a, b] *has a finite derivative almost everywhere on* [a, b]. Before proving this proposition, due to Lebesgue, we must first introduce some further definitions and then establish three preliminary lemmas.

The derivative of a function f at a point x_0 is defined in the familiar way as the limit of the ratio

$$\frac{f(x) - f(x_0)}{x - x_0} \tag{3}$$

as $x \to x_0$. Even if this limit fails to exist, the following four quantities (which may take infinite values) always exist:

1) The lower limit of (3) as $x \to x_0$ from the left, denoted by λ_L;
2) The upper limit of (3) as $x \to x_0$ from the left, denoted by Λ_L;[2]
3) The lower limit of (3) as $x \to x_0$ from the right, denoted by λ_R;
4) The upper limit of (3) as $x \to x_0$ from the right, denoted by Λ_R.

These four quantities, with the geometric meaning shown in Figure 17, are called the *derived numbers* of f at x_0.[3] It is clear that the inequalities

$$\lambda_L \leqslant \Lambda_L, \qquad \lambda_R \leqslant \Lambda_R \tag{4}$$

always hold. If λ_L and Λ_L exist and are equal, their common value is just the left-hand derivative of f at x_0. Similarly, if λ_R and Λ_R exist and are equal, their common value is just the right-hand derivative of f at x_0. Moreover, f has a derivative at x_0 if and only if all four derived numbers λ_L, Λ_L

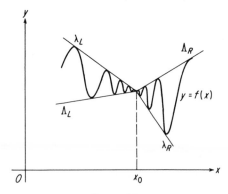

FIGURE 17

---

[2] Upper and lower limits are defined on p. 111.

[3] To distinguish these quantities further, we can call λ_L the *left-hand lower derived number*, Λ_R the *right-hand upper derived number*, and so on.

λ_R and Λ_R exist and are equal at x_0. Hence the italicized assertion at the beginning of this section can be restated as follows: *For a monotonic function defined on an interval* $[a, b]$, *the formula*

$$-\infty < \lambda_L = \Lambda_L = \lambda_R = \Lambda_R < +\infty$$

holds almost everywhere on $[a, b]$.

DEFINITION 5. *Let* f *be a continuous function defined on an interval* $[a, b]$. *Then a point* $x_0 \in [a, b]$ *is said to be **invisible from the right** (with respect to* f) *if there is a point* ξ *such that* $x_0 < \xi \leqslant b$ *and* $f(x_0) < f(\xi)$, *and **invisible from the left** if there is a point* ξ *such that* $a \leqslant \xi < x_0$ *and* $f(x_0) < f(\xi)$.

Example. In Figure 18, the points belonging to the intervals $[a_1, b_1)$ and (a_2, b_2) are invisible from the right (interpret the word "invisible").

LEMMA 1 (*F. Riesz*). *The set of all points invisible from the right with respect to a function* f *continuous on* $[a, b]$ *is the union of no more than countably many pairwise disjoint open intervals* (a_k, b_k),[4] *such that*

$$f(a_k) \leqslant f(b_k) \qquad (k = 1, 2, \ldots). \tag{5}$$

Proof. If x_0 is invisible from the right with respect to f, then the same is true of any point sufficiently close to x_0, by the continuity of f. Hence the set of all points invisible from the right is an open set G. It follows from Theorem 6, p. 51 that G is the union of a finite or countable system of pairwise disjoint open intervals. Let (a_k, b_k) be one of these intervals, and suppose

$$f(a_k) > f(b_k). \tag{6}$$

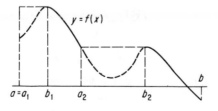

FIGURE 18

---

[4] However, if $a_1 = a$ (say), then in some cases (a_1, b_1) should be replaced by the half-open interval $[a_1, b_1)$, as in Figure 18. This is permissible, since $[a_1, b_1)$ is open relative to $[a, b]$.

Then there is an (interior) point $x_0 \in (a_k, b_k)$ such that $f(x_0) > f(b_k)$. Of the points $x \in (a_k, b_k)$ such that $f(x) = f(x_0)$, let x^* be the one with largest abscissa (x^* may coincide with x_0). Since x^* belongs to (a_k, b_k) and hence is invisible from the right, there is a point $\xi > x^*$ such that $f(\xi) > f(x^*)$. Clearly ξ cannot belong to (a_k, b_k), since x^* is the point x with largest abscissa for which $f(x) = f(x_0)$, while $f(b_k) < f(x_0)$, so that $\xi \in (a_k, b_k)$ would imply the existence of a point $x > x^*$ such that $f(x) = f(x_0)$. On the other hand, the inequality $\xi > b_k$ is also impossible, since it would imply $f(b_k) < f(x_0) < f(\xi)$ despite the fact that b_k is *not* invisible from the right. Thus (6) leads to a contradiction (obviously $\xi \neq b_k$). It follows that $f(a_k) \leqslant f(b_k)$. ∎

LEMMA 1′. *The set of all points invisible from the left with respect to a function f continuous on $[a, b]$ is the union of no more than countably many pairwise disjoint open intervals (a_k, b_k), such that*

$$f(a_k) \geqslant f(b_k) \qquad (k = 1, 2, \ldots).$$

Proof. Virtually the same as that of Lemma 1. ∎

LEMMA 2. *Let f be a continuous nondecreasing function on $[a, b]$, with λ_L and Λ_R as two of its derived numbers. Given any numbers c, C and ρ such that*

$$0 < c < C < \infty, \qquad \rho = \frac{c}{C},$$

let E_ρ be the set

$$E_\rho = \{x : \lambda_L < c, \Lambda_R > C\}.$$

Then

$$\mu\{x : x \in E \cap (\alpha, \beta)\} \leqslant \rho(\beta - \alpha)$$

for every open interval $(\alpha, \beta) \subset [a, b]$.

Proof. Let x_0 be a point of (α, β) for which $\lambda_L < c$. Then there is a point $\xi < x$ such that

$$\frac{f(\xi) - f(x_0)}{\xi - x_0} < c$$

i.e., such that

$$f(\xi) - c > f(x_0) - cx_0.$$

Therefore x_0 is invisible from the left with respect to the function $f(x) - cx$. Hence, by Lemma 1′, the set of all such x_0 is the union of no more than countably many pairwise disjoint open intervals $(\alpha_k, \beta_k) \subset (\alpha, \beta)$, where

$$f(\alpha_k) - c\alpha_k \geqslant f(\beta_k) - c\beta_k,$$

or equivalently

$$f(\beta_k) - f(\alpha_k) \leqslant c(\beta_k - \alpha_k). \tag{7}$$

Let G_k be the set of points in (α_k, β_k) for which $\Lambda_R > C$. Then, by virtually the same argument together with Lemma 1, G_k is the union of no more than countably many pairwise disjoint open intervals $(\alpha_{k_n}, \beta_{k_n})$, where

$$\beta_{k_n} - \alpha_{k_n} \leqslant \frac{1}{C} [f(\beta_{k_n}) - f(\alpha_{k_n})] \tag{8}$$

(why?). Clearly $E_\rho \cap (\alpha, \beta)$ is covered by the system of intervals $(\alpha_{k_n}, \beta_{k_n})$. Moreover, it follows from (7) and (8) that

$$\sum_{k,n} (\beta_{k_n} - \alpha_{k_n}) \leqslant \frac{1}{C} \sum_{k,n} [f(\beta_{k_n}) - f(\alpha_{k_n})]$$

$$\leqslant \frac{1}{C} \sum_{k} [f(\beta_k) - f(\alpha_k)] \leqslant \frac{c}{C} \sum_{k} (\beta_k - \alpha_k) \leqslant \rho(\beta - \alpha). \quad \blacksquare$$

We are now in a position to prove

THEOREM 6 (*Lebesgue*). *A monotonic function f defined on an interval* $[a, b]$ *has a finite derivative almost everywhere on* $[a, b]$.

Proof. There is no loss of generality in assuming that f is nondecreasing, since if f is nonincreasing, then obviously $-f$ is nondecreasing. But if $-f$ has a derivative almost everywhere, then so does f. We also assume that f is continuous, dropping this restriction at the end of the proof. It will be enough to show that the two inequalities

$$\Lambda_R < +\infty \tag{9}$$

and

$$\lambda_L \geqslant \Lambda_R \tag{10}$$

hold almost everywhere on $[a, b]$, for any continuous nondecreasing function. In fact, setting $f^*(x) = -f(-x)$, we see that f^* is continuous and nondecreasing, like f itself. Moreover, it is easily verified that

$$\lambda_L^* = \lambda_R, \qquad \Lambda_R^* = \Lambda_L,$$

where λ_L^* and Λ_R^* are the indicated derived numbers of f^*. Therefore, applying (10) to f^*, we get

$$\lambda_L^* \geqslant \Lambda_R^*$$

or

$$\lambda_R \geqslant \Lambda_L. \tag{11}$$

Combining the inequalities (10) and (11), we obtain

$$\Lambda_R \leqslant \lambda_L \leqslant \Lambda_L \leqslant \lambda_R \leqslant \Lambda_R,$$

after using (4). Thus if (9) and (10) hold almost everywhere, we have[5]

$$-\infty < \lambda_L = \Lambda_L = \lambda_R = \Lambda_R < +\infty$$

almost everywhere, and the theorem is proved.

To prove that $\Lambda_R < +\infty$ almost everywhere, we argue as follows: If $\Lambda_R = +\infty$ at some point x_0, then, given any constant $C > 0$, there is a point $\xi > x_0$ such that

$$\frac{f(\xi) - f(x_0)}{\xi - x_0} > C,$$

i.e.,

$$f(\xi) - f(x_0) > C(\xi - x_0),$$

or equivalently

$$f(\xi) - C\xi > f(x_0) - Cx_0.$$

Thus x_0 is invisible from the right with respect to the function $f(x) - Cx$. Hence, by Lemma 1, the set of all points x_0 at which $\Lambda_R = +\infty$ is the union of no more than countably many open intervals (a_k, b_k), whose end points satisfy the inequalities

$$f(a_k) - Ca_k \leqslant f(b_k) - Cb_k$$

or

$$f(b_k) - f(a_k) \geqslant C(b_k - a_k).$$

Dividing by C and summing over all the intervals (a_k, b_k), we get

$$\sum_k (b_k - a_k) \leqslant \sum_k \frac{f(b_k) - f(a_k)}{C} \leqslant \frac{f(b) - f(a)}{C}.$$

But C can be made arbitrarily large. Hence the set of points where $\Lambda_R = +\infty$ can be covered by a collection of intervals the sum of whose lengths is arbitrarily small. It follows that this set is of measure zero, i.e., that $\Lambda_R < +\infty$ almost everywhere.

To prove that $\lambda_L \geqslant \Lambda_R$ almost everywhere, let the numbers c, C, ρ and the set E_ρ be the same as in Lemma 2. It will then follow that $\lambda_L \geqslant \Lambda_R$ almost everywhere if we succeed in showing that $\mu(E_\rho) = 0$, since the set of points where $\lambda_L < \Lambda_R$ can clearly be represented as the union of no more than countably many sets of the form E_ρ (why?). Let $\mu(E_\rho) = t$. Then, given any $\varepsilon > 0$, there is an open set G, equal to the union of no more than countably many open intervals (a_k, b_k) such that $E_\rho \subset G$ and

$$\sum_k (b_k - a_k) < t + \varepsilon$$

---

[5] Note that Λ_R cannot equal $-\infty$, since the difference quotient (3) is inherently nonnegative if f is nondecreasing.

(this follows from the very definition of Lebesgue measure on the line).
If

$$t_k = \mu[E_\rho \cap (a_k, b_k)],$$

then

$$t = \sum_k t_k$$

But $t_k \leqslant \rho(b_k - a_k)$, by Lemma 2. Hence

$$t \leqslant \rho \sum_k (b_k - a_k) < \rho(t + \varepsilon),$$

which implies $t \leqslant \rho t$, since $\varepsilon > 0$ is arbitrary. This in turn implies $t = 0$, since $0 < \rho < 1$. Therefore $\lambda_L \geqslant \Lambda_R$ almost everywhere, as asserted.

Finally, to drop the requirement that f be continuous, we need only generalize Lemmas 1 and 1' in the way indicated in Problem 6, noting that the proof continues to go through (check details).[6] ∎

Remark. Despite its apparent complexity, the proof of Theorem 6 is based on simple intuitive ideas. For example, the finiteness of Λ_R (and Λ_L) almost everywhere is easily made plausible. In fact, let f be continuous and nondecreasing on $[a, b]$. Then f maps $[a, b]$ into the interval $[f(a), f(b)]$, at the same time subjecting a small interval $[x, \xi]$ at x to a "magnification" approximately equal to

$$\gamma(x) = \frac{f(\xi) - f(x)}{\xi - x}.$$

But the interval $[f(a), f(b)]$ is finite, and hence $\gamma(x)$ cannot be infinite on a set of positive measure. As for the part of the proof based on Lemma 2, it merely says that if the intersection of a subset $A \subseteq [a, b]$ with *every* interval (α, β) has measure no greater than $\rho(\beta - \alpha)$ for some fixed number $\rho < 1$, then A cannot have positive measure.

31.3. Differentiation of an integral with respect to its upper limit. Returning to the problem of differentiating the indefinite Lebesgue integral, we have

THEOREM 7. *Let f be any function summable on $[a, b]$. Then*

$$\frac{d}{dx} \int_a^x f(t) \, dt \tag{12}$$

exists and is finite for almost all x.

Proof. As noted at the beginning of Sec. 31.1

$$f(t) = f_+(t) - f_-(t),$$

---

[6] For an alternative proof, see Problems 7–9.

where f_+ and f_- are nonnegative summable functions, so that

$$F(x) = \int_a^x f(t)\, dt = \int_a^x f_+(t)\, dt - \int_a^x f_-(t)\, dt = F_1(x) - F_2(x)$$

is the difference between two nondecreasing functions F_1 and F_2. But F_1 and F_2 have finite derivatives almost everywhere, by Theorem 6, and hence so does F. ∎

We now *evaluate* the derivative (12), thereby giving an affirmative answer to the first of the two questions posed on p. 314:

THEOREM 8. *Let f be any function summable on* $[a, b]$. *Then*

$$\frac{d}{dx}\int_a^x f(t)\, dt = f(x)$$

almost everywhere.

Proof. Let

$$F(x) = \int_a^x f(t)\, dt.$$

Then it will be enough to show that

$$f(x) \geqslant F'(x) \tag{13}$$

almost everywhere for any summable function. In fact, changing $f(x)$ to $-f(x)$ in (13), we get

$$-f(x) \geqslant -F'(x)$$

and hence

$$f(x) \leqslant F'(x). \tag{14}$$

But (13) and (14) together imply the desired result

$$f(x) = F'(x) = \frac{d}{dx}\int_a^x f(t)\, dt$$

(almost everywhere).

To prove (13), we observe that if

$$f(x) < F'(x),$$

then there are rational numbers α and β such that

$$f(x) < \alpha < \beta < F'(x). \tag{15}$$

Let $E_{\alpha\beta}$ be the set of all x satisfying (15). Then, as we now show, $\mu(E_{\alpha\beta}) = 0$. Since the number of sets $E_{\alpha\beta}$ is countable, this will imply

$$\mu\{x : f(x) < F'(x)\} = 0$$

and hence that (13) holds almost everywhere.

To prove that $\mu(E_{\alpha\beta}) = 0$, we first note that, given any $\varepsilon > 0$, there is a $\delta > 0$ such that $\mu(E) < \delta$ implies

$$\left| \int_E f(t)\, dt \right| < \varepsilon$$

(the existence of such a number δ follows from the absolute continuity of the Lebesgue integral, proved in Theorem 6, p. 300).[7] Let $G \subset [a, b]$ be an open set, made up of no more than countably many pairwise disjoint open intervals (a_k, b_k), such that

$$E_{\alpha\beta} \subset G, \qquad \mu(G) < \mu(E_{\alpha\beta}) + \delta,$$

and let x_0 be any point in $G_k = E_{\alpha\beta} \cap (a_k, b_k)$. Then

$$\frac{F(\xi) - F(x_0)}{\xi - x_0} > \beta \qquad (16)$$

for any point $\xi > x_0$ sufficiently close to x_0. Writing (16) in the form

$$F(\xi) - \beta\xi > F(x_0) - \beta x_0,$$

we see that the point x_0 is invisible from the right with respect to the continuous function $F(x) - \beta x$. It follows from Lemma 1 that G_k is the union of no more than countably many pairwise disjoint open intervals (a_{k_n}, b_{k_n}), where

$$F(a_{k_n}) - \beta a_{k_n} \leqslant F(b_{k_n}) + \beta b_{k_n},$$

i.e.,

$$F(b_{k_n}) - F(a_{k_n}) \geqslant \beta(b_{k_n} - a_{k_n}),$$

or equivalently

$$\int_{a_{k_n}}^{b_{k_n}} f(t)\, dt \geqslant \beta(b_{k_n} - a_{k_n}). \qquad (17)$$

If

$$S = \bigcup_{k,n} (a_{k_n}, b_{k_n}),$$

then clearly

$$E_{\alpha\beta} \subset S \subset G, \qquad \mu(S) < \mu(E_{\alpha\beta}) + \delta.$$

Summing (17) over all the intervals (a_{k_n}, b_{k_n}), we get

$$\int_S f(t)\, dt = \sum_{k,n} \int_{a_{k_n}}^{b_{k_n}} f(t)\, dt \geqslant \beta \sum_{k,n} (b_{k_n} - a_{k_n}) = \beta\mu(S).$$

---

[7] In particular, $F(x)$ is continuous. In fact,

$$|F(x') - F(x)| = \left| \int_x^{x'} f(t)\, dt \right| < \varepsilon$$

if $|x' - x| < \delta$.

On the other hand,

$$\int_S f(t)\,dt = \int_{E_{\alpha\beta}} f(t)\,dt + \int_{S-E_{\alpha\beta}} f(t)\,dt$$

$$< \alpha\mu(E_{\alpha\beta}) + \varepsilon < \alpha\mu(S) + |\alpha|\,\delta + \varepsilon. \tag{18}$$

Comparing (17) and (18), we get

$$\alpha\mu(S) + |\alpha|\,\delta + \varepsilon > \beta\mu(S)$$

or

$$\mu(S) < \frac{|\alpha|\,\delta + \varepsilon}{\beta - \alpha}.$$

Therefore $E_{\alpha\beta}$ is contained in an open set of arbitrarily small measure (it can be assumed that $|\alpha|\,\delta < \varepsilon$). It follows that $\mu(E_{\alpha\beta}) = 0$. ∎

Problem 1. Let $x_1, x_2, \ldots, x_n, \ldots$ be the set of all rational points in $[a, b]$, enumerated in any way, and let $h_n = 1/2^n$. Prove that the jump function

$$f(x) = \sum_{x_n < x} h_n$$

is discontinuous at every rational point and continuous at every irrational point.

Problem 2. Suppose we define a jump function by the formula

$$f(x) = \sum_{x_n \leqslant x} h_n, \tag{19}$$

rather than by the formula (2). Prove that f is continuous from the right, rather than from the left as in Theorem 4.

Problem 3. Find the derived numbers of the function

$$f(x) = \begin{cases} x \sin \dfrac{1}{x} & \text{if } x > 0, \\ 0 & \text{if } x \leqslant 0 \end{cases}$$

at the point $x = 0$.

Problem 4. Find the points invisible from the left in Figure 18, p. 319.

Problem 5. In Lemma 1, show that $f(a_k) = f(b_k)$ if $a_k \neq a$.

Problem 6. Prove that the requirement that f be continuous on $[a, b]$ can be dropped in Lemma 1, provided that

1) The discontinuity points of f are all of the first kind;
2) A point $x_0 \in [a, b]$ is said to be invisible from the right (with respect to f) if there is a point ξ such that $x_0 < \xi < b$ and

$$\max \{f(x_0 - 0), f(x_0), f(x_0 + 0)\} < f(\xi);$$

3) The inequality (5) is replaced by

$$f(a_k + 0) \leqslant \max \{f(b_k - 0), f(b_k), f(b_k + 0)\}.$$

State and prove the corresponding generalization of Lemma 1'.

Problem 7. Let

$$\sum_{n=0}^{\infty} \varphi_n(x) = f(x) \tag{20}$$

be an everywhere convergent series, whose general term $\varphi_n(x)$ is nondecreasing (alternatively, nonincreasing) on $[a, b]$. Prove that (20) can be differentiated term by term almost everywhere, i.e., that

$$\sum_{n=0}^{\infty} \varphi'_n(x) = f'(x)$$

almost everywhere.

Problem 8. Prove that every jump function has a zero derivative almost everywhere.
Hint. Use Problem 7.

Problem 9. Prove that the assumption that f be continuous from the left in Theorem 5 can be dropped if we define a jump function as a sum of a "left jump function" like (2) and a "right jump function" like (19). Use this fact and Problem 8 to complete the proof of Theorem 6 without recourse to Problem 6.
Hint. Use Problem 8 and Theorem 5.

Problem 10. Following van der Waerden, let

$$\varphi_0(x) = \begin{cases} x & \text{if } 0 \leqslant x \leqslant \frac{1}{2}, \\ 1 - x & \text{if } \frac{1}{2} \leqslant x \leqslant 1, \end{cases}$$

and continue φ_0 by periodicity, with period 1, over the whole x-axis. Then let

$$\varphi_n(x) = \frac{1}{4^n} \varphi_0(4^n x) \qquad (n = 1, 2, \ldots),$$

$$f(x) = \sum_{n=0}^{\infty} \varphi_n(x)$$

Prove that
a) The function f is continuous everywhere;
b) The derivative of f fails to exist at every point $x_0 \in (-\infty, \infty)$.

Hint. Consider the increments

$$\frac{f\left(x_0 \pm \dfrac{1}{4^n}\right) - f(x_0)}{\pm \dfrac{1}{4^n}}.$$

32. Functions of Bounded Variation

The problem of differentiating a Lebesgue integral with respect to its upper limit has led us to consider functions that can be represented as differences between two monotonic functions. We now give a different description of such functions (independent of the notion of monotonicity), afterwards studying some of their properties.

DEFINITION 1. *A function f defined on an interval $[a, b]$ is said to be of bounded variation if there is a constant $C > 0$ such that*

$$\sum_{k=1}^{n} |f(x_k) - f(x_{k-1})| \leqslant C \tag{1}$$

for every partition

$$a = x_0 < x_1 < \cdots < x_n = b \tag{2}$$

of $[a, b]$ by points of subdivision $x_0, x_1, \ldots, x_n$.

Example. Every monotonic function is of bounded variation, since the left-hand side of (1) equals $|f(b) - f(a)|$ regardless of the choice of partition.

DEFINITION 2. *Let f be a function of bounded variation. Then by the total variation of f on $[a, b]$, denoted by $V_a^b(f)$, is meant the quantity*

$$V_a^b(f) \equiv \sup \sum_{k=1}^{n} |f(x_k) - f(x_{k-1})|, \tag{3}$$

where the least upper bound is taken over all (finite) partitions (2) of the interval $[a, b]$.

Remark 1. A function f defined on the whole real line $(-\infty, \infty)$ is said to be of bounded variation if there is a constant $C > 0$ such that

$$V_a^b(f) \leqslant C$$

for every pair of real numbers a and b $(a < b)$. The quantity

$$\lim_{\substack{a \to -\infty \\ b \to \infty}} V_a^b(f)$$

is then called the total variation of f on $(-\infty, \infty)$, denoted by $V_{-\infty}^{\infty}(f)$.

Remark 2. It is an immediate consequence of (3) that

$$V_a^b(\alpha f) = |\alpha| \, V_a^b(f) \tag{4}$$

for any constant α.

THEOREM 1. *If f and g are functions of bounded variation on $[a, b]$, then so is $f + g$ and*

$$V_a^b(f + g) \leqslant V_a^b(f) + V_a^b(g). \tag{5}$$

Proof. For any partition of the interval $[a, b]$, we have

$$\sum_k |f(x_k) + g(x_k) - f(x_{k-1}) - g(x_{k-1})|$$
$$\leqslant \sum_k |f(x_k) - f(x_{k-1})| + \sum_k |g(x_k) - g(x_{k-1})|.$$

Taking the least upper bound of both sides over all partitions of $[a, b]$, and noting that

$$\sup \{x + y : x \in A, y \in B\} \leqslant \sup \{x : x \in A\} + \sup \{y : y \in B\},$$

we immediately get (5). ∎

It follows from (4) and (5) that any linear combination of functions of bounded variation is itself a function of bounded variation. In other words, the set of all functions of bounded variation on a given interval is a linear space (unlike the set of all monotonic functions).

THEOREM 2. *If $a < b < c$, then*

$$V_a^c(f) = V_a^b(f) + V_b^c(f). \tag{6}$$

Proof. First we consider a partition of the interval $[a, c]$ such that b is one of the points of subdivision, say $x_r = b$. Then

$$\sum_{k=1}^n |f(x_k) - f(x_{k-1})|$$
$$= \sum_{k=1}^r |f(x_k) - f(x_{k-1})| + \sum_{k=r+1}^n |f(x_k) - f(x_{k-1})| \leqslant V_a^b(f) + V_b^c(f). \tag{7}$$

Now consider an *arbitrary* partition of $[a, c]$. It is clear that adding an extra point of subdivision to this partition can never decrease the sum

$$\sum_{k=1}^n |f(x_k) - f(x_{k-1})|.$$

Therefore (7) holds for *any* subdivision of $[a, c]$, and hence

$$V_a^b(f) \leqslant V_a^b(f) + V_b^c(f). \tag{8}$$

On the other hand, given any $\varepsilon > 0$, there are partitions of the intervals $[a, b]$ and $[b, c]$, respectively, such that

$$\sum_i |f(x_i') - f(x_{i-1}')| > V_a^b(f) - \frac{\varepsilon}{2},$$

$$\sum_j |f(x_j'') - f(x_{j-1}'')| > V_b^c(f) - \frac{\varepsilon}{2}.$$

Combining all points of subdivision x'_i, x''_k, we get a partition of the interval $[a, c]$, with points of subdivision x_k, such that

$$\sum_k |f(x_k) - f(x_{k-1})| = \sum_i |f(x'_i) - f(x'_{i-1})| + \sum_j |f(x''_j) - f(x''_{j-1})|$$
$$> V_a^b(f) + V_b^c(f) - \varepsilon.$$

Since $\varepsilon > 0$ is arbitrary, it follows that

$$V_a^c(f) \geqslant V_a^b(f) + V_b^c(f). \tag{9}$$

Comparing (8) and (9), we get (6). ∎

COROLLARY. *The function*

$$v(x) = V_a^x(f) \tag{10}$$

is nondecreasing.

Proof. An immediate consequence of (6), since the total variation of any function of bounded variation on any interval is nonnegative. ∎

THEOREM 3. *Let f be a function of bounded variation on* $[a, b]$, *and let v be the function* (10). *Then if f is continuous from the left at a point* x^*, *so is v.*

Proof. Given any $\varepsilon > 0$, use the fact that f is continuous from the left to choose a $\delta > 0$ such that

$$|f(x^*) - f(x)| < \frac{\varepsilon}{2} \tag{11}$$

whenever $x^* - x < \delta$. Then choose a partition

$$a = x_0 < x_1 < \cdots < x_n = x^*$$

such that

$$V_a^{x^*}(f) - \sum_{k=1}^n |f(x_k) - f(x_{k-1})| < \frac{\varepsilon}{2}. \tag{12}$$

Here it can be assumed that

$$x^* - x_{n-1} < \delta,$$

since otherwise we need only add an extra point of subdivision which can never increase the left-hand side of (12). It follows from (11) and (12) that

$$V_a^{x^*}(f) - \sum_{k=1}^{n-1} |f(x_k) - f(x_{k-1})| < \varepsilon,$$

and hence

$$V_a^{x^*}(f) - V_a^{x_{n-1}}(f) < \varepsilon$$

a fortiori, i.e.,

$$v(x^*) - v(x_{n-1}) < \varepsilon.$$

But then, since v is nondecreasing,

$$v(x^*) - v(x) < \varepsilon$$

for all x such that $x_{n-1} \leqslant x \leqslant x^*$. In other words, v is continuous from the left at x^*, ∎

Remark. Virtually the same argument shows that if f is continuous from the right at x^*, then so is v. Together with Theorem 3, this shows that if f is continuous at x^*, or on the whole interval $[a, b]$, then so is v.

THEOREM 4. *If f is of bounded variation on $[a, b]$, then f can be represented as the difference between two nondecreasing functions on $[a, b]$.*

Proof. Let

$$v(x) = V_a^x(f),$$

and consider the function

$$g = v - f.$$

Then g is nondecreasing. In fact, if $x' \leqslant x''$, then

$$g(x'') - g(x') = [v(x'') - v(x')] - [f(x'') - f(x')]. \tag{13}$$

But

$$|f(x'') - f(x')| \leqslant v(x'') - v(x'),$$

by the very definition of v, and hence the right-hand side of (13) is nonnegative. Writing

$$f = v - g,$$

we get the desired representation of f as the difference between two nondecreasing functions. ∎

COROLLARY 1. *Every function of bounded variation has a finite derivative almost everywhere.*

Proof. An immediate consequence of Theorem 6, p. 321. ∎

COROLLARY 2. *If f is summable on $[a, b]$, then the indefinite integral*

$$\Phi(x) = \int_a^x f(t)\, dt$$

is a function of bounded variation on $[a, b]$.

Proof. Recall the remarks at the beginning of Sec. 9.1. ∎

Problem 1. Prove that $V_b^a(f) = 0$ if and only if $f(x) \equiv$ const on $[a, b]$.

Problem 2. Prove that the function

$$f(x) = \begin{cases} x^\alpha \sin \dfrac{1}{x^\beta} & \text{if } 0 < x \leqslant 1, \\ 0 & \text{if } x = 0 \end{cases}$$

is of bounded variation on $[0, 1]$ if $\alpha > \beta$ but not if $\alpha \leqslant \beta$.

Problem 3. Suppose f has a bounded derivative on $[a, b]$, so that $f'(x)$ exists and satisfies an inequality $|f'(x)| \leqslant C$ at every point $x \in [a, b]$. Prove that f is of bounded variation and

$$V_a^b(f) \leqslant C(b - a).$$

Problem 4. Prove that if f and g are functions of bounded variation on $[a, b]$, then so is fg and

$$V_a^b(fg) \leqslant V_a^b(f) \sup_x |g(x)| + V_a^b(g) \sup_x |f(x)|.$$

Problem 5. Let f be a function of bounded variation on $[a, b]$ such that

$$f(x) \geqslant c > 0.$$

Prove that $1/f$ is also a function of bounded variation and

$$V_a^b\left(\frac{1}{f}\right) \leqslant \frac{1}{c^2} V_a^b(f).$$

Problem 6. Prove the converse of Theorem 4.

Problem 7. Prove that a curve

$$y = f(x) \qquad (a \leqslant x \leqslant b)$$

is *rectifiable*, i.e., has finite length, as defined in Problem 3, p. 114, if and only if f is of bounded variation on $[a, b]$.

Problem 8. Let f be a function of bounded variation on $[a, b]$. Prove that

$$\|f\| = V_a^b(f)$$

has all the properties of a norm (cf. p. 138) if we impose the extra condition $f(a) = 0$.

Comment. Thus the space $V_{[a,b]}^0$ of all functions of bounded variation on $[a, b]$ equipped with this norm and vanishing at $x = a$ is a normed linear space (addition of functions and multiplication of functions by numbers being defined in the usual way).

Problem 9. Prove that the space $V_{[a,b]}^0$ defined in the preceding comment is complete.

Problem 10. Does there exist a continuous function which is not of bounded variation on any interval?

Hint. Recall Problem 10, p. 327 and Corollary 1 above.

33. Reconstruction of a Function from Its Derivative

33.1. Statement of the problem. We now address ourselves to the second of the problems posed on p. 314, i.e., we look for the largest class of functions F such that

$$\int_a^x F'(t)\, dt = F(x) - F(a), \tag{1}$$

or equivalently

$$F(x) = F(a) + \int_a^x F'(t)\, dt. \tag{2}$$

(As we know from calculus, these formulas hold if F is continuously differentiable.) From the outset, we must restrict ourselves to functions F which are differentiable (i.e., have a finite derivative) almost everywhere, since otherwise (2) would be meaningless. Every function of bounded variation has this property (see Corollary 1, p. 331). Moreover, the right-hand side of (2) is a function of bounded variation (see Corollary 2, p. 331). It follows that the largest class of functions satisfying (2) must be some subset of the class of functions of bounded variation. Since every function of bounded variation is the difference between two nondecreasing functions (see Theorem 4, p. 331), we begin by studying nondecreasing functions from the standpoint of formula (1).

THEOREM 1. *Let F be a nondecreasing function on $[a, b]$. Then the derivative F' is summable on $[a, b]$ and*

$$\int_a^b F'(t)\, dt \leqslant F(b) - F(a). \tag{3}$$

Proof. Let

$$\Phi_n(t) = n\left[F\left(t + \frac{1}{n}\right) - F(t)\right] \qquad (n = 1, 2, \ldots),$$

where, to make $\Phi_n(t)$ meaningful for all $t \in [a, b]$, we get $F(t) = F(b)$ for $b < t \leqslant b + 1$, by definition.[8] Clearly

$$F'(t) = \lim_{n \to \infty} \frac{F\left(t + \dfrac{1}{n}\right) - F(t)}{\dfrac{1}{n}} = \lim_{n \to \infty} \Phi_n(t)$$

almost everywhere on $[a, b]$. Since F is summable on $[a, b]$, by Theorem

---

[8] Verify that this does not affect the validity of the proof.

1, p. 316, so is every Φ_n. Integrating Φ_n, we get

$$\int_a^b \Phi_n(t)\,dt = n\int_a^b\left[F\left(t+\frac{1}{n}\right) - F(t)\right]dt = n\left[\int_{a+(1/n)}^{b+(1/n)}F(t)\,dt - \int_a^b F(t)\,dt\right]$$

$$= n\left[\int_b^{b+(1/n)}F(t)\,dt - \int_a^{a+(1/n)}F(t)\,dt\right] \leqslant F(b) - F(a),$$

where in the last step we use the fact that F is nondecreasing. The summability of F' and the inequality (3) now follow at once from Fatou's theorem (Theorem 3, p. 307). $\blacksquare$

Example 1. It is easy to find nondecreasing functions F for which (3) becomes a strict inequality, i.e., such that

$$\int_a^b F'(t)\,dt < F(b) - F(a). \tag{4}$$

For example, let

$$F(t) = \begin{cases} 0 & \text{if } 0 \leqslant t < \frac{1}{2}, \\ 1 & \text{if } \frac{1}{2} \leqslant t \leqslant 1. \end{cases}$$

Then

$$0 = \int_0^1 F'(t)\,dt < F(1) - F(0) = 1.$$

Example 2 (The Cantor function). In the preceding example, F is discontinuous. However, it is also possible to find *continuous* nondecreasing functions satisfying the strict inequality (4). To this end, let

$$[a_1^{(1)}, b_1^{(1)}] = [\tfrac{1}{3}, \tfrac{2}{3}]$$

be the middle third of the interval $[0, 1]$, let

$$[a_1^{(2)}, b_1^{(2)}] = [\tfrac{1}{9}, \tfrac{2}{9}], \qquad [a_2^{(2)}, b_2^{(2)}] = [\tfrac{7}{9}, \tfrac{8}{9}]$$

be the middle thirds of the intervals remaining after deleting $[a_1^{(1)}, b_1^{(1)}]$ from $[0, 1]$, let

$$[a_1^{(2)}, b_1^{(2)}] = [\tfrac{2}{27}, \tfrac{3}{27}], \qquad [a_2^{(2)}, b_2^{(2)}] = [\tfrac{7}{27}, \tfrac{8}{27}],$$

$$[a_3^{(2)}, b_3^{(2)}] = [\tfrac{19}{27}, \tfrac{20}{27}], \qquad [a_4^{(2)}, b_4^{(2)}] = [\tfrac{25}{27}, \tfrac{26}{27}]$$

be the middle thirds of the intervals remaining after deleting $[a_1^{(1)}, b_1^{(1)}]$, $[a_1^{(2)}, b_1^{(2)}]$ and $[a_2^{(2)}b_2^{(2)}]$ from $[0, 1]$, and so on, with

$$[a_1^{(n)}, b_1^{(n)}], \ldots, [a_k^{(n)}, b_k^{(n)}], \ldots, [a_{2^{n-1}}^{(n)}, b_{2^{n-1}}^{(n)}]$$

being the 2^{n-1} intervals deleted at the nth stage. Note that the complement of union of all the intervals $[a_k^{(n)}, b_k^{(n)}]$ · is the set of all "points of the second kind" of the Cantor set constructed in Example 4, p. 52, i.e., all points of the

Cantor set except the end points

$$0, 1, \tfrac{1}{3}, \tfrac{2}{3}, \tfrac{1}{9}, \tfrac{2}{9}, \tfrac{7}{9}, \tfrac{8}{9}, \ldots \tag{5}$$

of the deleted intervals (together with the points 0 and 1).

Now define a function

$$F(t) = \frac{2k-1}{2^n} \quad \text{if} \quad t \in [a_k^{(n)}, b_k^{(n)}],$$

so that

$$F(t) = \tfrac{1}{2} \quad \text{if} \quad \tfrac{1}{3} \leqslant t \leqslant \tfrac{2}{3},$$

$$F(t) = \begin{cases} \tfrac{1}{4} & \text{if} \quad \tfrac{1}{9} \leqslant t \leqslant \tfrac{2}{9}, \\ \tfrac{3}{4} & \text{if} \quad \tfrac{7}{9} \leqslant t \leqslant \tfrac{8}{9}, \end{cases}$$

$$F(t) = \begin{cases} \tfrac{1}{8} & \text{if} \quad \tfrac{2}{27} \leqslant t \leqslant \tfrac{3}{27}, \\ \tfrac{3}{8} & \text{if} \quad \tfrac{7}{27} \leqslant t \leqslant \tfrac{8}{27}, \\ \tfrac{5}{8} & \text{if} \quad \tfrac{19}{27} \leqslant t \leqslant \tfrac{20}{27}, \\ \tfrac{7}{8} & \text{if} \quad \tfrac{25}{27} \leqslant t \leqslant \tfrac{26}{27}, \end{cases}$$

and so on, as shown schematically in Figure 19. Then F is defined everywhere on $[0, 1]$ except at points of the second kind of the Cantor set. Given any such point t^*, let $\{t_n\}$ be an increasing sequence of points of the type (5) converging to t^*, and let $\{t_n'\}$ be a decreasing sequence of points of the same type converging to t^* (why do such sequences exist?). Then let

$$F(t^*) = \lim_{n \to \infty} F(t_n) = \lim_{n \to \infty} F(t_n')$$

(justify the equality of the limits). Completing the definition of F in this way, we obtain a *continuous* nondecreasing function on the whole interval $[0, 1]$, known as the *Cantor function*. (Fill in some missing details.) The derivative F' obviously vanishes at every interior point of the intervals $[a_k^{(n)}, b_k^{(n)}]$, and

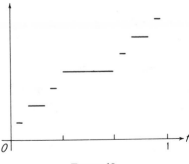

FIGURE 19

hence vanishes almost everywhere, since the sum of the lengths of these intervals equals

$$\tfrac{1}{3} + \tfrac{2}{9} + \tfrac{4}{27} + \cdots = 1$$

(the Cantor set is of measure zero). It follows that

$$0 = \int_0^1 F'(t)\,dt < F(1) - F(0) = 1.$$

33.2. Absolutely continuous functions. We have just given examples of functions for which formula (1) does not hold. To describe the class of functions satisfying (1), or equivalently (2), we will need the following

DEFINITION. *A function f defined on an interval $[a, b]$ is said to be* **absolutely continuous** *on $[a, b]$ if, given any $\varepsilon > 0$, there is a $\delta > 0$ such that*

$$\sum_{k=1}^n |f(b_k) - f(a_k)| < \varepsilon$$

for every finite system of pairwise disjoint subintervals

$$(a_k, b_k) \subset [a, b] \qquad (k = 1, \ldots, n)$$

of total length

$$\sum_{k=1}^n (b_k - a_k)$$

less than δ.

Remark 1. Clearly every absolutely continuous function is uniformly continuous, as we see by choosing a single subinterval $(a_1, b_1) \subset [a, b]$. However, a uniformly continuous function need not be absolutely continuous. For example, the Cantor function F constructed in Example 2 of the preceding section is continuous (and hence uniformly continuous) on $[0, 1]$, but not absolutely continuous on $[0, 1]$. In fact, the Cantor set can be covered by a *finite* system of subintervals (a_k, b_k) of arbitrarily small total length (why?). But obviously

$$\sum_{k=1}^n |F(b_k) - F(a_k)| = 1$$

for every such system. The same example shows that a function of bounded variation need not be absolutely continuous. On the other hand, an absolutely continuous function is necessarily of bounded variation (see Theorem 2).

Remark 2. In the definition, we can change "finite" to "finite or countable." In fact, suppose that given any $\varepsilon > 0$, there is a $\delta > 0$ such that

$$\sum_{k=1}^n |f(b_k) - f(a_k)| < \varepsilon' < \varepsilon$$

for every finite system of pairwise disjoint intervals $(a_k, b_k) \subset [a, b]$ of total length less than δ, and consider any *countable* system of pairwise disjoint intervals $(\alpha_k, \beta_k) \subset [a, b]$ of total length less than δ. Then obviously

$$\sum_{k=1}^{n} |f(\beta_k) - f(\alpha_k)| < \varepsilon'$$

for every n. Hence, taking the limit as $n \to \infty$, we get

$$\sum_{k=1}^{\infty} |f(\beta_k) - f(\alpha_k)| \leqslant \varepsilon' < \varepsilon.$$

THEOREM 2. *If f is absolutely continuous on $[a, b]$, then f is of bounded variation on $[a, b]$.*

Proof. Given any $\varepsilon > 0$, there is a $\delta > 0$ such that

$$\sum_{k=1}^{n} |f(b_k) - f(a_k)| < \varepsilon$$

for every system of pairwise disjoint intervals $(a_k, b_k) \subset [a, b]$ such that

$$\sum_{k=1}^{n} (b_k - a_k) < \delta.$$

Hence if $[\alpha, \beta]$ is any interval of length less than δ, we have

$$V_\alpha^\beta(f) \leqslant \varepsilon.$$

Let

$$a = x_0 < x_1 < \cdots < x_N = b$$

be a partition of $[a, b]$ into N subintervals $[x_{k-1}, x_k]$ all of length less than δ. Then, by Theorem 2, p. 329,

$$V_a^b(f) \leqslant N\varepsilon < \infty. \quad \blacksquare$$

THEOREM 3. *If f is absolutely continuous on $[a, b]$, then so is αf, where α is any constant. Moreover, if f and g are absolutely continuous on $[a, b]$, then so is $f + g$.*

Proof. An immediate consequence of the definition of absolute continuity and obvious properties of the absolute value. $\blacksquare$

It follows from Theorems 2 and 3 (together with Remark 1) that the set of all absolutely continuous functions on $[a, b]$ is a proper subspace of the linear space of all functions of bounded variation on $[a, b]$.

THEOREM 4. *If f is absolutely continuous on $[a, b]$, then f can be represented as the difference between two absolutely continuous nondecreasing functions on $[a, b]$.*

Proof. By Theorem 2, f is of bounded variation on $[a, b]$, and hence can be represented in the form

$$f = v - g,$$

where

$$v(x) = V_a^x(f), \qquad g = v - f$$

are the same nondecreasing functions as in Theorem 4, p. 331. We now verify that v and g are absolutely continuous. Given any $\varepsilon > 0$, let $\delta > 0$ be such that

$$\sum_{k=1}^{n} |f(b_k) - f(a_k)| < \varepsilon' < \varepsilon$$

for every finite system of pairwise disjoint subintervals $(a_k, b_k) \subset [a, b]$ of total length less than δ. Consider the sum

$$\sum_{k=1}^{n} |v(b_k) - v(a_k)| = \sum_{k=1}^{n} [v(b_k) - v(a_k)],$$

equal to the least upper bound of the sums

$$\sum_{k=1}^{n} \sum_{l=1}^{m_k} |f(x_{k,l}) - f(x_{k,l-1})| \tag{6}$$

taken over all possible finite partitions

$$a_1 = x_{1,0} < x_{1,1} < \cdots < x_{1,m_1} = b_1,$$
$$\cdots\cdots\cdots\cdots\cdots\cdots\cdots\cdots$$
$$a_k = x_{k,0} < x_{k,1} < \cdots < x_{k,m_k} = b_k,$$
$$\cdots\cdots\cdots\cdots\cdots\cdots\cdots\cdots$$
$$a_n = x_{n,0} < x_{n,1} < \cdots < x_{n,m_n} = b_n$$

of the intervals $(a_1, b_1), \ldots, (a_n, b_n)$. The total length of all the intervals $(x_{k,l-1}, x_{k,l})$ figuring in (6) is clearly less than δ, and hence the sum (6) is less than ε', by the absolute continuity of f. Therefore

$$\sum_{k=1}^{n} |v(b_k) - v(a_k)| \leqslant \varepsilon' < \varepsilon,$$

i.e., v is absolutely continuous on $[a, b]$. It follows from Theorem 3 that $g = v - f$ is also absolutely continuous on $[a, b]$. ∎

We now study the close connection between absolute continuity and the indefinite Lebesgue integral:

THEOREM 5. *The indefinite integral*

$$F(x) = \int_a^x f(t)\, dt$$

of a summable function f is absolutely continuous.

Proof. Given any finite collection of pairwise disjoint intervals (a_k, b_k), we have

$$\sum_{k=1}^{n} |F(b_k) - F(a_k)| = \sum_{k=1}^{n} \left| \int_{a_k}^{b_k} f(t) \, dt \right| \leqslant \sum_{k=1}^{n} \int_{a_k}^{b_k} |f(t)| \, dt = \int_{\bigcup_k (a_k, b_k)} |f(t)| \, dt.$$

But the last expression on the right approaches zero as the total length of the intervals (a_k, b_k) approaches zero, by the absolute continuity of the Lebesgue integral (Theorem 6, p. 300). ∎

LEMMA. *Let f be an absolutely continuous nondecreasing function on $[a, b]$ such that $f'(x) = 0$ almost everywhere. Then $f(x) = $ const.*

Proof. Since f is continuous and nondecreasing, its range is the closed interval $[f(a), f(b)]$. We will show that the length of this interval is zero if $f'(x) = 0$ almost everywhere, thereby proving the lemma. Let E be the set of points $x \in [a, b]$ such that $f'(x) = 0$, and let $Z = [a, b] - E$, where $\mu(Z) = 0$, by hypothesis. Given any $\varepsilon > 0$, we find $\delta > 0$ such that

$$\sum_k |f(b_k) - f(a_k)| < \varepsilon \tag{7}$$

for any finite or countable system of pairwise disjoint intervals $(a_k, b_k) \subset [a, b]$ of length less than δ (recall Remark 2, p. 336), and then cover Z by an open set of measure less than δ (this is possible, since Z is of measure zero). In other words, we cover Z by a finite or countable system of intervals (a_k, b_k) of total length less than δ. It then follows from (7) that the whole system of intervals, and hence (a fortiori) the set

$$Z \subset \bigcup_k (a_k, b_k),$$

is mapped into a set of measure less than ε. But then $\mu[f(Z)] = 0$, since $\varepsilon > 0$ is arbitrary.

Next consider the set $E = [a, b] - Z$, and let $x_0 \in E$. Then, since $f'(x_0) = 0$, we have

$$\frac{f(x) - f(x_0)}{x - x_0} < \varepsilon$$

for all $x > x_0$ sufficiently near x_0, i.e.,

$$f(x) - f(x_0) < \varepsilon(x - x_0)$$

or

$$\varepsilon x_0 - f(x_0) < \varepsilon x - f(x).$$

Therefore the point x_0 is invisible from the right with respect to the function $\varepsilon x - f(x)$. It follows from Lemma 1, p. 319 that E is the union of no more than countably many pairwise disjoint intervals (α_k, β_k), with end points satisfying the inequalities

$$\varepsilon\alpha_k - f(\alpha_k) \leqslant \varepsilon\beta_k - f(\beta_k)$$

or

$$f(\beta_k) - f(\alpha_k) \leqslant \varepsilon(\beta_k - \alpha_k).$$

But then

$$\sum_k [f(\beta_k) - f(\alpha_k)] \leqslant \varepsilon \sum_k (\beta_k - \alpha_k) \leqslant \varepsilon(b - a).$$

In other words, f maps E into a set covered by a system of intervals of total length less than $\varepsilon(b - a)$. Therefore $\mu[f(E)] = 0$, since $\varepsilon > 0$ is arbitrary.

We have just shown that the sets $f(Z)$ and $f(E)$ are both of measure zero. But the interval $[f(a), f(b)]$ is the union of $f(Z)$ and $f(E)$. It follows that $[f(a), f(b)]$ is of length zero, i.e., that $f(x) \equiv$ const. ∎

We are now in a position to prove

THEOREM 6 (*Lebesgue*). *If F is absolutely continuous on* $[a, b]$, *then the derivative* F' *is summable on* $[a, b]$ *and*

$$F(x) = F(a) + \int_a^x F'(t)\, dt. \tag{8}$$

Proof. We need only consider the case of nondecreasing F (why?). Then F' is summable, by Theorem 1, and the function

$$\Phi(x) = F(x) - \int_a^x F'(t)\, dt \tag{9}$$

is also nondecreasing. In fact, if $x'' > x'$, then

$$\Phi(x'') - \Phi(x') = F(x'') - F(x') - \int_{x'}^{x''} F'(t)\, dt \geqslant 0,$$

where we again use Theorem 1. Moreover, Φ is absolutely continuous, being the difference between two absolutely continuous functions (recall Theorems 3 and 5), and $\Phi'(x) = 0$ almost everywhere, by Theorem 8, p. 324. It follows from the lemma that $\Phi(x) \equiv$ const. Setting $x = a$, we find that this constant equals $F(a)$. Replacing $\Phi(x)$ by $F(a)$ in (9), we get (8). ∎

Remark. Combining Theorems 5 and 6, we can now give a definitive answer to the second of the questions posed on p. 314 (see also p. 333): *The formula*

$$\int_a^x F'(t)\, dt = F(x) - F(a),$$

or equivalently,

$$F(x) = F(a) + \int_a^x F'(t)\, dt,$$

holds for all $x \in [a, b]$ *if and only if F is absolutely continuous on* $[a, b]$.

33.3. The Lebesgue decomposition. Let f be a function of bounded variation on $[a, b]$. Then it follows from Theorem 4, p. 331 and Problem 9, p. 327 that f can (in general) be represented as a sum

$$f(x) = \varphi(x) + \psi(x), \tag{10}$$

where φ is a continuous function of bounded variation and ψ is a jump function.[9] Now let

$$\varphi_1(x) = \int_a^x \varphi'(t)\, dt, \tag{11}$$

$$\varphi_2(x) = \varphi(x) - \varphi_1(x).$$

Then φ_1 is absolutely continuous, while φ_2 is a continuous function of bounded variation such that

$$\varphi_2'(x) = \varphi'(x) - \frac{d}{dx} \int_a^x \varphi'(t)\, dt = 0$$

almost everywhere. A continuous function of bounded variation is said to be *singular* if its derivative vanishes almost everywhere. For example, the Cantor function F constructed in Example 2, p. 334 is singular. Combining (10) and (11), we find that a function f of bounded variation can (in general) be represented as a sum

$$f(x) = \varphi_1(x) + \varphi_2(x) + \psi(x) \tag{12}$$

of an absolutely continuous function φ_1, a singular function φ_2 and a jump function ψ. Formula (12) is known as the *Lebesgue decomposition*.

Remark. Differentiating (12), we get

$$f'(x) = \varphi_1'(x)$$

almost everywhere. Thus integration of the derivative of a function of bounded variation does not restore the function itself, but only its absolutely continuous "component," while the other two components, i.e., the singular function and the jump function, "disappear without a trace."

Problem 1. Prove that a function f is absolutely continuous on $[a, b]$ if and only if it is a continuous function of bounded variation mapping every subset $Z \subset [a, b]$ of measure zero into a set of measure zero.

---

[9] Generalizing Problem 9, p. 327, by a *jump function*, we now mean a function of the form

$$\sum_{x_n < x} h_n + \sum_{x_n' \leqslant x} h_n',$$

where the numbers $h_1, \ldots, h_n, \ldots$ and $h_1', \ldots, h_n', \ldots$ corresponding to the discontinuity points $x_1, \ldots, x_n, \ldots$ and $x_1', \ldots, x_n', \ldots$ satisfy the conditions

$$\sum_n |h_n| < \infty, \qquad \sum_n |h_n'| < \infty$$

(we now allow negative h_n, h_n').

Problem 2. Verify directly from the definition on p. 336 that the function

$$f(x) = \begin{cases} x \sin \dfrac{1}{x} & \text{if} \quad x \neq 0, \\ 0 & \text{if} \quad x = 0 \end{cases}$$

fails to be absolutely continuous on any interval $[a, b]$ containing the point $x = 0$.

Problem 3. Prove that if a function f satisfies a Lipschitz condition

$$|f(x') - f(x'')| \leqslant K |x' - x''|$$

for all $x', x'' \in [a, b]$, then f is absolutely continuous on $[a, b]$.

Problem 4. Prove that each of the terms φ_1, φ_2 and ψ in the Lebesgue decomposition (12) is unique to within an additive constant.

Comment. The stipulation "to within an additive constant" can be dropped if we require the function f and its "components" to vanish at $x = a$, say, or if we agree to regard all functions differing by a constant as equivalent.

Problem 5. Let $A^0_{[a,b]}$ be the space of all absolutely continuous functions f defined on $[a, b]$, satisfying the condition $f(a) = 0$. Prove that $A^0_{[a,b]}$ is a closed subspace of the space $V^0_{[a,b]}$ of all functions of bounded variation on $[a, b]$ satisfying the same condition, equipped with the norm $\|f\| = V^b_a(f)$.

Comment. There is no need for the condition $f(a) = 0$ if we regard all functions differing by a constant as equivalent. We then have $\|f\| = 0$ if and only if $f = \text{const}$.

Problem 6. Starting from a locally summable function f, i.e., a function summable on every finite interval, defined the corresponding generalized function f and generalized derivative f' by the formulas

$$(f, \varphi) = \int_{-\infty}^{\infty} f(x)\varphi(x)\, dx,$$

$$(f', \varphi) = -\int_{-\infty}^{\infty} f(x)\varphi'(x)\, dx$$

as in Sec. 21.2. (Here φ is any test function, i.e., any infinitely differentiable function of finite support.) Prove that the generalized derivative f' determines f to within an additive constant. Apply this to the case of the function

$$f(x) = \begin{cases} 0 & \text{if} \quad x < 0, \\ F(x) & \text{if} \quad 0 \leqslant x \leqslant 1, \\ 1 & \text{if} \quad x > 1, \end{cases}$$

where F is the Cantor function constructed in Example 2, p. 334.

Hint. See Theorem 1, p. 213.

Problem 7. Let f and f' be the same as in the preceding problem, and suppose f is of bounded variation on $(-\infty, \infty)$. Then f has an ordinary derivative almost everywhere. Let f_1 be the generalized function corresponding to df/dx, so that

$$(f_1, \varphi) = \int_{-\infty}^{\infty} \frac{df}{dx} \varphi(x) \, dx.$$

Prove that

a) In general, f_1 does not equal the generalized derivative f';
b) If f is absolutely continuous, then $f_1 = f'$;
c) If $f_1 = f'$, then f is equivalent to an absolutely continuous function[10] and, in particular, is absolutely continuous if it is continuous.

Hint. In a), consider the function

$$f(x) = \begin{cases} 0 & \text{if } x \leqslant 0, \\ 1 & \text{if } x > 0. \end{cases}$$

Comment. Problems 6 and 7 further illustrate the situation discussed on pp. 206–207. To carry out the operations of analysis (in this case, reconstruction of a function from its derivative), we can either restrict the class of admissible functions (by requiring them to be absolutely continuous) or else extend the notion of function itself (at the same time, extending the notion of a derivative).

34. The Lebesgue Integral as a Set Function

34.1. Charges. The Hahn and Jordan decompositions. As we now show, the theory developed in Secs. 31–33 for functions defined on the real line $(-\infty, \infty)$ continues to make sense in a much more general setting. Let X be a space (i.e., some "master set") equipped with a measure μ, and let f be a μ-summable function defined on X. Then f is summable on every measurable subset $E \subset X$, so that the integral

$$\Phi(E) = \int_E f(x) \, d\mu \tag{1}$$

(for fixed f) defines a set function on the system $\mathscr{S}_\mu$ of all μ-measurable subsets of X. By Theorem 4, p. 298, Φ is σ-additive, i.e., if a measurable set E is a finite or countable union

$$E = \bigcup_n E_n$$

---

[10] I.e., coincides almost everywhere with an absolutely continuous function.

of pairwise disjoint measurable sets E_n, then

$$\Phi(E) = \sum_n \Phi(E_n).$$

In other words, the set function (1) has all the properties of a σ-additive measure except that it may not be nonnegative in the case where f takes negative values. These considerations suggest

DEFINITION 1. *A σ-additive set function Φ defined on a σ-ring (in particular, a σ-algebra) of subsets of a space X and in general taking values of both signs is called a **signed measure** or **charge** (on X).*

Remark. Thus the notion of a measure is equivalent to that of a nonnegative charge.

In the case of electrical charge distributed on a surface, we can divide the surface into two regions, one carrying positive charge (i.e., such that every part of the region is positively charged) and one carrying negative charge. We will establish the mathematical equivalent of this fact in a moment, after first introducing

DEFINITION 2. *Let Φ be a charge defined on a σ-algebra $\mathscr{S}$ of subsets of a space X. Then a set $A \subset X$ is said to be **negative** with respect to Φ if $E \cap A \in \mathscr{S}$ and $\Phi(E \cap A) \leqslant 0$ for every $E \in \mathscr{S}$. Similarly, A is said to be **positive** with respect to Φ if $E \cap A \in \mathscr{S}$ and $\Phi(E \cap A) \geqslant 0$ for every $E \in \mathscr{S}$.*

THEOREM 1. *Given a charge Φ on a space X, there is a measurable set $A^- \subset X$ such that A^- is negative and $A^+ = X - A^-$ is positive with respect to Φ.*

Proof. Let

$$a = \inf \Phi(A),$$

where the greatest lower bound is taken over all measurable negative sets A. Let $\{A_n\}$ be a sequence of measurable negative sets such that

$$\lim_{n \to \infty} \Phi(A_n) = A.$$

Then

$$A^- = \bigcup_n A_n$$

is a measurable negative set such that

$$\Phi(A^-) = a$$

(why?). To show that A^- is the required set, we must now prove that $A^+ = X - A^-$ is positive. Suppose A^+ is not positive. Then A^+ contains a measurable subset B_0 such that $\Phi(B_0) < 0$. However, B_0 cannot be negative, since if it were, the set $\tilde{A} = A^- \cup B_0$ would be a negative set such that $\Phi(\tilde{A}) < a$, which is impossible. Hence there is a *least* positive integer k_1 such that B_0 contains a subset B_1 satisfying the condition

$$\Phi(B_1) \geqslant \frac{1}{k_1}.$$

Obviously $B_1 \neq B_0$. Applying the same argument to the set $B_0 - B_1$, we find a least positive integer k_2 such that $B_0 - B_1$ contains a subset B_2 satisfying the inequality

$$\Phi(B_2) \geqslant \frac{1}{k_2} \qquad (k_2 > k_1)$$

(explain why $k_2 > k_1$), a least positive integer k_3 such that $(B_0 - B_1) - B_2$ contains a subset B_3 satisfying the inequality

$$\Phi(B_3) \geqslant \frac{1}{k_3} \qquad (k_3 > k_2),$$

and so on. Now let

$$F = B_0 - \bigcup_{n=1}^{\infty} B_n.$$

Clearly F is nonempty, since $\Phi(B_0) < 0$ while $\Phi(B_n) > 0$ for all $n \geqslant 1$. Moreover, F is negative by construction (think things through). Hence the set $\tilde{A} = A^- \cup F$ is again negative and $\Phi(\tilde{A}) < a$, which is impossible. This contradiction shows that $A^+ = X - A^-$ must be positive.   ∎

Thus we can represent X as a union

$$X = A^+ \cup A^- \qquad (2)$$

of two disjoint measurable sets A^+ and A^-, where A^+ is positive and A^- is negative with respect to the charge Φ. The representation (2) is called the *Hahn decomposition* of X, and may not be unique. However, if

$$X = A_1^+ \cup A_1^-, \qquad X = A_2^+ \cup A_1^+$$

are two distinct Hahn decompositions of X, then

$$\Phi(E \cap A_1^-) = \Phi(E \cap A_2^-), \qquad \Phi(E \cap A_1^+) = \Phi(E \cap A_2^+) \qquad (3)$$

for every $E \in \mathscr{S}$. In fact,

$$E \cap (A_1^- - A_2^-) \subset E \cap A_1^- \qquad (4)$$

and at the same time

$$E \cap (A_1^- - A_2^-) \subset E \cap A_2^+. \qquad (5)$$

But (4) implies

$$\Phi(E \cap A(_1^- - A_2^-)) \leqslant 0,$$

while (5) implies

$$\Phi(E \cap (A_1^- - A_2^-)) \geqslant 0.$$

Therefore

$$\Phi(E \cap (A_1^- - A_2^-)) = 0, \tag{6}$$

and similarly

$$\Phi(E \cap (A_2^- - A_1^-)) = 0. \tag{7}$$

It follows from (6) and (7) that

$$\Phi(E \cap A_1^-) = \Phi(E \cap (A_1^- - A_2^-)) + \Phi(E \cap (A_1^- \cap A_2^-))$$
$$= \Phi(E \cap (A_2^- - A_1^-)) + \Phi(E \cap (A_1^- \cap A_2^-)) = \Phi(E \cap A_2^-),$$

which proves the first of the formulas (3). The second formula is proved in exactly the same way.

Thus a charge Φ on a space X uniquely determines two nonnegative set functions, namely

$$\Phi^+(E) = \Phi(E \cap A^+), \qquad \Phi^-(E) = -\Phi(E \cap A^-),$$

called the *positive variation* and *negative variation* of Φ, respectively. It is clear that

1) $\Phi = \Phi^+ - \Phi^-$;
2) Φ^+ and Φ^- are nonnegative σ-additive set functions, i.e., measures;
3) The set function $|\Phi| = \Phi^+ + \Phi^-$, called the *total variation* of Φ, is also a measure.

The representation

$$\Phi = \Phi^+ - \Phi^-$$

a charge Φ as the difference between its positive and negative variations is called the *Jordan decomposition* of Φ.

34.2. Classification of charges. The Radon-Nikodym theorem. We now classify charges on a space X equipped with a measure:

DEFINITION 3. *Let μ be a σ-additive measure on a σ-algebra $\mathscr{S}_\mu$ of (μ-measurable) subsets of a space X, and let Φ be a charge defined on $\mathscr{S}_\mu$. Then Φ is said to be **concentrated** on a set $A \in \mathscr{S}_\mu$ if $\Phi(E) = 0$ for every measurable set $E \subset X - A$.*

DEFINITION 4. *Let μ, $\mathscr{S}_\mu$, X and Φ be the same as in Definition 3. Then Φ is said to be*

1) **Continuous** *if $\Phi(E) = 0$ for every single-element set $E \subset X$ of measure zero;*

2) **Singular** if Φ is concentrated on a set of measure zero;

3) **Discrete** if Φ is concentrated on a finite or countable set of measure zero;

4) **Absolutely continuous** (with respect to μ) if $\Phi(E) = 0$ for every measurable set E such that $\mu(E) = 0$.

Clearly, the Lebesgue integral

$$\Phi(E) = \int_E \varphi(x) \, d\mu$$

of a fixed summable function φ is absolutely continuous with respect to the measure μ. As we will see in a moment, every absolutely continuous charge can be represented in this form. But first we need the following

LEMMA. Let μ be a σ-additive measure defined on a σ-algebra $\mathscr{S}_\mu$ of subsets of a space X, and let Φ be another such measure defined on $\mathscr{S}_\mu$. Suppose Φ is absolutely continuous with respect to μ and is not identically zero. Then there is a positive integer n and a set $A \in \mathscr{S}_\mu$ such that $\mu(A) > 0$ and A is positive with respect to the charge $\Phi - (1/n)\mu$.

Proof. Let

$$X = A_n^- \cup A_n^+ \qquad (n = 1, 2, \ldots)$$

be the Hahn decomposition corresponding to the charge $\Phi - (1/n)\mu$, and let

$$A_0^- = \bigcap_{n=1}^\infty A_n^-, \qquad A_0^+ = \bigcup_{n=1}^\infty A_n^+.$$

Then

$$\Phi(A_0^-) \leqslant \frac{1}{n}\,\mu(A_0^-)$$

for all $n = 1, 2, \ldots$, i.e., $\Phi(A_0^-) = 0$, and hence $\Phi(A_0^+) > 0$ since $X = A_0^- \cup A_0^+$ and Φ is not identically zero. But then $\mu(A_0^+) > 0$, by the absolute continuity of Φ. Hence there is an n such that $\mu(A_n^+) > 0$ (why?). This n and the set $A = A_n^+$ satisfy the conditions of the lemma.

THEOREM 2 (**Radon-Nikodým**). Let μ be a σ-additive measure defined on a σ-algebra $\mathscr{S}_\mu$ of subsets of a space X, and let Φ be a charge defined on $\mathscr{S}_\mu$. Suppose Φ is absolutely continuous with respect to μ. Then there is a μ-summable function φ on X such that

$$\Phi(E) = \int_E \varphi(x) \, d\mu \tag{8}$$

for every $E \in \mathscr{S}_\mu$. The function φ is unique to within its values on a set of μ-measure zero.

Proof. We can assume that Φ is not identically zero, since otherwise we need only choose φ to be any function equal to zero almost everywhere (discuss the uniqueness of φ in this case). Let K be the set of all μ-summable functions on X such that

$$\int_E f(x)\, d\mu \leqslant \Phi(E)$$

for every $E \in \mathscr{S}_\mu$, and let

$$M = \sup_{f \in K} \int_X f(x)\, d\mu.$$

Moreover, let $\{f_n\}$ be a sequence of functions in K such that

$$\lim_{n \to \infty} \int_X f_n(x)\, d\mu = M, \qquad (9)$$

and let

$$g_n(x) = \max \{f_1(x), \ldots, f_n(x)\}.$$

Then clearly

$$g_1(x) \leqslant g_2(x) \leqslant \cdots \leqslant g_n(x) \leqslant \cdots.$$

Moreover,

$$\int_E g_n(x)\, d\mu \leqslant \Phi(E) \qquad (10)$$

for every $E \in \mathscr{S}_\mu$. In fact, E can be written in the form

$$E = \bigcup_{k=1}^n E_k,$$

where the sets $E_1, \ldots, E_n$ are pairwise disjoint and $g_n(x) = f_k(x)$ on E_k, and hence

$$\int_E g_n(x)\, d\mu = \sum_{k=1}^n \int_E f_k(x)\, d\mu \leqslant \sum_{k=1}^n \Phi(E_k) = \Phi(E).$$

In particular, it follows from (10) that $g_n \in K$, so that

$$\int_X g_n(x)\, d\mu \leqslant M.$$

But then

$$\lim_{n \to \infty} \int_X g_n(x)\, d\mu = M,$$

since otherwise

$$\lim_{n \to \infty} \int_X f_n(x)\, d\mu \leqslant \lim_{n \to \infty} \int_X g_n(x)\, d\mu < M,$$

contrary to (10). Writing

$$\varphi(x) = \sup_n g_n(x),$$

we find that

$$\varphi(x) = \lim_{n \to \infty} g_n(x),$$

and hence, by Levi's theorem (Theorem 2, p. 305),

$$\int_X \varphi(x)\, d\mu = \lim_{n \to \infty} \int_X g_n(x)\, d\mu = M. \tag{11}$$

Next we show that φ is the required function, figuring in the representation (8). By construction, the set function

$$\lambda(E) = \Phi(E) - \int_E \varphi(x)\, d\mu$$

is nonnegative and in fact is a σ-additive measure. If $\lambda(E) \not\equiv 0$, then, by the lemma, there is an $\varepsilon > 0$ and a set $A \in \mathscr{S}_\mu$ such that $\mu(A) > 0$ and

$$\varepsilon\mu(E \cap A) \leqslant \lambda(E \cap A)$$

for every $E \in \mathscr{S}_\mu$. Let

$$h(x) = \varphi(x) + \varepsilon\chi_A(x),$$

where[11]

$$\chi_A(x) = \begin{cases} 1 & \text{if } x \in A, \\ 0 & \text{if } x \notin A. \end{cases}$$

Then

$$\int_E h(x)\, d\mu = \int_E \varphi(x)\, d\mu + \varepsilon\mu(E \cap A)$$

$$\leqslant \int_{E-A} \varphi(x)\, d\mu + \Phi(E \cap A) \leqslant \Phi(E),$$

so that h belongs to the set K introduced at the beginning of the proof. On the other hand, it follows from (11) that

$$\int_X h(x)\, d\mu = \int_X \varphi(x)\, d\mu + \varepsilon\mu(A) > M,$$

contrary to the definition of M. Therefore $\lambda(E) \equiv 0$, which is equivalent to (8).

Finally, to prove that φ is unique to within its values on a set of measure zero, suppose

$$\Phi(E) = \int_E \varphi(x)\, d\mu = \int_E \varphi^*(x)\, d\mu$$

for all $E \in \mathscr{S}_\mu$. Then, by Chebyshev's inequality (Theorem 5, p. 299), we have

$$\mu(A_m) \leqslant m\int_{A_m} [\varphi(x) - \varphi^*(x)]\, d\mu = 0$$

---

[11] χ_A is called the *characteristic function* of the set A.

for every set

$$A_m = \left\{ x : \varphi(x) - \varphi^*(x) > \frac{1}{m} \right\} \qquad (m = 1, 2, \ldots),$$

and similarly

$$\mu(B_n) = 0$$

for every set

$$B_n = \left\{ x : \varphi^*(x) - \varphi(x) > \frac{1}{n} \right\} \qquad (n = 1, 2, \ldots).$$

But

$$\{ x : \varphi(x) \neq \varphi^*(x) \} = \left(\bigcup_m A_m \right) \cup \left(\bigcup_n B_n \right),$$

and hence

$$\mu \{ x : \varphi(x) \neq \varphi^*(x) \} = 0,$$

i.e., $\varphi(x) = \varphi^*(x)$ almost everywhere. ∎

Remark 1. The function φ figuring in the representation (8) is called the *Radon-Nikodým derivative* (or simply the *density*) of the charge Φ with respect to the measure μ, and is denoted

$$\frac{d\Phi}{d\mu}.$$

Clearly, Theorem 2 is the natural generalization of Lebesgue's theorem (Theorem 6, p. 340), which states that an absolutely continuous function F is the integral of its own derivative F'. However, in the case of a function F defined on the real line there is an explicit procedure for finding the derivative of F at a point x_0, namely evaluation of the limit

$$\lim_{\Delta x \to 0} \frac{\Delta F}{\Delta x} = \lim_{\Delta x \to 0} \frac{F(x_0 + \Delta x) - F(x_0)}{\Delta x},$$

whereas the Radon-Nikodým theorem only establishes the existence of the derivative $d\Phi/d\mu$, without telling how to find it. However, an explicit procedure can be given for evaluating $d\Phi/d\mu$ at a point $x_0 \in X$ by calculating the limit

$$\lim_{\varepsilon \to 0} \frac{\Phi(A_\varepsilon)}{\mu(A_\varepsilon)},$$

where $\{A_\varepsilon\}$ is a system of sets "converging to the point x_0" as $\varepsilon \to 0$, in a suitably defined sense.[12]

---

[12] For the details, see G. E. Shilov and B. L. Gurevich, *Integral, Measure and Derivative: A Unified Approach* (translated by R. A. Silverman), Prentice-Hall, Inc., Englewood Cliffs, N.J. (1966), Chap. 10.

Remark 2. It can also be shown[13] that an arbitrary charge Φ has a unique representation as the sum

$$\Phi(E) = A(E) + S(E) + D(E)$$

of an absolutely continuous charge A, a singular charge S and a discrete charge D. This is the exact analogue of the Lebesgue decomposition on p. 341.

Problem 1. Given any charge Φ defined on a σ-algebra $\mathscr{S}$, prove that there is a constant $M > 0$ such that $|\Phi(E)| \leqslant M$ for all $E \in \mathscr{S}$.

Problem 2. Give an example of two distinct Hahn decompositions of a space X.

Problem 3. Prove that a charge Φ vanishes identically if it is both absolutely continuous and singular with respect to a measure μ.

Problem 4. Prove that if a charge Φ is concentrated on a set A_0, then so are its positive, negative and total variations.

Problem 5. Prove that

a) Every absolutely continuous charge is continuous;
b) Every discrete charge is singular.

Problem 6. Prove that if a charge Φ is absolutely continuous (with respect to a measure μ), then so are its positive, negative and total variations.

Problem 7. Prove that if a charge Φ is discrete, then there are no more than countably many points $x_1, x_2, \ldots, x_n, \ldots$ and corresponding real numbers $h_1, h_2, \ldots, h_n, \ldots$ such that $\mu(\{x_n\}) = 0$ and

$$\Phi(E) = \sum_{x_n \in E} h_n.$$

Write expressions for the positive, negative and total variations of Φ.

Problem 8. Let X be the square $0 \leqslant x \leqslant 1$, $0 \leqslant y \leqslant 1$ equipped with ordinary two-dimensional Lebesgue measure μ, and let $\Phi(E)$ be the ordinary one-dimensional Lebesgue measure of the intersection of E with the interval $0 \leqslant x \leqslant 1$. Prove that Φ is continuous and singular, but not absolutely continuous.

---

[13] G. E. Shilov and B. L. Gurevich, *op. cit.*, Chap. 9.

10

MORE ON INTEGRATION

35. Product Measures. Fubini's Theorem

The problem of reducing double (or multiple) integrals to iterated integrals plays an important role in classical analysis. In the Lebesgue theory, the key result along these lines is Fubini's theorem, proved in Sec. 35.3. En route to Fubini's theorem we will need the preliminary topics treated in Secs. 35.1 and 35.2, which are also of interest in their own right.

35.1. Direct products of sets and measures. By the *direct* (or *Cartesian*) *product* of two sets X and Y, denoted by $X \times Y$, we mean the set of all ordered pairs (x, y) where $x \in X$, $y \in Y$. Similarly, by the direct product of n sets $X_1, X_2, \ldots, X_n$, denoted by

$$X_1 \times X_2 \times \cdots \times X_n, \tag{1}$$

we mean the set of all ordered n-tuples $(x_1, x_2, \ldots, x_n)$, where $x_1 \in X_1$, $x_2 \in X_2, \ldots, x_n \in X_n$. In particular, if

$$X_1 = X_2 = \cdots = X_n = X,$$

we write (1) simply as X^n, the "nth power of X."

Example 1. Real n-space R^n is the nth power of the real line R^1, as anticipated by the notation.

Example 2. The unit cube I^n in n-space, i.e., the set of all elements of R^n with coordinates satisfying the inequalities

$$0 \leqslant x_k \leqslant 1 \qquad (k = 1, 2, \ldots, n),$$

is the nth power of the closed unit interval $I^1 = [0, 1]$.

Now let $\mathscr{S}_1, \mathscr{S}_2, \ldots, \mathscr{S}_n$ be systems of subsets of the sets $X_1, X_2, \ldots, X_n$, respectively. Then by

$$\mathfrak{S} = \mathscr{S}_1 \times \mathscr{S}_2 \times \cdots \times \mathscr{S}_n$$

we mean the system of subsets of the direct product (1) which can be represented in the form

$$A = A_1 \times A_2 \times \cdots \times A_n,$$

where

$$A_k \in \mathscr{S}_k \qquad (k = 1, 2, \ldots, n).$$

If

$$\mathscr{S}_1 = \mathscr{S}_2 = \cdots = \mathscr{S}_n = \mathscr{S},$$

then $\mathfrak{S}$ is the "nth power of $\mathscr{S}$," written

$$\mathfrak{S} = \mathscr{S}^n.$$

For example, the system of all closed rectangular parallelepipeds in R^n is the nth power of the system of all closed intervals in R^1.

THEOREM 1. *If $\mathscr{S}_1, \mathscr{S}_2, \ldots, \mathscr{S}_n$ are semirings, then so is the set* $\mathfrak{S} = \mathscr{S}_1 \times \mathscr{S}_2 \times \cdots \times \mathscr{S}_n$.

Proof. By the definition of a semiring (see p. 32), we must show that[1]

a) If $A, B \in \mathfrak{S}$, then $A \cap B \in \mathfrak{S}$;
b) If $A, B \in \mathfrak{S}$ and $B \subset A$, then A can be represented as a finite union

$$A = \bigcup_{k=1}^n C^{(k)}$$

of pairwise disjoint sets $C^{(k)} \in \mathfrak{S}$, with $B = C^{(1)}$.

It is clearly enough to prove these assertions for the case $n = 2$. Thus suppose $A \in \mathscr{S}_1 \times \mathscr{S}_2$, $B \subset \mathscr{S}_1 \times \mathscr{S}_2$. Then

$$\begin{aligned} A &= A_1 \times A_2 \qquad (A_1 \in \mathscr{S}_1, A_2 \in \mathscr{S}_2) \\ B &= B_1 \times B_2 \qquad (B_1 \in \mathscr{S}_1, B_2 \in \mathscr{S}_2), \end{aligned} \qquad (2)$$

and hence

$$A \cap B = (A_1 \times A_2) \cap (B_1 \times B_2) = (A_1 \cap A_2) \times (A_2 \cap B_2).$$

But $A_1 \cap B_1 \in \mathscr{S}_1$, $A_2 \cap B_2 \in \mathscr{S}_2$, since $\mathscr{S}_1$ and $\mathscr{S}_2$ are semirings. It follows that $A \cap B \in \mathscr{S}_1 \times \mathscr{S}_2$. This proves a).

To prove b), suppose that

$$B_1 \subset A_1, \qquad B_2 \subset A_2,$$

---

[1] Note that the empty set $\varnothing$ belongs to $\mathfrak{S}$, since $\varnothing = \varnothing \times \varnothing \times \cdots \times \varnothing$ (why?).

in addition to (2). Then, since $\mathcal{S}_1$ and $\mathcal{S}_2$ are semirings, there are finite expansions

$$A_1 = B_1 \cup B_1^{(1)} \cup \cdots \cup B_1^{(i)},$$

$$A_2 = B_2 \cup B_2^{(1)} \cup \cdots \cup B_2^{(j)},$$

where the sets $B_1, B_1^{(1)}, \ldots, B_1^{(i)}$ are pairwise disjoint and belong to $\mathcal{S}_1$, while the sets $B_2, B_2^{(1)}, \ldots, B_2^{(j)}$ are pairwise disjoint and belong to $\mathcal{S}_2$. Therefore

$$\begin{aligned}
A = A_1 \times A_2 = &(B_1 \times B_2) \cup (B_1 \times B_2^{(1)}) \cup \cdots \cup (B_1 \times B_2^{(j)}) \\
&\cup (B_1^{(1)} \times B_2) \cup (B_1^{(1)} \times B_2^{(1)}) \cup \cdots \cup (B_1^{(1)} \times B_2^{(j)}) \\
&\cup \cdots \cup (B_1^{(i)} \times B_2) \cup (B_1^{(i)} \times B_2^{(1)}) \cup \cdots \cup (B_1^{(i)} \times B_2^{(j)})
\end{aligned}$$

is the desired finite expansion of $A_1 \times A_2$, where $B_1 \times B_2$ is the first term and the other terms are pairwise disjoint and belong to $\mathfrak{S} = \mathcal{S}_1 \times \mathcal{S}_2$. ∎

Now let $\mathcal{S}_1, \mathcal{S}_2, \ldots, \mathcal{S}_n$ be n semirings, equipped with measures

$$\mu_1(A_1), \mu_2(A_2), \ldots, \mu_n(A_n) \qquad (A_k \in \mathcal{S}_k), \tag{3}$$

and let μ be the measure on the semiring $\mathfrak{S} = \mathcal{S}_1 \times \mathcal{S}_2 \times \cdots \times \mathcal{S}_n$ defined by the formula

$$\mu(A) = \mu_1(A_1)\mu_2(A) \cdots \mu_n(A_n)$$

for every $A = A_1 \times A_2 \times \cdots \times A_n$. Then μ is called the *direct* (or *Cartesian*) *product*[2] of the measures (3), and is denoted by

$$\mu = \mu_1 \times \mu_2 \times \cdots \times \mu_n.$$

To confirm that μ is indeed a measure, we now show that μ is additive (μ is obviously real and nonnegative). It will again be enough to consider the case $n = 2$. Suppose

$$A = A_1 \times A_2 = \bigcup_{k=1}^{t} B^{(k)}, \tag{4}$$

where

$$B^{(i)} \cap B^{(j)} = \varnothing \qquad (i \neq j)$$

and

$$B^{(k)} = B_1^{(k)} \times B_2^{(k)}.$$

According to Lemma 2, p. 33, there are finite expansions

$$A_1 = \bigcup_{m=1}^{r} C_1^{(m)}, \qquad A_2 = \bigcup_{n=1}^{s} C_2^{(n)},$$

_____

[2] The term *product measure* will be used with a different meaning below.

each involving pairwise disjoint sets, such that each $B_1^{(k)}$ is a finite union

$$B_1^{(k)} = \bigcup_{m \in M_k} C_1^{(m)}$$

of certain of the sets $C_1^{(m)}$, while each $B_2^{(k)}$ is a finite union

$$B_2^{(k)} = \bigcup_{n \in N_k} C_2^{(n)}$$

of certain of the sets $C_2^{(n)}$ (here M_k denotes some subset of the set $\{1, 2, \ldots, r\}$ and N_k some subset of the set $\{1, 2, \ldots, s\}$). But then, by the additivity of μ_1 and μ_2, we have

$$\mu(A) = \mu_1(A_1)\mu_2(A_2) = \sum_{m=1}^{r} \mu_1(C_1^{(m)}) \sum_{n=1}^{s} \mu_2(C_2^{(n)})$$

$$= \sum_{k=1}^{t} \sum_{m \in M_k} \mu_1(C_1^{(m)}) \sum_{n \in N_k} \mu_2(C_2^{(n)})$$

$$= \sum_{k=1}^{t} \mu_1(B_1^{(k)})\mu_2(B_2^{(k)}) = \sum_{k=1}^{t} \mu(B_k),$$

which, when compared with (4), shows the additivity of $\mu = \mu_1 \times \mu_2$.

Example 3. Thus the additivity of area of rectangles in the plane follows from the additivity of length of intervals on the line.

THEOREM 2. *If the measures $\mu_1, \mu_2, \ldots, \mu_n$ are σ-additive, then so is the measure $\mu = \mu_1 \times \mu_2 \times \cdots \times \mu_n$.*

Proof. Again we need only consider the case $n = 2$. Let λ_1 denote the Lebesgue extension of the measure μ_1, and suppose

$$C = \bigcup_{n=1}^{\infty} C_n,$$

where the sets C_n are pairwise disjoint and the sets C, C_n belong to $\mathscr{S}_1 \times \mathscr{S}_2$, i.e.,

$$C = A \times B \qquad (A \in \mathscr{S}_1, B \in \mathscr{S}_2),$$

$$C_n = A_n \times B_n \qquad (A_n \in \mathscr{S}_1, B_n \in \mathscr{S}_2).$$

Moreover, let

$$f_n(x) = \begin{cases} \mu_2(B_n) & \text{if } x \in A_n, \\ 0 & \text{if } x \notin A_n. \end{cases}$$

We then have

$$\sum_{n=1}^{\infty} f_n(x) = \mu_2(B) \quad \text{if } x \in A,$$

and hence, by the corollary on p. 307,

$$\sum_{n=1}^{\infty} \int_A f_n(x)\, d\lambda_1 = \int_A \mu_2(B)\, d\lambda_1 = \lambda_1(A)\mu_2(B)$$

$$= \mu_1(A)\mu_2(B) = \mu(C). \tag{5}$$

But

$$\int_A f_n(x) \, d\lambda_1 = \mu_1(A_n)\mu_2(B_n) = \mu(C_n). \tag{6}$$

Substituting (6) into (5), we get

$$\mu(C) = \sum_{n=1}^{\infty} \mu(C_n). \quad \blacksquare$$

Again let $\mathscr{S}_1, \mathscr{S}_2, \ldots, \mathscr{S}_n$ be n semirings, this time equipped with σ-*additive* measures (3). Then it follows from Theorem 2 that the measure[3]

$$m = \mu_1 \times \mu_2 \times \cdots \times \mu_n \tag{7}$$

is σ-additive on the semiring

$$\mathfrak{S} = \mathscr{S}_1 \times \mathscr{S}_2 \times \cdots \times \mathscr{S}_n.$$

Therefore, as in Sec. 27, m has a Lebesgue extension μ defined on a σ-ring $\mathscr{S}_\mu \supset \mathfrak{S}$. This measure μ is called the *product measure* of the measures (3), and is denoted by

$$\mu = \mu_1 \otimes \mu_2 \otimes \cdots \otimes \mu_n. \tag{8}$$

The distinction between the meaning of the symbols $\times$ and $\otimes$ in (7) and (8) is crucial.

Example 4. Let

$$\mu_1 = \mu_2 = \cdots = \mu_n = \mu^1,$$

where μ^1 is ordinary Lebesgue measure on the line. Then the product measure (8) is ordinary Lebesgue measure in n-space.

35.2. Evaluation of a product measure. Let G be a region in the xy-plane bounded by the vertical lines $x = a$, $x = b$ $(a < b)$ and the curves $y = f(x)$, $y = g(x)$, where $f(x) \leqslant g(x)$. Then it will be recalled from calculus that the area of G is given by the integral

$$\int_a^b [g(x) - f(x)] \, dx,$$

where the difference $g(x_0) - f(x_0)$ is just the length of the segment in which the vertical line $x = x_0$ intersects the region G. As we now show, the natural generalization of this method can be used to evaluate an arbitrary product measure:

THEOREM 3. *Let μ be the product measure*

$$\mu = \mu_x \otimes \mu_y,$$

---

[3] We change to the symbol m here, to "free" μ for use in formula (8).

of two measures μ_x *and* μ_y *such that*

1) μ_x *is* σ-*additive on a Borel algebra* $\mathscr{S}_{\mu_x}$ *of subsets of a set* X;
2) μ_y *is* σ-*additive on a Borel algebra* $\mathscr{S}_{\mu_y}$ *of subsets of a set* Y;
3) μ_x *and* μ_y *are complete, in the sense that* $B \subset A$ *and* $\mu_x(A) = 0$
   implies that B *is measurable* (*with measure zero*), *and similarly for*
   μ_y.[4]

Then

$$\mu(A) = \int_X \mu_y(A_x)\, d\mu_x = \int_Y \mu_x(A_y)\, d\mu_y \tag{9}$$

for every μ-*measurable set* A, *where*[5]

$$A_x = \{y : (x, y) \in A\} \qquad (x \text{ fixed}),$$
$$A_y = \{x : (x, y) \in A\} \qquad (y \text{ fixed}).$$

Proof. We note in passing that the integral over X in (9) reduces to an integral over the set of the form

$$\bigcup_y A_y \subset X$$

outside which $\mu_y(A_x)$ vanishes (and similarly for the integral over Y). It will be enough to prove that

$$\mu(A) = \int_X \varphi_A(x)\, d\mu_x, \tag{10}$$

where

$$\varphi_A(x) = \mu_y(A_x),$$

since the other part of (9) is proved in exactly the same way. Observe that implicit in the theorem is the conclusion that the set A_x is μ_x-measurable for almost all x (in the sense of the measure μ_x) and that the function $\varphi_A(x)$ is μ_x-measurable, since otherwise (10) would be meaningless.

The measure μ is the Lebesgue extension of the measure

$$m = \mu_x \times \mu_y$$

defined on the semiring $\mathscr{S}_m$ of all sets of the form

$$A = A_{y_0} \times A_{x_0} \qquad (A \in \mathscr{S}_\mu),$$

where $\mathscr{S}_\mu$ is the Borel algebra of μ-measurable subsets of $X \times Y$. But (10) obviously holds for all such sets, since for them

$$\varphi_A(x) = \begin{cases} \mu_y(A_{x_0}) & \text{if} \quad x \in A_{y_0}, \\ 0 & \text{if} \quad x \notin A_{y_0}. \end{cases}$$

---

[4] The Lebesgue extension of any measure is complete (see Problem 7, p. 280).

[5] If X is the x-axis and Y the y-axis (so that $X \times Y$ is the xy-plane), then A_{x_0} is the projection onto the y-axis of the set in which the vertical line $x = x_0$ intersects the set A (and similarly for A_{y_0}).

Moreover, (10) carries over at once to the ring $\mathscr{R}(\mathscr{S}_m)$ generated by $\mathscr{S}_m$, since $\mathscr{R}(\mathscr{S}_m)$ is just the system of all sets which can be represented as finite unions of pairwise disjoint sets of $\mathscr{S}_m$ (recall Theorem 3, p. 34).

To prove (10) for an arbitrary set $A \in \mathscr{S}_\mu$, we recall from Theorem 8, p. 277 that there are sets

$$B_{nk} \in \mathscr{R}(\mathscr{S}_m) \qquad (B_{n1} \subset B_{n2} \subset \cdots \subset B_{nk} \subset \cdots)$$

and corresponding sets

$$B_n = \bigcup_k B_{nk} \in \mathscr{S}_\mu \qquad (B_1 \supset B_2 \supset \cdots \supset B_n \supset \cdots)$$

such that

$$A \subset B = \bigcap_n B_n,$$
$$\mu(A) = \mu(B). \tag{11}$$

Clearly,

$$\varphi_{B_n}(x) = \lim_{k \to \infty} \varphi_{B_{nk}}(x), \qquad \varphi_{B_{n1}}(x) \leqslant \varphi_{B_{n2}}(x) \leqslant \cdots \leqslant \varphi_{B_{nk}}(x) \leqslant \cdots,$$

$$\varphi_B(x) = \lim_{n \to \infty} \varphi_{B_n}(x), \qquad \varphi_{B_1}(x) \geqslant \varphi_{B_2}(x) \geqslant \cdots \geqslant \varphi_{B_n}(x) \geqslant \cdots.$$

Hence we can invoke Levi's theorem[6] to extend (10) from the ring $\mathscr{R}(\mathscr{S}_m)$ to the system of all sets $B \in \mathscr{S}_\mu$ of the form

$$\bigcap_n \bigcup_k B_{nk} \qquad (B_{nk} \in \mathscr{S}_m). \tag{12}$$

Moreover if $\mu(A) = 0$, then $\mu(B) = 0$, because of (11), and hence

$$\varphi_B(x) = \mu_y(B_x) = 0$$

almost everywhere. Therefore A_x is measurable and

$$\varphi_A(x) = \mu_y(A_x) = 0$$

for almost all x, since $A_x \subset B_x$. But then

$$\int_X \varphi_A(x) \, d\mu_x = 0 = \mu(A).$$

In other words, (10) holds for all sets of measure zero, as well as for all sets of the form (12). But, according to (11), an arbitrary set $A \in \mathscr{S}_\mu$ can be represented as

$$A = B - Z,$$

where B is of the form (12) and Z is of measure zero. Therefore

$$B = A \cup Z \qquad (A \cap Z = \varnothing).$$

---

[6] See Theorem 2, p. 305 and Problem 2, p. 311.

It follows that

$$\mu(A) = \mu(B) = \int_X \varphi_B(x)\, d\mu_x$$

$$= \int_X \varphi_A(x)\, d\mu_x + \int_X \varphi_Z(x)\, d\mu_x = \int_X \varphi_A(x)\, d\mu_x.$$

i.e., (10) holds for every $A \in \mathscr{S}_\mu$. ∎

Example 1. Let M be any μ_x-measurable set, and let f be an integrable nonnegative function. Moreover, let Y be the y-axis, and let μ_y be ordinary Lebesgue measure on the line. Consider the set

$$A = \{(x, y) : x \in M, 0 \leqslant y \leqslant f(x)\}. \tag{13}$$

Then

$$\varphi_A(x) = \mu_y(A_x) = \begin{cases} f(x) & \text{if } x \in A, \\ 0 & \text{if } x \notin A, \end{cases}$$

and hence, by Theorem 3,

$$\mu(A) = \int_X \varphi_A(x)\, d\mu_x = \int_M f(x)\, d\mu_x. \tag{14}$$

This allows us to interpret the Lebesgue integral of a nonnegative function over a set $M \subset X$ in terms of the μ-measure of the set (13), where $\mu = \mu_x \otimes \mu_y$.

Example 2. In the preceding example, let X be the x-axis and let M be a closed interval $[a, b]$. Moreover, suppose f is nonnegative and Riemann-integrable on $[a, b]$. Then (14) reduces to the familiar formula

$$\mu(A) = \int_a^b f(x)\, dx$$

for the area under the graph of the function $y = f(x)$ between $x = a$ and $x = b$.

35.3. Fubini's theorem. The next theorem is basic in the theory of multiple integration:

THEOREM 4 (*Fubini*). *Let μ_x and μ_y be the same as in Theorem 3, let μ be the product measure $\mu_x \otimes \mu_y$, and let $f(x, y)$ be μ-integrable on the set $A \subset X \times Y$. Then*

$$\int_A f(x, y)\, d\mu = \int_X \left(\int_{A_x} f(x, y)\, d\mu_y \right) d\mu_x = \int_Y \left(\int_{A_y} f(x, y)\, d\mu_x \right) d\mu_y. \tag{15}$$

Proof. Note that implicit in the theorem is the conclusion that the "inner integrals" in parentheses exist for almost all values of the variable over which they are integrated (x in the first case, y in the second). We begin by assuming temporarily that $f(x, y) \geqslant 0$. Consider the triple Cartesian product

$$U = X \times Y \times Z,$$

where Z is the real line, equipped with the product measure

$$\mu_u = \mu_x \otimes \mu_y \otimes \mu^1 = \mu \otimes \mu^1 = \mu_x \otimes (\mu_y \otimes \mu^1)$$

(see Problem 3), where μ^1 is ordinary Lebesgue measure on the line. Moreover, consider the set $W \subset U$ defined by

$$W = \{(x, y, z) : x \in A_y, y \in A_x, 0 \leqslant z \leqslant f(x, y)\}.$$

By (14),

$$\mu_u(W) = \int_A f(x, y) \, d\mu. \tag{16}$$

On the other hand, by Theorem 3,

$$\mu_u(W) = \int_X \lambda(W_x) \, d\mu_x, \tag{17}$$

where

$$\lambda = \mu_y \otimes \mu^1,$$
$$W_x = \{(y, z) : (x, y, z) \in W\} \qquad (x \text{ fixed}).$$

Using (14) again, we obtain

$$\lambda(W_x) = \int_{A_x} f(x, y) \, d\mu_y. \tag{18}$$

Comparing (16)–(18), we get part of (15). The rest of (15) is proved in exactly the same way. To remove the restriction that $f(x, y)$ be nonnegative, we merely note that

$$f(x, y) = f^+(x, y) - f^-(x, y),$$

where the functions

$$f^+(x, y) = \frac{|f(x, y)| + f(x, y)}{2},$$

$$f^-(x, y) = \frac{|f(x, y)| - f(x, y)}{2}$$

are both nonnegative. ∎

Remark. Thus Fubini's theorem asserts that if the "double integral"

$$I = \int_A f(x, y) \, d\mu \tag{19}$$

exists, then so do the "iterated integrals"

$$I_{xy} = \int_X \left(\int_{A_x} f(x, y) \, d\mu_y \right) d\mu_x, \qquad I_{yx} = \int_Y \left(\int_{A_y} f(x, y) \, d\mu_x \right) d\mu_y, \tag{20}$$

and moreover $I = I_{xy} = I_{yx}$.

Problem 1. Give an example of a set in R^2 which is not a direct product of any two sets in R^1.

Problem 2. Prove that the direct product of two rings (or σ-rings) need not be a ring (or σ-ring).

Problem 3. Given three spaces X, Y and Z, equipped with measures μ_x, μ_y and μ_z, respectively, prove that $(\mu_x \otimes \mu_y) \otimes \mu_z$ and $\mu_x \otimes (\mu_y \otimes \mu_z)$ are identical measures on $X \times Y \times Z$.

Problem 4. Let $A = [-1, 1] \times [-1, 1]$ and

$$f(x, y) = \frac{xy}{(x^2 + y^2)^2}.$$

Prove that
  a) The iterated integrals (20) exist and are equal;
  b) The double integral (19) fails to exist.

Hint. Since

$$\int_{-1}^{1} f(x, y)\, dx = \int_{-1}^{1} f(x, y)\, dy = 0,$$

we have

$$\int_{-1}^{1} \left(\int_{-1}^{1} f(x, y)\, dx \right) dy = \int_{-1}^{1} \left(\int_{-1}^{1} f(x, y)\, dy \right) dx = 0.$$

On the other hand, the double integral fails to exist, since

$$\int_{-1}^{1} \int_{-1}^{1} |f(x, y)|\, dx\, dy > \int_{0}^{1} dr \int_{0}^{2\pi} \frac{|\sin \theta \cos \theta|}{r}\, d\theta = 2 \int_{0}^{1} \frac{dr}{r} = \infty,$$

after transforming to polar coordinates.

Problem 5. Let $A = [0, 1] \times [0, 1]$ and

$$f(x, y) = \begin{cases} 2^{2n} & \text{if } \dfrac{1}{2^n} \leqslant x < \dfrac{1}{2^{n-1}}, \quad \dfrac{1}{2^n} \leqslant y < \dfrac{1}{2^{n-1}}, \\ -2^{2n+1} & \text{if } \dfrac{1}{2^{n+1}} \leqslant x < \dfrac{1}{2^n}, \quad \dfrac{1}{2^n} \leqslant y < \dfrac{1}{2^{n-1}}, \\ 0 & \text{otherwise.} \end{cases}$$

Prove that the iterated integrals (20) exist but are unequal.

Ans. $\displaystyle\int_{0}^{1} \left(\int_{0}^{1} f(x, y)\, dx \right) dy = 0$, $\displaystyle\int_{0}^{1} \left(\int_{0}^{1} f(x, y)\, dy \right) dx = 1$.

Problem 6. The preceding two problems show that the existence of the iterated integrals (20) does not imply either the existence of the double integral (19) or the validity of formula (15). However, show that the existence of either of the integrals

$$\int_X \left(\int_{A_x} |f(x, y)| \, d\mu_y \right) d\mu_x, \qquad \int_Y \left(\int_{A_y} |f(x, y)| \, d\mu_x \right) d\mu_y \qquad (21)$$

implies both the existence of (19) and the validity of (15).

Hint. Suppose the first of the integrals (21) exists and equals M. The function

$$f_n(x, y) = \min \{|f(x, y)|, n\}$$

is measurable and bounded, and hence summable on A. By Fubini's theorem,

$$\int_A f_n(x, y) \, d\mu = \int_X \left(\int_{A_x} f_n(x, y) \, d\mu_y \right) d\mu_x \leqslant M.$$

Moreover, $\{f_n(x, y)\}$ is a nondecreasing sequence of functions converging to $|f(x, y)|$. Use Levi's theorem to deduce the summability of $|f(x, y)|$ and hence that of $f(x, y)$ on A.

Problem 7. Show that Fubini's theorem continues to hold for the case of σ-finite measures (cf. Sec. 30.2).

36. The Stieltjes Integral

36.1. Stieltjes measures. Let F be a nondecreasing function defined on a closed interval $[a, b]$, and suppose F is continuous from the left at every point of $(a, b]$. Let $\mathscr{S}$ be the semiring of all subintervals (open, closed or half-open) of $[a, b)$, and let m be the measure on $\mathscr{S}$ defined by the formulas[7]

$$\begin{aligned}
m(\alpha, \beta) &= F(\beta) - F(\alpha + 0), \\
m[\alpha, \beta] &= F(\beta + 0) - F(\alpha), \\
m(\alpha, \beta] &= F(\beta + 0) - F(\alpha + 0), \\
m[\alpha, \beta) &= F(\beta) - F(\alpha).
\end{aligned} \qquad (1)$$

Finally, let μ_F be the Lebesgue extension of m, defined on the σ-algebra $\mathscr{S}_{\mu_F}$ of μ_F-measurable sets. In particular, $\mathscr{S}_{\mu_F}$ contains all subintervals of $[a, b)$ and hence all Borel subsets of $[a, b)$. Then μ_F is called the (*Lebesgue-*) *Stieltjes measure* corresponding to the function F, and the function F itself is called the *generating function* of μ_F.

Example 1. The Stieltjes measure corresponding to the generating function $F(x) = x$ is just ordinary Lebesgue measure on the line.

---

[7] To avoid confusion, we omit "outer parentheses," writing $\mu(\alpha, \beta)$ instead of $\mu((\alpha, \beta))$, and similarly in the rest of the formulas (1). Moreover, in $m[\alpha, \beta]$, we allow the case $\alpha = \beta$.

Example 2. Let F be a jump function, with discontinuity points $x_1, x_2, \ldots, x_n, \ldots$ and corresponding jumps $h_1, h_2, \ldots, h_n, \ldots$. Then every subset $A \subset [a, b)$ is μ_F-measurable, with measure

$$\mu_F(A) = \sum_{x_n \in A} h_n. \tag{2}$$

In fact, according to (1), every single-element set $\{x_n\}$ has measure h_n, and moreover it is clear that the measure of the complement of the set $\{x_1, x_2, \ldots, x_n, \ldots\}$ is zero. But then (2) holds, by the σ-additivity of μ_F. A Stieltjes measure μ_F of this type, generated by a jump function, is said to be *discrete*.

Example 3. Let F be an absolutely continuous nondecreasing function on $[a, b)$, with derivative $f = F'$. Then the Stieltjes measure μ_F is defined on all Lebesgue-measurable subsets $A \subset [a, b)$ and

$$\mu_F(A) = \int_A f(x)\, dx. \tag{3}$$

In fact, by Theorem 6, p. 340,

$$\mu_F(\alpha, \beta) = F(\beta) - F(\alpha) = \int_\alpha^\beta f(x)\, dx \tag{4}$$

for every open interval (α, β). But then (3) holds for every Lebesgue-measurable set $A \subset [a, b)$ since a Lebesgue extension of a σ-additive measure is uniquely determined by its values on the original semiring.[8] A Stieltjes measure μ_F of this type, with an absolutely continuous generating function, is itself said to be *absolutely continuous*.

Example 4. Let F be singular (and continuous) as on p. 341. Then the corresponding Stieltjes measure μ_F is concentrated on the set of Lebesgue measure zero where the derivative F' is nonzero or fails to exist. A Stieltjes measure of this type is said to be *singular*.

Example 5. By the Lebesgue decomposition (p. 341), an arbitrary generating function F can be represented as a sum

$$F(x) = D(x) + A(x) + S(x) \tag{5}$$

of a jump function D, an absolutely continuous function A and a singular function S (verify that D, A and S are themselves generating functions). Moreover, each of the "components" D, A and S is uniquely determined to within an additive constant (see Problem 4, p. 342). But clearly

$$\mu_F = \mu_D + \mu_A + \mu_S.$$

---

[8] Give a more detailed argument, recalling Problem 1, p. 279. Note that in this case $m(\alpha, \beta) = m[\alpha, \beta] = m(\alpha, \beta] = m[\alpha, \beta)$.

It follows that an arbitrary Lebesgue-Stieltjes measure can be represented as a sum of a discrete measure μ_D, an absolutely continuous measure μ_A and a singular measure μ_S. Moreover, this representation is unique (why?).

Remark. We can easily extend the notion of a Stieltjes measure on a (finite) interval $[a, b)$ to that of a Stieltjes measure on the whole line $(-\infty, \infty)$. Let F be a bounded nondecreasing function on $(-\infty, \infty)$, so that

$$m \leqslant F(x) \leqslant M \qquad (-\infty < x < \infty).$$

Using the formulas (1) to define the measure of *arbitrary* intervals (open, closed or half-open), not just subintervals of a fixed interval $[a, b)$, we get a finite measure μ_F on the whole line, called a (*Lebesgue-*) *Stieltjes measure*, as before. In particular, we have

$$\mu(-\infty, \infty) = F(\infty) - F(-\infty)$$

for the measure of the whole line, where

$$F(\infty) = \lim_{x \to \infty} F(x), \qquad F(-\infty) = \lim_{\alpha \to -\infty} F(x)$$

(the existence of the limits follows from the fact that F is bounded and monotonic).

36.2. The Lebesgue-Stieltjes integral. Let μ_F be a Stieltjes measure on the interval $[a, b)$, corresponding to the generating function F, and let f be a μ_F-summable function. Then by the *Lebesgue-Stieltjes integral* of f (with respect to F), denoted by

$$\int_a^b f(x)\, dF(x), \tag{6}$$

we simply mean the Lebesgue integral

$$\int_{[a,b)} f(x)\, d\mu_F.$$

Example 1. Let F be the jump function

$$F(x) = \sum_{x_n < x} h_n,$$

so that μ_F is a discrete measure. Then (6) reduces to the sum

$$\sum_n f(x_n) h_n.$$

Example 2. If F is absolutely continuous, then

$$\int_a^b f(x)\, dF(x) = \int_a^b f(x) F'(x)\, dx, \tag{7}$$

where the right-hand side is the integral of fF' with respect to ordinary Lebesgue measure on the line. In the case where $f(x) \equiv$ const, this is an immediate consequence of (4). Moreover, by the σ-additivity of integrals, (7) can be extended to the case of any simple function f which is μ_F-summable. More generally, let $\{f_n\}$ be a sequence of such simple functions converging uniformly to f, so that $\{f_n F'\}$ converges uniformly to fF'. It can be assumed without loss of generality that

$$f_1(x) \leqslant f_2(x) \leqslant \cdots \leqslant f_n(x) \leqslant \cdots ,$$

and hence that

$$f_1(x)F(x) \leqslant f_2(x)F(x) \leqslant \cdots \leqslant f_n(x)F(x) \leqslant \cdots .$$

Therefore, applying Levi's theorem (Theorem 2, p. 305) to both sequences $\{f_n\}$ and $\{f_n F'\}$, we get

$$\int_a^b f(x)\, dF(x) = \lim_{n \to \infty} \int_a^b f_n(x)\, dF(x) = \lim_{n \to \infty} \int_a^b f_n(x)F'(x)\, dx = \int_a^b f(x)F'(x)\, dx.$$

Example 3. Suppose

$$F(x) = D(x) + A(x),$$

where D is the jump function

$$D(x) = \sum_{x_n < x} h_n$$

and A is absolutely continuous. Then it follows from Examples 1 and 2 that

$$\int_a^b f(x)\, dF(x) = \sum_n f(x_n)h_n + \int_a^b f(x)A'(x)\, dx.$$

In the case where F also contains a singular component, as in (5), there is no such representation of the Lebesgue-Stieltjes integral (6) as the sum of a series and an ordinary Lebesgue integral.

Remark. We can easily extend the notion of a Lebesgue-Stieltjes integral with respect to a nondecreasing function F to that of a Lebesgue-Stieltjes integral with respect to an arbitrary function of bounded variation Φ. In fact, as in Theorem 4, p. 331, let

$$\Phi = v - g,$$

where v, the total variation of Φ on the interval $[a, x]$, and $g = v - \Phi$ are both nondecreasing. We then set

$$\int_a^b f(x)\, d\Phi(x) = \int_a^b f(x)\, dv(x) - \int_a^b f(x)\, dg(x) \tag{8}$$

by definition (see Problem 2).

36.3. Applications to probability theory. The Lebesgue-Stieltjes integral is widely used in mathematical analysis and its applications. The concept plays a particularly important role in probability theory. Given a random variable ξ,[9] let

$$F(x) = \mathbf{P}\{\xi < x\},$$

i.e., let $F(x)$ be the probability that ξ takes a value less than x. Then F is clearly nondecreasing and continuous from the left. Moreover, F satisfies the conditions

$$F(-\infty) = 0, \qquad F(\infty) = 1$$

(why?). Conversely, every such function f can be represented as the probability distribution of some random variable ξ.

Two basic numerical characteristics of a random variable ξ are its *mathematical expectation* or *mean* (*value*)

$$\mathbf{E}\xi = \int_{-\infty}^{\infty} x \, dF(x), \tag{9}$$

and *variance*

$$\mathbf{D}\xi = \int_{-\infty}^{\infty} (x - \mathbf{E}\xi)^2 \, dF(x) \tag{10}$$

(however, see Problem 5).

Example 1. A random variable ξ is said to be *discrete* if it can take no more than countably many values $x_1, x_2, \ldots, x_n, \ldots$. For example, the number of calls received on a given telephone line during a given time interval is a discrete random variable. Let

$$p_n = \mathbf{P}\{\xi = x_n\} \qquad (n = 1, 2, \ldots)$$

be the probability of the random variable ξ taking the value x_n. Then the distribution function of ξ is just the jump function

$$F(x) = \sum_{x_n < x} p_n.$$

In this case, the integrals (9) and (10) for the mean and variance of ξ reduce to the sums

$$\mathbf{E}\xi = \sum_n x_n p_n,$$

$$\mathbf{D}\xi = \sum_n (x_n - a)^2 p_n \qquad (a = \mathbf{E}\xi).$$

Example 2. A random variable ξ is said to be *continuous* if its distribution function F is absolutely continuous. The derivative

$$p(x) = F'(x)$$

---

[9] We presuppose familiarity with the rudiments of probability theory. See e.g., Y. A. Rozanov, *Introductory Probability Theory* (translated by R. A. Silverman), Prentice-Hall, Inc., Englewood Cliffs, N.J. (1969).

of the distribution function is then called the *probability density* of ξ. It follows from Example 2, p. 364 that in this case the integrals (9) and (10) for the mean and variance of ξ reduce to the following integrals with respect to ordinary Lebesgue measure on the line:

$$\mathbf{E}\xi = \int_{-\infty}^{\infty} xp(x)\, dx,$$

$$\mathbf{D}\xi = \int_{-\infty}^{\infty} (x-a)^2 p(x)\, dx \qquad (a = \mathbf{E}\xi).$$

36.4. The Riemann-Stieltjes integral. Besides the Lebesgue-Stieltjes integral introduced in Sec. 36.2 (which is in effect nothing but the difference between two ordinary Lebesgue integrals with respect to two measures on the real line[10]), we can also introduce the Riemann-Stieltjes integral, defined as a limit of certain approximating sums, analogous to those used to define the ordinary Riemann integral. To this end, let f and Φ be two functions on $[a, b]$, where Φ is of bounded variation and continuous from the left, and let

$$a = x_0 < x_1 < x_2 < \cdots < x_n = b$$

be a partition of the interval $[a, b]$ by points of subdivision $x_0, x_1, x_2, \ldots, x_n$. Choosing an arbitrary point ξ_k in each subinterval $[x_{k-1}, x_k]$, we form the sum

$$\sum_{k=1}^{n} f(\xi_k)[\Phi(x_k) - \Phi(x_{k-1})]. \tag{11}$$

Suppose that as the partition is "refined," i.e., as the quantity

$$\max \{x_1 - x_0, x_2 - x_1, \ldots, x_n - x_{n-1}\} \tag{12}$$

(equal to the maximum length of the subintervals) approaches zero, the sum (11) approaches a limit independent of the choice of both the points of subdivision x_k and the "intermediate points" ξ_k. Then this limit is called the *Riemann-Stieltjes integral* of f with respect to Φ, and is denoted by

$$\int_a^b f(x)\, d\Phi(x)$$

(just as in the case of the Lebesgue-Stieltjes integral).

Remark. If $\Phi = \Phi_1 + \Phi_2$, then

$$\int_a^b f(x)\, d\Phi(x) = \int_a^b f(x)\, d\Phi_1(x) + \int_a^b f(x)\, d\Phi_2(x) \tag{13}$$

(provided the integrals on the right exist). In fact, we need only write the

---

[10] Recall formula (8).

identity

$$\sum_{k=1}^{n} f(\xi_k)[\Phi(x_k) - \Phi(x_{k-1})]$$

$$= \sum_{k=1}^{n} f(\xi_k)[\Phi_1(x_k) - \Phi_1(x_{k-1})] + \sum_{k=1}^{n} f(\xi_k)[\Phi_2(x_k) - \Phi_2(x_{k-1})],$$

and then pass to the limit as the quantity (12) approaches zero.

THEOREM 1. *If f is continuous on $[a, b]$, then its Riemann-Stieltjes integral exists and coincides with its Lebesgue-Stieltjes integral.*

Proof. The sum (11) can be regarded as the Lebesgue-Stieltjes integral of the step function

$$f_n(x) = \xi_k \quad \text{if} \quad x_{k-1} \leqslant x < x_k \quad (k = 1, \ldots, n).$$

As the partition of $[a, b]$ is refined, the sequence $\{f_n\}$ converges uniformly to f (why?). Hence, by the very definition of the Lebesgue integral (recall p. 294),

$$\lim_{n \to \infty} \int_a^b f_n(x)\, dx = I,$$

where I is the Lebesgue-Stieltjes integral of f over $[a, b)$. But then

$$\lim_{n \to \infty} \sum_{i=1}^{n} f(x_k)[\Phi(x_k) - \Phi(x_{k-1})] = I,$$

where the limit on the left is the Riemann-Stieltjes integral of f over $[a, b]$. ∎

THEOREM 2. *If f is continuous on $[a, b]$, then*

$$\left| \int_a^b f(x)\, d\Phi(x) \right| \leqslant V_a^b(f) \max_{a \leqslant x \leqslant b} |f(x)|, \qquad (14)$$

where $V_a^b(\Phi)$ is the total variation of Φ on $[a, b]$.

Proof. The inequality

$$\left| \sum_{k=1}^{n} f(\xi_k)[\Phi(x_k) - \Phi(x_{k-1})] \right| \leqslant \sum_{k=1}^{n} |f(\xi_k)|\, |\Phi(x_k) - \Phi(x_{k-1})|$$

$$\leqslant \max_{a \leqslant x \leqslant b} |f(x)| \sum_{k=1}^{n} |\Phi(x_k) - \Phi(x_{k-1})| \leqslant V_a^b(f) \max_{a \leqslant x \leqslant b} |f(x)|$$

holds for any partition of the interval $[a, b]$. Taking the limit of the left-hand side as $\max\{x_1 - x_0, \ldots, x_n - x_{n-1}\} \to 0$, we get (14). ∎

Remark. If $\Phi(x) = x$, (14) reduces to the familiar estimate

$$\left| \int_a^b f(x)\, dx \right| \leqslant (b - a) \max_{a \leqslant x \leqslant b} |f(x)|$$

for the ordinary Riemann integral.

THEOREM 3. *Let Φ be a function of bounded variation on $[a, b]$, different from zero at no more than countably many points $c_1, c_2, \ldots, c_n, \ldots$ in (a, b). Then*

$$\int_a^b f(x) \, d\Phi(x) = 0 \tag{15}$$

for any function f continuous on $[a, b]$.

Proof. The assertion is obvious if Φ is nonzero at only a single point $c_1 \in (a, b)$, since then

$$\sum_{k=1}^n f(x_k)[\Phi(x_k) - \Phi(x_{k-1})] = 0$$

for an "arbitrarily fine" partition

$$a = x_0 < x_1 < \cdots < x_n = b,$$

i.e., a partition for which the quantity (12) is arbitrarily small, provided we make sure that c_1 is not one of the points of subdivision $x_0, x_1, \ldots, x_n$.[11] Hence, by (13), the assertion is also true if Φ is nonzero at only finitely many points in (a, b). Now suppose Φ is different from zero at countably many points

$$c_1, c_2, \ldots, c_n, \ldots$$

in (a, b), and let

$$y_n = \Phi(c_n).$$

Then

$$\sum_{n=1}^\infty |y_n| < \infty,$$

since Φ is of bounded variation. Given any $\varepsilon > 0$, we choose N such that

$$\sum_{n=N+1}^\infty |y_n| < \varepsilon,$$

and write Φ in the form

$$\Phi = \Phi_N + \Phi^*, \tag{16}$$

where Φ_N takes the values $y_1, \ldots, y_N$ at the points $c_1, \ldots, c_N$ and is zero elsewhere, while Φ^* takes the values $y_{N+1}, y_{N+2}, \ldots$ at the points $c_{N+1}, c_{N+2}, \ldots$ and is zero elsewhere. Then, as just shown,

$$\int_a^b f(x) \, d\Phi_N(x) = 0. \tag{17}$$

Moreover

$$\left| \sum_{k=1}^m f(\xi_k)[\Phi^*(x_k) - \Phi^*(x_{k-1})] \right| \leqslant 2M \sum_{n=N+1}^\infty |y_n| < 2M\varepsilon,$$

---

[11] Note that here we rely on the fact that c_1 is not an end point of $[a, b]$.

where
$$M = \max_{a \leqslant x \leqslant b} |f(x)|,$$
or
$$\left| \int_a^b f(x) \, d\Phi^*(x) \right| < 2M\varepsilon$$

after taking the limit as $m \to \infty$. This in turn implies
$$\int_a^b f(x) \, d\Phi(x) = 0, \tag{18}$$

since $\varepsilon > 0$ is arbitrary. Formula (15) now follows at once from (13) and (16)–(18). ∎

36.5. Helly's theorems. In Sec. 30.1 we found conditions insuring the validity of passing to the limit in Lebesgue integrals, i.e., conditions under which
$$\lim_{n \to \infty} \int_A f_n(x) \, d\mu = \int_A f(x) \, d\mu, \tag{19}$$

where $\{f_n\}$ is a sequence of functions converging (almost everywhere) to a function f and the integrals are all with respect to a fixed measure μ. In the case of Stieltjes integrals, we now ask a closely related but somewhat different question: Under what conditions does the formula
$$\lim_{n \to \infty} \int_a^b f(x) \, d\Phi_n(x) = \int_a^b f(x) \, d\Phi(x) \tag{20}$$

hold, where f is continuous and $\{\Phi_n\}$ is a sequence of functions of bounded variation converging (everywhere) to a function Φ? (Note that here, unlike (19), the function f is fixed, and it is the function Φ_n, or the corresponding Stieltjes measure, which varies.) The answer to this question is given by

THEOREM 4 (*Helly's convergence theorem*). *Let $\{\Phi_n\}$ be a sequence of functions of bounded variation on $[a, b]$, converging to a function Φ at every point of $[a, b]$. Suppose the sequence of total variations $\{V_a^b(\Phi_n)\}$ is bounded, so that*
$$V_a^b(\Phi_n) \leqslant C \qquad (n = 1, 2, \ldots) \tag{21}$$

for some constant $C > 0$. Then Φ is also of bounded variation on $[a, b]$, and (20) holds for every function f continuous on $[a, b]$.

Proof. Let
$$a = x_0 < x_1 < \cdots < x_m = b$$

be any partition of the interval $[a, b]$ by points of subdivision $x_0, x_1, \ldots, x_m$. Then
$$\sum_{k=1}^m |\Phi(x_k) - \Phi(x_{k-1})| = \lim_{n \to \infty} \sum_{k=1}^m |\Phi_n(x_k) - \Phi_n(x_{k-1})| \leqslant C,$$

and hence

$$V_a^b(\Phi) \leqslant C, \tag{22}$$

i.e., Φ is of bounded variation on $[a, b]$, as asserted.

Next we show that (20) holds if f is a step function. Suppose

$$f(x) = h_k \qquad \text{if} \quad x_{k-1} \leqslant x < x_k.$$

Then

$$\int_a^b f(x)\, d\Phi_n(x) = \sum_k h_k [\Phi_n(x_k) - \Phi_n(x_{k-1})] \tag{23}$$

and[12]

$$\int_a^b f(x)\, d\Phi(x) = \sum_k h_k [\Phi(x_k) - \Phi(x_{k-1})], \tag{24}$$

where obviously (23) approaches (24) as $n \to \infty$. Now let f be continuous on $[a, b]$. Given any $\varepsilon > 0$, choose a step function f_ε such that

$$|f(x) - f_\varepsilon(x)| < \frac{\varepsilon}{3C} \qquad (a \leqslant x < b) \tag{25}$$

(why is this possible?). Then

$$\left| \int_a^b f(x)\, d\Phi(x) - \int_a^b f(x)\, d\Phi_n(x) \right| \leqslant |I_1| + |I_2| + |I_3|, \tag{26}$$

where

$$I_1 = \int_a^b f(x)\, d\Phi(x) - \int_a^b f_\varepsilon(x)\, d\Phi(x),$$

$$I_2 = \int_a^b f_\varepsilon(x)\, d\Phi(x) - \int_a^b f_\varepsilon(x)\, d\Phi_n(x),$$

$$I_3 = \int_a^b f_\varepsilon(x)\, d\Phi_n(x) - \int_a^b f(x)\, d\Phi_n(x).$$

By the inequality (14), which clearly holds for Lebesgue-Stieltjes integrals as well as for Riemann-Stieltjes integrals (why?), we have

$$\begin{aligned}
|I_1| &\leqslant \int_a^b |f(x) - f_\varepsilon(x)|\, d\Phi(x) < \frac{\varepsilon}{3C} V_a^b(\Phi) \leqslant \frac{\varepsilon}{3}, \\
|I_3| &\leqslant \int_a^b |f_\varepsilon(x) - f(x)|\, d\Phi_n(x) < \frac{\varepsilon}{3C} V_a^b(\Phi_n) \leqslant \frac{\varepsilon}{3},
\end{aligned} \tag{27}$$

after using (21), (22) and (25). Moreover, as just shown,

$$|I_2| < \frac{\varepsilon}{3} \tag{28}$$

---

[12] Think of (23) and (24) as Lebesgue-Stieltjes integrals.

for sufficiently large n. It follows from (26)–(28) that

$$\left| \int_a^b f(x)\, d\Phi(x) - \int_a^b f(x)\, d\Phi_n(x) \right| < \varepsilon,$$

which implies (20), since $\varepsilon > 0$ is arbitrary. ∎

Theorem 1 gives conditions under which we can take the limit of a sequence $\{\Phi_n\}$ of functions of bounded variation inside a Stieltjes integral. The next theorem gives conditions guaranteeing the existence of a sequence $\{\Phi_n\}$ meeting the requirements of Theorem 4.

THEOREM 5 (*Helly's selection principle*). *Let Φ be a family of functions defined on an interval $[a, b]$ and satisfying the conditions*

$$V_a^b(\varphi) \leqslant C, \qquad \sup_{a \leqslant x \leqslant b} |\varphi(x)| \leqslant M \tag{29}$$

for suitable C and M. Then Φ contains a sequence which converges for every $x \in [a, b]$.

Proof. It is enough to prove the theorem for nondecreasing functions. In fact, let

$$\varphi = v - g,$$

where v is the total variation of φ on $[a, x]$. Then the functions v corresponding to all $\varphi \in \Phi$ are nondecreasing and satisfy the conditions of the theorem, since

$$V_a^b(v) = V_a^b(\varphi) \leqslant C, \qquad \sup_{a \leqslant x \leqslant b} |v(x)| \leqslant C.$$

Assuming that the theorem holds for nondecreasing functions, we choose a sequence $\{\varphi_n\}$ from Φ such that v_n converges to a limit v^* on $[a, b]$. Then the functions

$$g_n = v_n - \varphi_n$$

are also nondecreasing and satisfy the conditions of the theorem (why?). Therefore $\{\varphi_n\}$ contains a subsequence $\{\varphi_{n_k}\}$ such that $\{g_{n_k}\}$ converges to a limit g^* on $[a, b]$. But then

$$\lim_{n \to \infty} \varphi_{n_k}(x) = \varphi^*(x),$$

where

$$\varphi^*(x) = v^*(x) - g^*(x).$$

Thus we now proceed to prove the theorem for nondecreasing functions. Let $r_1, r_2, \ldots, r_n, \ldots$ be the rational points of $[a, b]$. It follows from (29) that the set of numbers

$$\varphi(r_1) \qquad (\varphi \in \Phi)$$

is bounded. Hence there is a sequence of functions $\{\varphi_n^{(1)}\}$ converging at the point r_1. Similarly, $\{\varphi_n^{(1)}\}$ contains a subsequence $\{\varphi_n^{(2)}\}$ converging at the point r_2 as well as at r_1, $\{\varphi_n^{(2)}\}$ contains a subsequence $\{\varphi_n^{(3)}\}$ converging at the point r_3 as well as at r_1 and r_2, and so on. The "diagonal sequence"

$$\{\psi_n\} = \{\varphi_n^{(n)}\}$$

will then converge at every rational point of $[a, b]$. The limit of this sequence is a nondecreasing function ψ, defined only at the points $r_1, r_2, \ldots, r_n, \ldots$. We complete the definition of ψ at the remaining points of $[a, b]$ by setting

$$\psi(x) = \lim_{\substack{r \to x-0 \\ r \text{ rational}}} \psi(r) \text{ if } x \text{ is irrational.}$$

The resulting function ψ is then the limit of $\{\psi_n\}$ at every continuity point of ψ. In fact, let x^* be such a point. Then, given any $\varepsilon > 0$, there is a $\delta > 0$ such that

$$|\psi(x^*) - \psi(x)| < \frac{\varepsilon}{6} \tag{30}$$

if

$$|x^* - x| < \delta.$$

Let r and r' be rational numbers such that

$$x^* - \delta < r' < x^* < r'' < x^* + \delta,$$

and let n be so large that

$$|\psi_n(r') - \psi(r')| < \frac{\varepsilon}{6}, \qquad |\psi_n(r'') - \psi(r'')| < \frac{\varepsilon}{6}. \tag{31}$$

It follows from (30) and (31) that

$$|\psi_n(r') - \psi_n(r'')| < \frac{2\varepsilon}{3}.$$

Since ψ_n is a nondecreasing function, we have

$$\psi_n(r') \leqslant \psi_n(x^*) \leqslant \psi_n(r''),$$

and hence

$$|\psi(x^*) - \psi_n(x^*)| \leqslant |\psi(x^*) - \psi(r')| + |\psi(r') - \psi_n(r')|$$
$$+ |\psi_n(r') - \psi_n(x^*)| \leqslant \frac{\varepsilon}{6} + \frac{\varepsilon}{6} + \frac{2\varepsilon}{3} = \varepsilon.$$

Therefore

$$\lim_{n \to \infty} \psi_n(x^*) = \psi(x^*),$$

since $\varepsilon > 0$ is arbitrary.

Thus we have constructed a sequence $\{\psi_n\}$ of functions in Φ converging to a limit function ψ everywhere except possibly at discontinuity points of ψ. Since there are no more than countably many such points (why?), we can again use the "diagonal process" to find a subsequence of $\{\psi_n\}$ which converges at these points as well, and hence converges everywhere on $[a, b]$. ∎

36.6. The Riesz representation theorem. Next we show how Stieltjes integrals can be used to represent the general linear functional on the space $C_{[a,b]}$ of all functions continuous on the interval $[a, b]$:

·THEOREM 6 (**F. Riesz**). *Every continuous linear functional φ on the space $C_{[a,b]}$ can be represented in the form*

$$\varphi(f) = \int_a^b f(x)\, d\Phi(x), \tag{32}$$

where Φ is a function of bounded variation on $[a, b]$, and moreover

$$\|\varphi\| = V_a^b(\Phi). \tag{33}$$

Proof. The space $C_{[a,b]}$ can be regarded as a subspace of the space $M_{[a,b]}$ of all bounded functions on $[a, b]$, with the same norm

$$\|f\| = \sup_{a \leqslant x \leqslant b} |f(x)|$$

as in $C_{[a,b]}$. Let φ be a continuous linear functional on $C_{[a,b]}$. By the Hahn-Banach theorem (Theorem 5, p. 180), φ can be extended without changing its norm from $C_{[a,b]}$ onto the whole space $M_{[a,b]}$. In particular, this extended functional will be defined on all functions of the form

$$f_\tau(x) = \begin{cases} 1 & \text{if } x < \tau, \\ 0 & \text{if } x \geqslant \tau \end{cases} \qquad (a \leqslant \tau \leqslant b). \tag{34}$$

Let

$$\Phi(\tau) = \varphi(f_\tau). \tag{35}$$

Then Φ is of bounded variation on $[a, b]$. In fact, given any partition

$$a = x_0 < x_1 < \cdots < x_n = b \tag{36}$$

of $[a, b]$, let

$$\alpha_k = \text{sgn}\,[\Phi(x_k) - \Phi(x_{k-1})] \qquad (k = 1, \ldots, n),$$

where

$$\text{sgn}\, x = \begin{cases} 1 & \text{if } x > 0, \\ 0 & \text{if } x = 0, \\ -1 & \text{if } x < 0. \end{cases}$$

Then

$$\sum_{k=1}^{n} |\Phi(x_k) - \Phi(x_{k-1})| = \sum_{k=1}^{n} \alpha_k [\Phi(x_k) - \Phi(x_{k-1})]$$

$$= \sum_{k=1}^{n} \alpha_k \varphi(f_{x_k} - f_{x_{k-1}}) = \varphi\left(\sum_{k=1}^{n} \alpha_k(f_{x_k} - f_{x_{k-1}})\right)$$

$$\leqslant \|\varphi\| \left\| \sum_{k=1}^{n} \alpha_k(f_{x_k} - f_{x_{k-1}}) \right\|.$$

But the function

$$\sum_{k=1}^{n} \alpha_k(f_{x_k} - f_{x_{k-1}})$$

can only take the values $0, \pm 1$, and hence its norm equals 1. Therefore

$$\sum_{k=1}^{n} |\Phi(x_k) - \Phi(x_{k-1})| \leqslant \|\varphi\|.$$

Since this is true for any partition of $[a, b]$, we have

$$V_a^b(\Phi) \leqslant \|\varphi\|, \tag{37}$$

i.e., Φ is of bounded variation on $[a, b]$, as asserted.

We now show that the functional φ can be represented in the form of a Stieltjes integral with respect to the function Φ just constructed. Let f be any function continuous on $[a, b]$. Given any $\varepsilon > 0$, let $\delta > 0$ be such that $|x' - x''| < \delta$ implies $|f(x') - f(x'')| < \varepsilon$. Suppose the partition (36) is such that each subinterval $[x_{k-1}, x_k]$ is of length less than δ, and consider the step function

$$f^{(\varepsilon)}(x) = f(x_k) \quad \text{if} \quad x_{k-1} \leqslant x < x_k \qquad (k = 1, \ldots, n),$$

which can obviously be written in the form

$$f^{(\varepsilon)}(x) = \sum_{k=1}^{n} f(x_k)[f_{x_k}(x) - f_{x_{k-1}}(x)], \tag{38}$$

where f_τ is the function defined by (34). Clearly,

$$|f(x) - f^{(\varepsilon)}(x)| < \varepsilon$$

for all $x \in [a, b]$,[13] i.e.,

$$\|f - f^{(\varepsilon)}\| < \varepsilon. \tag{39}$$

It follows from (35) and (38) that

$$\varphi(f^{(\varepsilon)}) = \sum_{k=1}^{n} f(x_k)[\varphi(f_{x_k}) - \varphi(f_{x_{k-1}})] = \sum_{k=1}^{n} f(x_k)[\Phi(x_k) - \Phi(x_{k-1})],$$

---

[13] We complete the definition of $f^{(\varepsilon)}$ by setting $f^{(\varepsilon)}(b) = f(x_n) = f(b)$ for every $\varepsilon > 0$.

i.e., $\varphi(f^{(\varepsilon)})$ is an "approximating sum" of the Riemann-Stieltjes integral

$$\int_a^b f(x)\, d\Phi(x).$$

Therefore

$$\left| \varphi(f^{(\varepsilon)}) - \int_a^b f(x)\, d\Phi(x) \right| < \varepsilon$$

for a "sufficiently fine" partition of the interval $[a, b]$. On the other hand,

$$|\varphi(f) - \varphi(f^{(\varepsilon)})| \leqslant \|\varphi\| \, \|f - f^{(\varepsilon)}\| \leqslant \|\varphi\| \, \varepsilon$$

because of (39). But then

$$\left| \varphi(f) - \int_a^b f(x)\, d\Phi(x) \right| < (\|\varphi\| + 1)\varepsilon,$$

which implies (32), since $\varepsilon > 0$ is arbitrary. To prove (33), we merely combine (37) with the opposite inequality

$$\|\varphi\| \leqslant V_a^b(\Phi),$$

which is an immediate consequence of Theorem 2 and the representation (32). ∎

Problem 1. Let μ be an arbitrary finite σ-additive measure on the real line $(-\infty, \infty)$. Represent μ as the Stieltjes measure corresponding to some generating function F.

Hint. Let $F(x) = \mu(-\infty, x)$.

Comment. Thus the term "Stieltjes measure" does not refer to a special kind of measure, but rather to a special way of constructing a measure (by using a generating function).

Problem 2. Let Φ be a function of bounded variation with two distinct representations $\Phi = v - g$, $\Phi = v^* - g^*$ in terms of nondecreasing functions v, g, v^* and g^* (give an example). Prove that

$$\int_a^b f(x)\, dv(x) - \int_a^b f(x)\, dg(x) = \int_a^b f(x)\, dv^*(x) - \int_a^b f(x)\, dg^*(x).$$

Comment. Thus in the definition (8) of the Lebesgue-Stieltjes integral with respect to a function of bounded variation Φ, the particular representation of Φ as a difference between two nondecreasing functions does not matter, i.e., v need not be the total variation of Φ on $[a, x]$.

Problem 3. Let ξ be the number of spots obtained in throwing an unbiased die. Find the mean and variance of ξ.

Ans. $\mathbf{E}\xi = \frac{7}{2}$, $\mathbf{D}\xi = \frac{35}{12}$.

Problem 4. Find the mean and variance of the random variable ξ with probability density

$$p(x) = \tfrac{1}{2}e^{-|x|} \qquad (-\infty < x < \infty).$$

Problem 5. Let ξ be the random variable with probability density

$$p(x) = \frac{1}{\pi(1 + x^2)} \qquad (-\infty < x < \infty).$$

Prove that $\mathbf{E}\xi$ and $\mathbf{D}\xi$ fail to exist.

Problem 6. Discuss random variables which are neither discrete nor continuous.

Problem 7. Given a random variable ξ with distribution function F, consider the new random variable $\eta = \varphi(\xi)$, where φ is a function summable with respect to the Stieltjes measure μ_F generated by F. Express $\mathbf{E}\xi$ and $\mathbf{D}\xi$ in terms of F.

Hint. Consider the problem of changing variables in a Lebesgue integral.

Ans. For example, $\mathbf{E}\xi = \int_{-\infty}^{\infty} \varphi(x)\, dF(x)$.

Problem 8. Prove that if f is continuous on $[a, b]$, then the Riemann-Stieltjes integral

$$\int_a^b f(x)\, d\Phi(x) \tag{40}$$

does not depend on the values taken by Φ at its discontinuity points in (a, b).

Hint. Use Theorem 3 and formula (13).

Comment. Hence if f is continuous, we need not insist that Φ be continuous from the left at its discontinuity points in (a, b). In fact, Φ can be assigned arbitrary values at these points.

Problem 9. Write formulas for the Riemann-Stieltjes integral (40) in the case where f is continuous and

a) Φ is a jump function;
b) Φ is an absolutely continuous function with a Riemann-integrable derivative.

Problem 10. Evaluate the following Riemann-Stieltjes integrals:

a) $\int_{-1}^{3} x \, dF(x)$, where $F(x) = \begin{cases} 0 & \text{if } x = -1, \\ 1 & \text{if } -1 < x < 2, \\ -1 & \text{if } 2 \leqslant x \leqslant 3; \end{cases}$

b) $\int_{0}^{2} x^2 \, dF(x)$, where $F(x) = \begin{cases} -1 & \text{if } 0 \leqslant x < \frac{1}{2}, \\ 0 & \text{if } \frac{1}{2} \leqslant x < \frac{3}{2}, \\ 2 & \text{if } x = \frac{3}{2}, \\ -2 & \text{if } \frac{3}{2} < x \leqslant 2; \end{cases}$

c) $\int_{0}^{1} x^2 \, dF(x)$, where $F(x) = \begin{cases} x & \text{if } 0 \leqslant x \leqslant \frac{1}{2}, \\ \dfrac{1}{x} & \text{if } \frac{1}{2} < x \leqslant 1. \end{cases}$

Problem 11. Develop a theory of Riemann-Stieltjes integration on the whole real line $(-\infty, \infty)$.

Problem 12. Extend Theorem 4 to the case where $a = -\infty$ or $b = \infty$ (or both), assuming that $f(x)$ approaches a limit as $x \to \pm\infty$.

Problem 13. Let $\{\Phi_n\}$ be the same as in Theorem 4, and let $\{f_n\}$ be a sequence of continuous functions on $[a, b]$ converging uniformly to a limit f. Prove that

$$\lim_{n \to \infty} \int_a^b f_n(x) \, d\Phi_n(x) = \int_a^b f(x) \, d\Phi(x).$$

Problem 14. Prove that there is a one-to-one correspondence between the set of all continuous linear functionals φ on $C_{[a,b]}$ and the space $V_{[a,b]}^0$ of Problem 8, p. 332, provided we identify any two elements of $V_{[a,b]}^0$ which coincide at all their continuity points. Prove that the inequality

$$V_a^b(\Phi) \leqslant \|\varphi\|$$

need not hold for every $\Phi \in V_{[a,b]}^0$ corresponding to a given functional $\varphi \in C_{[a,b]}$, but that there is always at least one such element Φ for which the inequality holds.

37. The Spaces L_1 and L_2

37.1. Definition and basic properties of L_1. Let X be a space equipped with a measure μ, where the measure of X itself may be either finite or infinite. Then by $L_1(X, \mu)$, or simply L_1, we mean the set of all real functions

f summable on X (however, see Problem 1). Clearly L_1 is a linear space (with addition of functions and multiplication of functions by numbers defined in the usual way), since a linear combination of summable functions is again a summable function. To introduce a norm in L_1, we define

$$\|f\| = \int |f(x)|\, d\mu, \tag{1}$$

where, as in the rest of this section, the symbol $\int$ by itself denotes integration over the whole space X. Of the various properties of a norm (see p. 138), it follows at once from (1) that

$$\|f\| \geqslant 0,$$

$$\|\alpha f\| = |\alpha|\, \|f\|.$$

$$\|f_1 + f_2\| \leqslant \|f_1\| + \|f_2\|,$$

and we need only verify that $\|f\| = 0$ if and only if $f = 0$. To insure this, we agree to regard equivalent functions (i.e., functions differing only on a set of measure zero) as identical elements of the space L_1. Thus the elements of L_1 are, to be perfectly exact, classes of equivalent summable functions.[14] In particular, the zero element of L_1 is the class consisting of all functions vanishing almost everywhere. With this understanding, we will continue to talk (more casually) about "functions in L_1."

In L_1, as in any normed linear space, we can use the formula

$$\rho(f, g) = \|f - g\|$$

to define a distance. Let $\{f_n\}$ be a sequence of functions in L_1. Then $\{f_n\}$ is said to *converge in the mean* to a function $f \in L_1$ if $\rho(f_n, f) \to 0$ as $n \to \infty$.

THEOREM 1. *The space L_1 is complete.*

Proof. Let $\{f_n\}$ be a Cauchy sequence in L_1, so that

$$\|f_m - f_n\| \to 0 \text{ as } m, n \to \infty.$$

Then we can find a sequence of indices $\{n_k\}$ (where $n_1 < n_2 < \cdots < n_k < \cdots$) such that

$$\|f_{n_k} - f_{n_{k+1}}\| = \int |f_{n_k}(x) - f_{n_{k+1}}(x)|\, d\mu < \frac{1}{2^k} \qquad (k = 1, 2, \ldots).$$

It follows from the corollary to Levi's theorem (see p. 307) that the series

$$|f_{n_1}| + |f_{n_2} - f_{n_1}| + \cdots$$

---

[14] Thus the precise definition of addition of two elements $\varphi_1, \varphi_2 \in L_1$ is the following: Let f_1 and f_2 be "representatives" of φ_1 and φ_2, respectively, i.e., let $f_1 \in \varphi_1, f_2 \in \varphi_2$. Then $\varphi_1 + \varphi_2$ is the class containing $f_1 + f_2$ (this class clearly does not depend on the particular choice of f_1 and f_2).

converges almost everywhere on X. Therefore the series

$$f_{n_1} + f_{n_2} - f_{n_1} + \cdots$$

also converges almost everywhere on X to some function

$$f(x) = \lim_{k \to \infty} f_{n_k}(x).$$

But $\{f_{n_k}\}$ converges in the mean to the same function f. In fact, given any $\varepsilon > 0$,

$$\int |f_{n_k}(x) - f_{n_l}(x)| \, d\mu < \varepsilon \qquad (2)$$

for sufficiently large k and l, since $\{f_n\}$ is a Cauchy sequence. Hence, by Fatou's theorem (Theorem 3, p. 307), we can take the limit as $l \to \infty$ behind the integral sign in (2), obtaining

$$\int |f_{n_k}(x) - f(x)| \, d\mu \leqslant \varepsilon.$$

It follows that $f \in L_1$ (why?) and that $f_{n_k} \to f$ in the mean. But if a Cauchy sequence contains a subsequence converging to a limit, then the sequence itself must converge to the same limit. Hence $f_n \to f$ in the mean. ∎

According to the definition of the Lebesgue integral (see p. 296), given any function f summable on X and any $\varepsilon > 0$, there is a summable simple function $\varphi(x)$ such that

$$\int |f(x) - \varphi(x)| < \varepsilon.$$

Moreover, the Lebesgue integral of a summable simple function φ taking values $y_1, y_2, \ldots$ on sets $E_1, E_2, \ldots$ is defined as the sum of the series

$$\sum_{n=1}^{\infty} y_n \mu(E_n)$$

(assumed to converge absolutely). Therefore every summable simple function can be represented as the limit in the mean (i.e., as the limit in the sense of convergence in the mean) of a sequence of summable simple functions, each taking only *finitely* many values. In fact, given any $\varepsilon > 0$, let N be such that

$$\sum_{n=N+1}^{\infty} |y_n| \, \mu(E_n) < \varepsilon,$$

and let[15]

$$\varphi_N(x) = \begin{cases} y_k & \text{if } x \in E_k, \, 1 \leqslant k \leqslant N, \\ 0 & \text{otherwise.} \end{cases}$$

---

[15] Note that φ_N is a finite linear combination of characteristic functions, namely

$$\varphi_N(x) = y_1 \chi_{E_1}(x) + \cdots + y_N \chi_{E_N}(x)$$

(see footnote 11, p. 349).

Then

$$\int |\varphi(x) - \varphi_N(x)| \, d\mu \leqslant \sum_{n=N+1}^{\infty} |y_n| \, \mu(E_n) < \varepsilon.$$

In other words, the set of all simple functions taking only finitely many values is everywhere dense in the space L_1.

THEOREM 2. *Let X be a metric space equipped with a measure μ such that*[16]

1) *Every open set and every closed set in X is measurable*;
2) *If a set $M \subset X$ is measurable, then*

$$\mu(M) = \inf_{M \subset G} \mu(G), \tag{3}$$

where the greatest lower bound is taken over all open sets $G \subset X$ containing M.

Then the set of all continuous functions on X is everywhere dense in

$$L_1(X, \mu).$$

Proof. We need only show that every simple function taking only finitely many values is the limit in the mean of a sequence of continuous functions. But every simple function taking only finitely many values is a finite linear combination of characteristic functions of measurable sets, and hence we need only show that every such characteristic function $\chi_M(x)$ is the limit in the mean of a sequence of continuous functions. If $M \subset X$ is measurable, then (3) implies that given any $\varepsilon > 0$, there is a closed set F_M and an open set G_M such that

$$F_M \subset M \subset G_M, \qquad \mu(G_M) - \mu(F_M) < \varepsilon. \tag{4}$$

Now let[17]

$$\varphi_\varepsilon(x) = \frac{\rho(X - G_M, x)}{\rho(X - G_M, x) + \rho(F_M, x)}$$

Then

$$\varphi_\varepsilon(x) = \begin{cases} 0 & \text{if} \quad x \in X - G_M, \\ 1 & \text{if} \quad x \in F_M. \end{cases}$$

Moreover, φ_ε is continuous, since $\rho(F_M, x)$ and $\rho(X - G_M, x)$ are both continuous functions, with a nonvanishing sum. But $|\chi_M - \varphi_\varepsilon|$ does not exceed 1 on $G_M - F_M$, and vanishes outside this set. Using (4), we find that

$$\int |\chi_M(x) - \varphi_\varepsilon(x)| \, d\mu < \varepsilon. \quad \blacksquare$$

---

[16] These conditions are satisfied by ordinary Lebesgue measure in n-space, and in many other cases of practical interest.

[17] As usual, $\rho(A, x)$ denotes the distance between the set A and the point x (see Problem 9, p. 54).

The space $L_1(X, \mu)$ depends on the choice of both the space X and the measure μ. For example, $L_1(X, \mu)$ is essentially a finite-dimensional space if μ is concentrated on a finite set of points (why?). In analysis, we are mainly interested in the case where L_1 is infinite-dimensional but has a countable everywhere dense subset.[18] To characterize such spaces, we introduce the following concept, stemming from general measure theory:

DEFINITION. *Suppose a space X equipped with a measure μ has a countable system $\mathscr{A}$ of measurable subsets $A_1, A_2, \ldots$ such that given any $\varepsilon > 0$ and any measurable subset $M \subset X$, there is a set $A_k \in \mathscr{A}$ satisfying the inequality*

$$\mu(M \triangle A_k) < \varepsilon.$$

*Then μ is said to have a **countable base**, consisting of the sets $A_1, A_2, \ldots$*

Example. Let μ be a Lebesgue extension of a measure m originally defined on a countable semiring $\mathscr{S}_m$. Then the ring $\mathscr{R}(\mathscr{S}_m)$ is obviously itself countable, and hence, by Theorem 3, p. 277, is a countable base for μ. In particular, ordinary Lebesgue measure on the line has a countable base, since we can choose the original semiring $\mathscr{S}_m$ to consist of all intervals (open, closed and half-open) with rational end points.

THEOREM 3. *Let X be a space equipped with a measure μ, and suppose μ has a countable base $A_1, A_2, \ldots$. Then $L_1(X, \mu)$ has a countable everywhere dense subset.*

Proof. We will show that the set M of all finite linear combinations of the form

$$\sum_{k=1}^{n} c_k f_k(x), \tag{5}$$

where f_k is the characteristic function of A_k and the numbers $c_1, \ldots, c_n$ are rational, forms a countable everywhere dense subset of $L_1 = L_1(X, \mu)$. The countability of M is obvious, and we need only show that M is everywhere dense in L_1. As already noted, the set of all simple functions taking only finitely many values is everywhere dense in L_1. But every such function can be approximated arbitrarily closely by a function of the same type taking only rational values. Hence we need only show that every function f taking rational values $y_1, \ldots, y_n$ on pairwise disjoint sets $E_1, \ldots, E_n$ (with X as their union) can be approximated arbitrarily closely in the L_1-metric by functions of the form (5). Clearly, there is no loss of generality in assuming that the base $A_1, A_2, \ldots$ is closed under the operations of taking differences and forming finite unions and intersections (why?).

---

[18] So that L_1 is separable, as defined on p. 48.

Now, according to the definition, given any $\varepsilon > 0$, there are sets $A_1, \ldots, A_n$ such that

$$\mu[(E_k - A_k) \cup (A_k - E_k)] < \varepsilon \qquad (k = 1, \ldots, n).$$

Let

$$A_k' = A_k - \bigcup_{j < k} A_j \qquad (k = 1, \ldots, n),$$

and define a function

$$f^*(x) = \begin{cases} y_k & \text{if } x \in A_k', \\ 0 & \text{if } x \in X - \bigcup_{k=1}^n A_k'. \end{cases}$$

Then clearly

$$\mu\{x : f(x) \neq f^*(x)\},$$

and hence the left-hand side of

$$\int |f(x) - f^*(x)| \, d\mu < 2 \left(\max_n |y_n|\right) \mu\{x : f(x) \neq f^*(x)\},$$

can be made arbitrarily small by choosing $\varepsilon > 0$ sufficiently small. This proves the theorem, since f^* is a function of the form (5). ∎

37.2. Definition and basic properties of L_2. As we have seen, the space $L_1 = L_1(X, \mu)$ is a Banach space, i.e., a complete normed linear space. However, L_1 is not Euclidean, since its norm cannot be derived from any scalar product. This follows from the "parallelogram theorem" (Theorem 15, p. 160). For example, if $X = [0, 2\pi]$ and μ is ordinary Lebesgue measure on the line, then the condition

$$\|f + g\|^2 + \|f - g\|^2 = 2(\|f\|^2 + \|g\|^2)$$

fails for the summable functions $f(x) \equiv 1$, $g(x) = \sin x$.[19] To get a function space which is not only a normed linear space but also a Euclidean space, we now consider the set of functions whose *squares* are summable.

Thus let X be a space equipped with a measure μ, where we temporarily assume that $\mu(X) < \infty$. Then by $L_2(X, \mu)$, or simply L_2, we mean the set of all real functions f whose squares are summable on X, i.e., which satisfy the condition

$$\int f^2(x) \, d\mu < \infty$$

(however, see Problem 6). As in the case of L_1, we do not distinguish between equivalent functions (i.e., functions differing only on a set of measure zero).

---

[19] As an exercise, show that the same kind of counterexample works quite generally.

THEOREM 4. *If f and g belong to L_2, then so do $\alpha f, f + g$, and fg, where α is an arbitrary constant. In particular, L_2 is a linear space.*

Proof. Obviously $\alpha f \in L_2$, since

$$\int [\alpha f(x)]^2 \, d\mu = \alpha^2 \int f^2(x) \, d\mu < \infty.$$

The fact that $fg \in L_2$ follows from the inequality

$$|f(x)g(x)| \leqslant \tfrac{1}{2}[f^2(x) + g^2(x)] \tag{6}$$

and Theorem 3, p. 297.[20] But then $f + g \in L_2$, since

$$[f(x) + g(x)]^2 \leqslant f^2(x) + 2\,|f(x)g(x)| + g^2(x),$$

where each term on the right is summable. ∎

Next we define a scalar product in L_2, setting

$$(f, g) = \int f(x)g(x) \, d\mu.$$

This choice obviously has all the properties of a scalar product listed on p. 142:

1) $(f, f) \geqslant 0$ where $(f, f) = 0$ if and only if $f = 0$;
2) $(f, g) = (g, f)$;
3) $(\lambda f, g) = \lambda(f, g)$;
4) $(f, g_1 + g_2) = (f, g_1) + (f, g_2)$.

(In asserting that $(f, f) = 0$ if and only if $f = 0$, we rely on the fact that every function vanishing almost everywhere is identified with the zero element of L_2.) Thus L_2 is a Euclidean space, with the norm defined by the usual formula

$$\|f\| = \sqrt{(f, f)} \tag{7}$$

(recall Theorem 1, p. 142). In the case of L_2, (7) takes the form

$$\|f\| = \sqrt{\int f^2(x) \, d\mu}.$$

By the same token, the distance between two elements $f, g \in L_2$ is just

$$(f, g) = \|f - g\| = \sqrt{\int [f(x) - g(x)]^2 \, d\mu}.$$

The quantity

$$\int [f(x) - g(x)]^2 \, d\mu = \|f - g\|^2$$

is called the *mean square deviation* of the functions f and g (from each other).

---

[20] Setting $g(x) \equiv 1$ in (6), we find that $f \in L_2$ implies $f \in L_1$ (provided that X is of finite measure).

Let $\{f_n\}$ be a sequence of functions in L_2. Then $\{f_n\}$ is said to *converge in the mean square* to a function $f \in L_2$ if $\rho(f_n, f) \to 0$ as $n \to \infty$.

In L_2, as in any other Euclidean space, we have the Schwarz inequality

$$|(f, g)| \leqslant \|f\| \, \|g\|,$$

which here takes the form

$$\left| \int f(x)g(x) \, d\mu \right| \leqslant \sqrt{\int f^2(x) \, d\mu} \, \sqrt{\int g^2(x) \, d\mu}. \tag{8}$$

The L_2-version of the triangle inequality

$$\|f + g\| \leqslant \|f\| + \|g\|$$

is clearly

$$\sqrt{\int [f(x) + g(x)]^2 \, d\mu} \leqslant \sqrt{\int f^2(x) \, d\mu} + \sqrt{\int g^2(x) \, d\mu}.$$

In particular, replacing f by $|f|$ and setting $g(x) \equiv 1$ in (8), we get

$$\int |f(x)| \, d\mu \leqslant \sqrt{\mu(K)} \sqrt{\int f^2(x) \, d\mu}, \tag{9}$$

from which it is again apparent (cf. footnote 20) that $f \in L_2$ implies $f \in L_1$ if $\mu(X) < \infty$.

THEOREM 5. *The space L_2 is complete.*

Proof. Let $\{f_n\}$ be a Cauchy sequence in L_2, so that

$$\|f_m - f_n\| \to 0 \quad \text{as} \quad m, n \to \infty.$$

Then, by (9), given any $\varepsilon > 0$, we have

$$\int |f_m(x) - f_n(x)| \, d\mu \leqslant \sqrt{\mu(X)} \sqrt{\int [f_m(x) - f_n(x)]^2 \, d\mu} < \varepsilon \sqrt{\mu(X)}$$

for sufficiently large m and n, i.e., $\{f_n\}$ is also a Cauchy sequence in the L_1-metric. Repeating the argument given in the proof of the completeness of L_1, we choose a subsequence $\{f_{n_k}\}$ from $\{f_n\}$ converging almost everywhere to some function f. Clearly, given any $\varepsilon > 0$, we have

$$\int [f_{n_k}(x) - f_{n_l}(x)]^2 \, d\mu < \varepsilon \tag{10}$$

for sufficiently large k and l. Hence, by Fatou's theorem (Theorem 3, p. 307), we can take the limit as $l \to \infty$ behind the integral sign in (10), obtaining

$$\int [f_{n_k}(x) - f(x)]^2 \, d\mu \leqslant \varepsilon.$$

It follows that $f \in L_2$ (why?) and that $f_{n_k} \to f$ in the mean square. But if a Cauchy sequence contains a subsequence converging to a limit, then the sequence itself must converge to the same limit. Hence $f_n \to f$ in the mean square. ∎

We now drop the restriction $\mu(X) < \infty$, allowing X to have infinite measure. In the case $\mu(X) = \infty$, it is no longer true that $f \in L_2$ implies $f \in L_1$, a fact deduced from (6) or (9) in the case $\mu(X) < \infty$. For example, let X be the real line equipped with ordinary Lebesgue measure, and let

$$f(x) = \frac{1}{\sqrt{1 + x^2}}.$$

Then f belongs to L_2 but not to L_1, since

$$\int_{-\infty}^{\infty} \frac{dx}{\sqrt{1 + x^2}} = \infty, \qquad \int_{-\infty}^{\infty} \frac{dx}{1 + x^2} = \pi < \infty.$$

Moreover, if a sequence $\{f_n\}$ converges to a limit f in the L_2-metric, it follows from (9) that $\{f_n\}$ also converges to f in the L_1-metric if $\mu(X) < \infty$. However, this conclusion fails if $\mu(X) = \infty$, as shown by the example

$$f_n(x) = \begin{cases} \dfrac{1}{n} & \text{if } |x| \leqslant n, \\ 0 & \text{if } |x| > n, \end{cases}$$

where $\{f_n\}$ approaches no limit in L_1 but approaches the zero function in L_2 (give the details). Despite all this, we have[21]

THEOREM 5′. *The space L_2 is complete even if $\mu(X) = \infty$, provided that μ is σ-finite.*

Proof. As in Sec. 30.2, let

$$X = \bigcup_n X_n, \qquad \mu(X_n) < \infty,$$

where

$$X_1 \subset X_2 \subset \cdots \subset X_n \subset \cdots .$$

Moreover, given any function φ on X, let

$$\varphi^{(n)}(x) = \begin{cases} \varphi(x) & \text{if } x \in X_n, \\ 0 & \text{if } x \notin X_n, \end{cases}$$

---

[21] Note that in the proof of the completeness of L_1 (Theorem 1), X can have either finite or infinite measure.

so that

$$\int \varphi(x)\, d\mu = \int_X \varphi(x)\, d\mu = \lim_{n \to \infty} \int_{X_n} \varphi(x)\, d\mu = \lim_{n \to \infty} \int_{X_n} \varphi^{(n)}(x)\, d\mu,$$

if φ is summable on X. Let $\{f_n\}$ be a Cauchy sequence in L_2, so that, given any $\varepsilon > 0$,

$$\int [f_k(x) - f_l(x)]^2\, d\mu < \varepsilon$$

for all sufficiently large k and l. Then

$$\lim_{n \to \infty} \int_{X_n} [f_k^{(n)}(x) - f^{(n)}(x)]^2\, d\mu = \int [f_k(x) - f_l(x)]^2\, d\mu < \varepsilon,$$

and hence, a fortiori,

$$\int_{X_n} [f_k^{(n)}(x) - f_l^{(n)}(x)]^2\, d\mu < \varepsilon. \tag{11}$$

But $L_2(X_n, \mu)$ is complete, by Theorem 5, since $\mu(X_n) < \infty$. Therefore $\{f_k^{(n)}\}$ converges in the metric of $L_2(X_n, \mu)$ to a function $f^{(n)} \in L_2(X_n, \mu)$. Taking the limit as $l \to \infty$ behind the integral sign in (11), we get

$$\int_{X_n} [f_k^{(n)}(x) - f^{(n)}(x)]^2\, d\mu \leqslant \varepsilon \tag{12}$$

(why is this justified?). Since (12) holds for every n, we can now take the limit as $n \to \infty$, obtaining

$$\lim_{n \to \infty} \int_{X_n} [f_k^{(n)}(x) - f^{(n)}(x)]^2\, d\mu \leqslant \varepsilon. \tag{13}$$

Now let

$$f(x) = f^{(n)}(x) \qquad \text{if} \quad x \in X_n.$$

Then (13) implies

$$\int [f_k(x) - f(x)]^2\, d\mu \leqslant \varepsilon.$$

It follows that $f \in L_2(X, \mu)$ and $f_k \to f$ in the mean square. ∎

Problem 1. A complex function is said to be *summable* if its real and imaginary parts are summable. Show that the considerations of Sec. 37.1 carry over verbatim to the case where L_1 consists of all complex summable functions (defined on X).

Problem 2. Prove that if each of the measures μ_1 and μ_2 has a countable base, then so does their direct product $\mu = \mu_1 \times \mu_2$.

Comment. In particular, Lebesgue measure in the plane (or more generally in n-space) has a countable base.

Problem 3. Let X be the interval $[a, b]$, and let μ be ordinary Lebesgue measure on the line. Prove that the set $\mathscr{P}$ of all polynomials on $[a, b]$ with rational coefficients is everywhere dense in $L_1(X, \mu)$.

Hint. Use Theorem 2 and the fact that every function continuous on $[a, b]$ can be approximated in the mean (or even uniformly) by elements of $\mathscr{P}$.

Problem 4. Prove that $L_2(X, \mu)$ is separable, i.e., has a countable everywhere dense subset, if μ has a countable base.

Comment. Thus $L_2(X, \mu)$ is a Hilbert space if μ has a countable base (we disregard the case where $L_2(X, \mu)$ is finite-dimensional). It follows from Theorem 11, p. 155 that all such spaces are isomorphic, in particular, that $L_2(X, \mu)$ is isomorphic to the space l_2 of all sequences $(x_1, x_2, \ldots, x_n, \ldots)$ such that

$$\sum_{n=1}^{\infty} x_n^2 < \infty.$$

(in fact, l_2 corresponds to the case where the measure μ is concentrated on a countable set of points).

Problem 5. Prove that every continuous linear functional φ on $L_2(X, \mu)$, where μ has a countable base, can be represented in the form

$$\varphi(f) = \int f(x)g(x) \, d\mu,$$

where g is a fixed element of $L_2(X, \mu)$.

Hint. Recall Theorem 2, p. 188.

Problem 6. Show that the considerations of Sec. 37.2 carry over verbatim to the case where L_2 consists of all complex functions f satisfying the condition

$$\int |f(x)|^2 \, d\mu < \infty,$$

provided the scalar product of two such functions f and g is now defined as

$$(f, g) = \int f(x)\overline{g(x)} \, d\mu.$$

Show that the resulting space L_2 is a complex Hilbert space if the measure μ has a countable base (again disregard the finite-dimensional case).

Problem 7. Let $\{f_n\}$ be a sequence of functions defined on a space X equipped with a measure μ such that $\mu(X) < \infty$. Prove that

a) If $\{f_n\}$ converges uniformly, then $\{f_n\}$ converges in the mean and in the mean square;

b) If $\{f_n\}$ converges in the mean or in the mean square, then $\{f_n\}$ converges in measure (as defined in Problem 6, p. 292);

c) If $\{f_n\}$ converges in the mean or in the mean square, then $\{f_n\}$ contains a subsequence $\{f_{n_k}\}$ which converges almost everywhere.

Hint. See Problem 9, p. 292. Alternatively, recall the proof of Theorem 1.

Problem 8. Prove that the sequence of functions constructed in Problem 8, p. 292 converges to $f(x) \equiv 0$ in the mean and in the mean square, without converging at a single point.

Problem 9. Give an example of a sequence of functions $\{f_n\}$ which converges everywhere on $[0, 1]$, but does not converge in the mean.

Hint. Let

$$f_n(x) = \begin{cases} n & \text{if} \quad x \in (0, 1/n), \\ 0 & \text{otherwise.} \end{cases}$$

Problem 10. Give an example of a sequence of functions $\{f_n\}$ which converges uniformly, but does not converge in the mean or in the mean square.

Hint. According to Problem 7a, we must have $\mu(X) = \infty$. Let

$$f_n(x) = \begin{cases} \dfrac{1}{\sqrt{n}} & \text{if} \quad |x| \leqslant n, \\ 0 & \text{if} \quad |x| > n. \end{cases}$$

Problem 11. Show that convergence in the mean need not imply convergence in the mean square, whether or not $\mu(X) < \infty$.

Problem 12. Let $L_p(X, \mu)$ be the set of all classes of equivalent (real or complex) functions f such that

$$\int |f|^p \, d\mu < \infty \qquad (1 \leqslant p < \infty),$$

equipped with the norm

$$\|f\| = \left(\int |f|^p \, d\mu \right)^{1/p}$$

Prove that $L_p(X, \mu)$ is a Banach space.

BIBLIOGRAPHY

Akhiezer, N. I. and I. M. Glazman, *Theory of Linear Operators in Hilbert Space* (translated by M. Nestell), Frederick Ungar Publishing Co., New York, *Volume I* (1961), *Volume II* (1963).

Berberian, S. K., *Measure and Integration*, The Macmillan Co., New York (1965).

Burkill, J. C., *The Lebesgue Integral*, Cambridge University Press, New York (1953).

Day, M. M., *Normed Linear Spaces*, Springer-Verlag, New York (1963).

Dieudonné, J , *Foundations of Modern Analysis*, Academic Press, Inc., New York (1960).

Dunford, N. and J. T Schwartz, *Linear Operators*, Interscience Publishers, Inc., New York, *Part I: General Theory* (1958), *Part II: Spectral Theory* (1963).

Edwards, R. E., *Functional Analysis, Theory and Applications*, Holt, Rinehart and Winston, New York (1965).

Fraenkel, A. A., *Abstract Set Theory*, third edition, North-Holland Publishing Co., Amsterdam (1966).

Friedman, A A., *Generalized Functions and Partial Differential Equations*, Prentice-Hall, Inc., Englewood Cliffs, N.J. (1963).

Gelfand, I. M. and G. E Shilov, *Generalized Functions* (translated by E. Saletan et al.), Academic Press, Inc., New York, *Volume 1: Properties and Operations* (1964), *Volume 2: Spaces of Fundamental and Generalized Functions* (1968).

Hahn, H. and A. Rosenthal, *Set Functions*, University of New Mexico Press, Albuquerque, New Mexico (1948).

Halmos, P. R., *Measure Theory*, D. Van Nostrand Co., Inc., Princeton, N.J. (1950).

Halmos, P. R., *Introduction to Hilbert Space*, second edition, Chelsea Publishing Co., New York (1957).

Halmos, P. R., *Naive Set Theory*, D. Van Nostrand Co., Inc., Princeton, N.J. (1960).

Halmos, P. R., *A Hilbert Space Problem Book*, D. Van Nostrand Co., Inc., Princeton, N.J. (1967).

Hewitt, E. and K. Stromberg, *Real and Abstract Analysis*, Springer-Verlag, New York (1965).

Hildebrandt, T. H., *Introduction to the Theory of Integration*, Academic Press, Inc., New York (1963).

Kelley, J. L., *General Topology*, D. Van Nostrand Co., Inc., Princeton, N.J. (1955).

Kelley, J. L., I. Namioka et al., *Linear Topological Spaces*, D. Van Nostrand Co., Princeton, N.J. (1963).

Kuratowski, K., *Introduction to Set Theory and Topology* (translated by L. F. Boron), Pergamon Press, Inc., New York (1961).

Liusternik, L. A. and V. I. Sobolev, *Elements of Functional Analysis* (translated by A. E. Labarre, Jr. et al.), Frederick Ungar Publishing Co., Inc., New York (1961).

Loève, M., *Probability Theory*, third edition, D. Van Nostrand Co., Inc., Princeton, N.J. (1963).

McShane, E. J., *Integration*, Princeton University Press, Princeton, N.J. (1944).

Natanson, I. P., Theory of Functions of a Real Variable (translated by L. F. Boron, with the collaboration of E. Hewitt), Frederick Ungar Publishing Co., New York, *Volume I* (1955), *Volume II* (1960).

Riesz, F. and B. Sz.-Nagy, *Functional Analysis* (translated by L. F. Boron), Frederick Ungar Publishing Co., New York (1955).

Royden, H. L., *Real Analysis*, second edition, The Macmillan Co., New York (1968).

Rudin, W., *Principles of Mathematical Analysis*, second edition, McGraw-Hill Book Co., Inc., New York (1964).

Saks, S., *Theory of the Integral* (translated by L. C. Young, with two notes by S. Banach), second edition, Dover Publications, Inc., New York (1964).

Schaefer, H. H., *Topological Vector Spaces*, The Macmillan Co., New York (1966)·

Shilov, G. E., *Generalized Functions and Partial Differential Equations* (translated by B. D. Seckler), Gordon and Breach Science Publishers, Inc., New York (1968).

Shilov, G. E. and B. L. Gurevich, *Integral, Measure and Derivative: A Unified Approach* (translated by R. A. Silverman), Prentice-Hall, Inc., Englewood Cliffs, N.J. (1966).

Taylor, A. E., *Introduction to Functional Analysis*, J. Wiley and Sons, Inc., New York (1958).

Titchmarsh, E. C., *The Theory of Functions*, second edition, Oxford University Press, New York (1939).

Yosida, K., *Functional Analysis*, second edition, Springer-Verlag, New York (1968).

Zaanen, A. C., *An Introduction to the Theory of Integration*, Interscience Publishers, Inc., New York (1958).

INDEX